*Edited by Monique A.V. Axelos and
Marcel Van de Voorde*

**Nanotechnology in Agriculture
and Food Science**

Further Volumes of the Series "Nanotechnology Innovation & Applications"

Cornier, J., Kwade, A., Owen, A., Van de Voorde, M. (eds.)

Pharmaceutical Nanotechnology

Innovation and Production

2017
Print ISBN: 9783527340545

Fermon, C. and Van de Voorde, M. (eds.)

Nanomagnetism

Applications and Perspectives

2017
Print ISBN: 9783527339853

Mansfield, E., Kaiser, D. L., Fujita, D., Van de Voorde, M. (eds.)

Metrology and Standardization for Nanotechnology

Protocols and Industrial Innovations

2017
Print ISBN: 9783527340392

Meyrueis, P., Sakoda, K., Van de Voorde, M. (eds.)

Micro- and Nanophotonic Technologies

2017
Print ISBN: 9783527340378

Müller, B. and Van de Voorde, M. (eds.)

Nanoscience and Nanotechnology for Human Health

2017
Print ISBN: 978-3-527-33860-3

Puers, R., Baldi, L., Van de Voorde, M., van Nooten, S. E. (eds.)

Nanoelectronics

Materials, Devices, Applications

2017
Print ISBN: 9783527340538

Raj, B., Van de Voorde, M., Mahajan, Y. (eds.)

Nanotechnology for Energy Sustainability

2017
Print ISBN: 9783527340149

Sels, B. and Van de Voorde, M. (eds.)

Nanotechnology in Catalysis

Applications in the Chemical Industry, Energy Development, and Environment Protection

2017
Print ISBN: 9783527339143

Edited by Monique A.V. Axelos and Marcel Van de Voorde

Nanotechnology in Agriculture and Food Science

Verlag GmbH & Co. KGaA

Volume Editors

Dr. Monique A.V. Axelos
INRA
Science & Engineering of Agricultural
Products
Rue de la Géraudière
44316 Nantes cedex 03
France

Prof. Dr. Dr. h.c. Marcel H. Van de Voorde
Member of the Science Council
of the French Senate and National
Assembly, Paris
Rue du Rhodania, 5
BRISTOL A, Appartement 31
3963 Crans-Montana
Switzerland

Series Editor

Prof. Dr. Dr. h.c. Marcel H. Van de Voorde
Member of the Science Council
of the French Senate and National
Assembly, Paris
Rue du Rhodania, 5
BRISTOL A, Appartement 31
3963 Crans-Montana
Switzerland

Cover credits: Glasware: fotolia_© kasto
Orange: fotolia_© Olga Kovalenko
Nanostructure: fotolia_© psdesign1

All books published by **Wiley-VCH** are carefully produced. Nevertheless, authors, editors, and publisher do not warrant the information contained in these books, including this book, to be free of errors. Readers are advised to keep in mind that statements, data, illustrations, procedural details or other items may inadvertently be inaccurate.

Library of Congress Card No.: applied for

British Library Cataloguing-in-Publication Data
A catalogue record for this book is available from the British Library.

Bibliographic information published by the Deutsche Nationalbibliothek
The Deutsche Nationalbibliothek lists this publication in the Deutsche Nationalbibliografie; detailed bibliographic data are available on the Internet at http://dnb.d-nb.de.

© 2017 Wiley-VCH Verlag GmbH & Co. KGaA, Boschstr. 12, 69469 Weinheim, Germany

All rights reserved (including those of translation into other languages). No part of this book may be reproduced in any form – by photoprinting, microfilm, or any other means – nor transmitted or translated into a machine language without written permission from the publishers. Registered names, trademarks, etc. used in this book, even when not specifically marked as such, are not to be considered unprotected by law.

Print ISBN: 978-3-527-33989-1
ePDF ISBN: 978-3-527-69771-7
ePub ISBN: 978-3-527-69773-1
Mobi ISBN: 978-3-527-69774-8
oBook ISBN: 978-3-527-69772-4

Cover Design   Adam Design
Typesetting   Thomson Digital, Noida, India
Printing and Binding   Markono Print Media Pte Ltd, Singapore

Printed on acid-free paper

*Thanks to my wife for her patience with me spending
many hours working on the book series through
the nights and over weekends.
The assistance of my son Marc Philip related to the complex
and large computer files with many sophisticated scientific
figures is also greatly appreciated.*

Marcel Van de Voorde

Series Editor Preface

Since years, nanoscience and nanotechnology have become particularly an important technology areas worldwide. As a result, there are many universities that offer courses as well as degrees in nanotechnology. Many governments including European institutions and research agencies have vast nanotechnology programmes and many companies file nanotechnology-related patents to protect their innovations. In short, nanoscience is a hot topic!

Nanoscience started in the physics field with electronics as a forerunner, quickly followed by the chemical and pharmacy industries. Today, nanotechnology finds interests in all branches of research and industry worldwide. In addition, governments and consumers are also keen to follow the developments, particularly from a safety and security point of view.

This books series fills the gap between books that are available on various specific topics and the encyclopedias on nanoscience. This well-selected series of books consists of volumes that are all edited by experts in the field from all over the world and assemble top-class contributions. The topical scope of the book is broad, ranging from nanoelectronics and nanocatalysis to nanometrology. Common to all the books in the series is that they represent top-notch research and are highly application-oriented, innovative, and relevant for industry. Finally they collect a valuable source of information on safety aspects for governments, consumer agencies and the society.

The titles of the volumes in the series are as follows:

Human-related nanoscience and nanotechnology

- *Nanoscience and Nanotechnology for Human Health*
- *Pharmaceutical Nanotechnology*
- *Nanotechnology in Agriculture and Food Science*

Nanoscience and nanotechnology in information and communication

- *Nanoelectronics*
- *Micro- and Nanophotonic Technologies*
- *Nanomagnetism: Perspectives and Applications*

Nanoscience and nanotechnology in industry

- Nanotechnology for Energy Sustainability
- Metrology and Standardization of Nanomaterials
- Nanotechnology in Catalysis: Applications in the Chemical Industry, Energy Development, and Environmental Protection

The book series appeals to a wide range of readers with backgrounds in physics, chemistry, biology, and medicine, from students at universities to scientists at institutes, in industrial companies and government agencies and ministries.

Ever since nanoscience was introduced many years ago, it has greatly changed our lives – and will continue to do so!

March 2016 *Marcel Van de Voorde*

About the Series Editor

Marcel Van de Voorde, Prof. Dr. ir. Ing. Dr. h.c., has 40 years' experience in European Research Organisations, including CERN-Geneva and the European Commission, with 10 years at the Max Planck Institute for Metals Research, Stuttgart. For many years, he was involved in research and research strategies, policy, and management, especially in European research institutions.

He has been a member of many Research Councils and Governing Boards of research institutions across Europe, the United States, and Japan. In addition to his Professorship at the University of Technology in Delft, the Netherlands, he holds multiple visiting professorships in Europe and worldwide. He holds a doctor honoris causa and various honorary professorships.

He is a senator of the European Academy for Sciences and Arts, Salzburg, and Fellow of the World Academy for Sciences. He is a member of the Science Council of the French Senate/National Assembly in Paris. He has also provided executive advisory services to presidents, ministers of science policy, rectors of Universities, and CEOs of technology institutions, for example, to the president and CEO of IMEC, Technology Centre in Leuven, Belgium. He is also a Fellow of various scientific societies. He has been honored by the Belgian King and European authorities, for example, he received an award for European merits in Luxemburg given by the former President of the European Commission. He is author of multiple scientific and technical publications and has coedited multiple books, especially in the field of nanoscience and nanotechnology.

Contents

Series Editor Preface *VII*
About the Series Editor *IX*
Foreword *XXI*
Introduction *XXV*

Part One Basic Elements of Nanofunctional Agriculture and Food Science *1*

1 Nanotechnologies for Agriculture and Foods: Past and Future *3*
Cecilia Bartolucci
References *13*

2 Nanoscience: Relevance for Agriculture and the Food Sector *15*
Shahin Roohinejad and Ralf Greiner
2.1 Introduction *15*
2.2 Fundamental of Nanoscience *16*
2.3 Applications of Nanotechnology in the Agriculture Sector *18*
2.3.1 Delivery of Agriculture Chemicals *19*
2.3.2 Nanosensors/Nanobiosensors *21*
2.3.3 Diagnosis and Control of Plant Diseases *22*
2.3.4 Waste Reduction and Production of High-Value Added Products *22*
2.4 Applications of Nanotechnology in the Food Sector *23*
2.4.1 Delivery of Active Compounds *24*
2.4.2 Food Packaging *25*
2.4.3 Other Applications *26*
2.5 Challenges of Using Nanotechnology in Agriculture and Food Sectors *27*
2.6 Conclusions *28*
Acknowledgment *28*
References *28*

3	**Naturally Occurring Nanostructures in Food** *33*
	Saïd Bouhallab, Christelle Lopez, and Monique A.V. Axelos
3.1	Introduction *33*
3.2	Protein-based Nanostructures *34*
3.2.1	Examples of Protein Nanostructures Present in Foods *35*
3.2.1.1	β-Lactoglobulin *35*
3.2.1.2	Serum Albumin *36*
3.2.1.3	α-Lactalbumin and Lysozyme *37*
3.2.1.4	Ovalbumin and Avidin *38*
3.2.1.5	Transferrins *38*
3.2.1.6	Osteopontin and Lactoperoxydase *39*
3.2.2	Formation of Natural Nanostructure Subsequently to Molecular Interaction/Complexation *40*
3.2.3	Special Case: Casein Micelles *41*
3.2.3.1	Casein Micelle Composition *42*
3.2.3.2	Casein Micelle Structure *42*
3.3	Lipid-Based Nanostructures *44*
3.3.1	Lipid Nanodroplets *44*
3.3.2	Special Case: Milk Fat Globules *45*
3.4	Concluding Remarks and Future Prospects *46*
	References *47*
4	**Artificial Nanostructures in Food** *49*
	Jared K. Raynes, Sally L. Gras, John A. Carver, and Juliet A. Gerrard
4.1	Introduction *49*
4.2	Types and Uses of Artificial Organic Nanostructures Found in Food *52*
4.2.1	Protein Nanostructures *52*
4.2.2	Polysaccharide Nanostructures *55*
4.2.3	Lipid Nanostructures *60*
4.3	Conclusion *62*
	References *63*
5	**Engineered Inorganic Nanoparticles in Food** *69*
	Marie-Hélène Ropers and Hélène Terrisse
5.1	Introduction *69*
5.2	Engineered Inorganic Materials Containing Nanoparticles *69*
5.2.1	Silica (SiO_2) and Silicates *72*
5.2.2	Titania or Titanium Dioxide (TiO_2) *74*
5.2.3	Iron Oxides and Hydroxides *75*
5.2.4	Silver (Ag) *75*
5.2.5	Miscellaneous *76*
5.2.5.1	Other Metals (Fe, Se, Ca, etc.) *76*
5.2.5.2	Calcium Carbonate ($CaCO_3$) *76*

5.2.5.3	Calcium Chloride (CaCl$_2$)	77
5.2.6	Knowledge Gaps	77
5.2.6.1	Gold (Au)	77
5.2.6.2	Aluminum (Al)	77
5.2.6.3	Zinc Oxide (ZnO)	78
5.3	Characterization of Engineered Inorganic Nanomaterials	78
5.3.1	Characterization of Engineered Inorganic Nanomaterials as Manufactured	79
5.3.2	Characterization of Engineered Inorganic Nanomaterials as Present in the Food Matrices	80
5.4	Conclusion and Perspectives	81
	References	82

6 Nanostructure Characterization Using Synchrotron Radiation and Neutrons 87
Francois Boué

6.1	Introduction	87
6.1.1	Observing at Nanosizes *In Situ*	87
6.1.2	Nanoparticles in Food and Agricultural Products: What is Here, What Can Be Seen	89
6.2	Principles	89
6.2.1	Scattering Process	89
6.2.2	q and r: Orders of Magnitude	91
6.2.3	Binary System: Contrast	91
6.2.4	Contrast Strategies	92
6.3	The Basic Information from a SAS Profile	93
6.3.1	Form Factors	93
6.3.2	Structure Factors: Interactions between Objects	97
6.4	A Few Examples: From Soft Matter to Agrofood	100
6.4.1	Proteins/Polymer: Opposite Architectures of Complexes in Mixed Systems	100
6.4.2	Lipids: Micelles, Bilayers, Crystalline Phases	102
6.4.3	A Complex but Model Structure: Casein Micelle in Cow Milk	105
6.4.4	Foams	105
6.5	Other Scattering Techniques	106
6.6	Recommendation and Practical: A Checklist for Scattering	107
6.6.1	Requirements for Sample Composition and Preparation	107
6.6.2	Sample Sizes, Volumes, and Quantities	108
6.6.3	Sample Damage	109
6.6.4	Spectrometer Setups: Sample Environment	109
6.6.5	Before and After: Proposal, Data Treatment, and Fitting	109
6.7	Summary and Conclusion	110
	References	110

Part Two Opportunities, Innovations, and New Applications in Agriculture and Food Systems *113*

7 Nanomaterials in Plant Protection *115*
Angelo Mazzaglia, Elena Fortunati, Josè Maria Kenny, Luigi Torre, and Giorgio Mariano Balestra
7.1 Introduction *115*
7.2 Nanotechnology and Agricultural Sector *117*
7.2.1 Nanomaterials *119*
7.2.1.1 Organic Nanomaterials *120*
7.2.1.2 Inorganic Nanomaterials *120*
7.2.1.3 Combined Organic/Inorganic Nanomaterials *122*
7.2.2 Functionalization of Nanomaterials (NMs): Development of Novel Nanoformulations for Pests and Plant Pathogens Control *123*
7.3 Applications of Nanomaterials against Plant Pathogens and Pests *125*
7.3.1 Bacteria *125*
7.3.2 Fungi *126*
7.3.3 Insects *127*
7.3.4 Virus *128*
7.4 Conclusions *129*
References *130*

8 Nanoparticle-Based Delivery Systems for Nutraceuticals: Trojan Horse Hydrogel Beads *135*
Benjamin Zeeb and David Julian McClements
8.1 Introduction *135*
8.2 Overview of Nanoparticles-Based Colloidal Delivery Systems *136*
8.2.1 Microemulsions *136*
8.2.2 Nanoliposomes *137*
8.2.3 Nanoemulsions *137*
8.2.4 Solid Lipid Nanoparticles *137*
8.2.5 Biopolymer Nanoparticles and Nanogels *138*
8.3 Designing Particle Characteristics *138*
8.3.1 Composition *138*
8.3.2 Particle Size *139*
8.3.3 Particle Charge *139*
8.3.4 Particle Structure *140*
8.4 Trojan Horse Nanoparticle Delivery Systems *140*
8.4.1 Biopolymers as Building Blocks to Form Hydrogel Beads *140*
8.4.2 Fabrication Methods for Hydrogel Beads *143*
8.4.3 Thermodynamic Incompatibility *143*
8.4.4 Complex Coacervation *144*
8.4.5 Antisolvent Precipitation *144*
8.4.6 Electrospinning *145*
8.4.7 Extrusion Techniques *145*
8.4.8 Fibril Formation *146*

8.5	Case Study: Alginate Hydrogel Beads as Trojan Horse Nanoparticle Delivery Systems for Curcumin *146*
8.6	Conclusions *149*
	References *149*

9	**Bottom-Up Approaches in the Design of Soft Foods for the Elderly** *153*
	José Miguel Aguilera and Dong June Park
9.1	Foods and the Elderly *153*
9.1.1	An Aging Society *153*
9.1.2	The Elderly and Food-Related Issues *153*
9.1.3	Special Foods for the Elderly: Texture-Modified Foods *155*
9.2	Rational Design of Soft and Nutritious Gel Particles *155*
9.2.1	Structure and Food Properties *155*
9.2.2	Molecular Gastronomy: An Example of Food Design *156*
9.2.3	Nanotechnology and Foods for the Elderly *157*
9.2.4	Building-Up Healthy Gels with Soft Textures *158*
9.3	Technological Alternatives for the Design of TM Foods *160*
9.4	Conclusions *162*
	Acknowledgments *163*
	References *163*

10	**Barrier Nanomaterials and Nanocomposites for Food Packaging** *167*
	Jose M. Lagaron, Luis Cabedo, and Maria J. Fabra
10.1	Introduction *167*
10.2	Nanocomposites *168*
10.3	Nanostructured Layers *172*
10.4	Conclusion and Future Prospects *174*
	References *174*

11	**Nanotechnologies for Active and Intelligent Food Packaging: Opportunities and Risks** *177*
	Nathalie Gontard, Stéphane Peyron, Jose M. Lagaron, Yolanda Echegoyen, and Carole Guillaume
11.1	Introduction and Definitions *177*
11.2	Nanomaterials in Active Packaging for Food Preservation *178*
11.2.1	Nanocomposites with Antioxidant Properties *178*
11.2.2	Nanocomposites with Antimicrobial Properties *179*
11.3	Nanotechnology for Intelligent Packaging as Food Freshness and Safety Monitoring Solution *181*
11.3.1	Stakes and Challenges of Nano-Enabled Intelligent Packaging *181*
11.3.2	Main Principles of Involved Nano-Enabled Sensing *183*
11.3.3	Indirect Nano-Enabled Indicators of Food Quality and Safety *184*
11.3.4	Direct Nano-Enabled Indicators of Food Quality and Safety *186*
11.4	Potential Safety Issues and Current Legislation *187*
11.5	Conclusions and Perspectives *190*
	References *191*

12	**Overview of Inorganic Nanoparticles for Food Science Applications** *197*	
	Xavier Le Guével	
12.1	Introduction *197*	
12.2	Food Packaging, Processing, and Storage *197*	
12.2.1	Antimicrobial Activities *198*	
12.2.2	Physical Barrier *199*	
12.3	Supplements/Additives *199*	
12.4	Food Analysis *200*	
12.4.1	NP Detection in Food *200*	
12.4.2	Nanoparticle-Based Sensors *201*	
12.4.2.1	Optical Detection *201*	
12.4.2.2	Electrochemical Sensing *202*	
12.5	Conclusion and Perspective *202*	
	Acknowledgment *203*	
	References *203*	
13	**Nanotechnology for Synthetic Biology: Crossroads Throughout Spatial Confinement** *209*	
	Denis Pompon, Luis F. Garcia-Alles, and Gilles Truan	
13.1	Convergence Between Nanotechnologies and Synthetic Biology *209*	
13.2	Spatially Constrained Functional Coupling in Biosystems *210*	
13.3	Functional Coupling Through Scaffold-Independent Structures *211*	
13.3.1	Functional Assembly Through Natural or Synthetic Fusions of Protein Domains *212*	
13.3.2	Functional Assembly Through Engineering of Natural or Synthetic Complexes *212*	
13.4	Spatial Confinement Mediated by Natural and Synthetic Scaffolds *213*	
13.4.1	Protein-Based Scaffolds *213*	
13.4.2	Nucleic Acids-Based Scaffolds *215*	
13.5	Encapsulated Biosystems Involving Natural or Engineered Nanocompartments *216*	
13.5.1	Lipid-Based Compartments *216*	
13.5.2	Protein-Based Nanocompartments *216*	
13.5.2.1	Shell-Independent Nanocompartments *217*	
13.5.2.2	Shell-Dependent Nanocompartments *221*	
13.5.2.3	Bacterial Microcompartments: Framework for Enzymatic Nanoreactors *223*	
13.5.2.4	Engineering of Natural BMC *224*	
13.6	Synthetically Designed Structures for Protein Coupling and Organization *225*	
13.7	Future Directions *226*	
	References *227*	

14	Modeling and Simulation of Bacterial Biofilm Treatment with Applications to Food Science *235*
	Jia Zhao, Tianyu Zhang, and Qi Wang
14.1	Introduction *235*
14.2	Review of Biofilm Models *237*
14.2.1	Hybrid Discrete–Continuum Models *239*
14.2.2	Multidimensional Continuum Models *240*
14.2.3	Individual-Based Modeling (IbM) *242*
14.2.4	Other Models Related to Biofilm Properties *243*
14.3	Biofilm Dynamics Near Antimicrobial Surfaces *244*
14.4	Antimicrobial Treatment of Biofilms by Targeted Drug Release *246*
14.5	Models for Intercellular and Surface Delivery by Nanoparticles *248*
14.6	Conclusion *250*
	Acknowledgments *251*
	References *251*

Part Three Technical Challenges of Nanoscale Detection Systems *257*

15	Smart Systems for Food Quality and Safety *259*
	Mark Bücking, Andreas Hengse, Heinrich Grüger, and Henning Schulte
15.1	Introduction *259*
15.2	Overview *260*
15.3	Roadmapping of Microsystem Technologies Toward Food Applications *261*
15.3.1	Implementation of Microsystems in the Dairy Sector *261*
15.3.1.1	Measurement of Contamination *262*
15.3.1.2	List of Specific Items to be Tested in the Dairy Industry *262*
15.3.2	Implementation of Microsystems in the Meat Sector *264*
15.3.3	Implementation of Microsystems in the Food and Beverage Sector *265*
15.4	Microsystem Technology Areas *266*
15.4.1	Detection Methods *266*
15.4.1.1	Near-Infrared Spectroscopy (NIRS) *266*
15.4.1.2	Mid-Infrared Spectroscopy *266*
15.4.1.3	Imaging Techniques *266*
15.4.1.4	Hyperspectral Imaging *266*
15.4.1.5	Ultrasound Imaging *267*
15.4.1.6	Magnetic Resonance Imaging (MRI) and X-ray scanning *267*
15.4.1.7	Dielectric sensor *267*
15.4.1.8	Process Viscometer *267*
15.4.1.9	Direct Sensing with Electronic Nose Technology *267*
15.4.1.10	Chemical and Biochemical Electronic sensors and systems *268*
15.4.2	Gas Sensing Devices and Systems *268*
15.4.2.1	E-Nose Instruments *269*

15.4.2.2	Microchromatographers *270*
15.4.3	NIR-Spectroscopy *270*
15.4.4	Biochemical Sensors *272*
15.4.4.1	E-Tongue Systems *272*
15.4.5	Microorganism Detection *273*
15.4.6	Tracking and Tracing *275*
	References *275*
16	**Nanoelectronics: Technological Opportunities for the Management of the Food Chain** *277*
	Kris Van De Voorde, Steven Van Campenhout, Veerle De Graef, Bart De Ketelaere, and Steven Vermeir
16.1	Technological Needs and Trends in the Food Industry *277*
16.2	Cooperation Model to Stimulate "The Introduction of New Nanoelectronics-Based Technologies in Food Industry": An Engine for Innovation and Bridging the Gap *279*
16.2.1	Awareness *279*
16.2.2	Platform Creation *280*
16.2.3	Validation of New Technologies *281*
16.2.4	Implementation of New Technologies *282*
16.3	Existing Technologies That Can Be Used in a Wide Range of Applications: The Present *282*
16.3.1	Characteristics *282*
16.3.1.1	Compact, State-of-the-Art Technology *282*
16.3.1.2	User-Friendly Technology *282*
16.3.1.3	Standardization *282*
16.3.1.4	Integration *283*
16.3.2	Some Examples of Existing Technologies and Suppliers *283*
16.3.2.1	Spectral Systems *283*
16.3.2.2	Portable Aroma Systems *284*
16.3.2.3	Biosensor Technologies *285*
16.3.2.4	Lab-on-a-Chip Systems *285*
16.4	New Technology Developments: The Future *285*
16.4.1	Short-Term New Technologies: Recently Validated *285*
16.4.1.1	New Reflection-Based Camera Technologies: Hyperspectral Imaging *285*
16.4.1.2	Optical Fiber Biosensor Technology *288*
16.4.1.3	New Transmission-Based Technology: Millimeter Wave Sensors (GHz–THz Sensor) *288*
16.4.2	Long-Term New/Future Technologies: To Be Validated *290*
16.4.2.1	Portable Hyperspectral Camera Technology *290*
16.4.2.2	NMR *291*
16.4.2.3	3D X-Ray *292*
16.4.3	The IoT (R)evolution and Big Data *293*
	References *295*

Part Four Nanotechnology: Toxicology Aspects and Regulatory Issues *297*

17   Quality and Safety of Nanofood *299*
Oluwatosin Ademola Ijabadeniyi
17.1   Introduction *299*
17.1.1   Nanotechnology and Nanofood: Background and Definition *299*
17.2   Current and Future Application of Nanotechnology in the Food Industry *300*
17.3   Food Quality and Food Safety *304*
17.4   How Safe is Nanofood? *304*
17.5   The Need for Risk Assessment *306*
17.6   Regulations for Food Nanotechnology *306*
17.7   Conclusion *307*
             References *307*

18   Interaction between Ingested-Engineered Nanomaterials and the Gastrointestinal Tract: *In Vitro* Toxicology Aspects *311*
Laurie Laloux, Madeleine Polet, and Yves-Jacques Schneider
18.1   Introduction *311*
18.2   Influence of the Gastrointestinal Tract on the Ingested Nanomaterials Characteristics *314*
18.2.1   *In vitro* Models of the Gastrointestinal Tract *315*
18.2.2   Influence of pH and Ionic Strength on Ingested Nanomaterials *316*
18.2.3   Influence of Digestive Enzymes and Food Matrices on Ingested Nanomaterials *316*
18.2.4   Characterization Techniques of Ingested Nanomaterials *318*
18.3   *In Vitro* Models of the Intestinal Barrier *318*
18.4   Cytotoxicity Assessment and Application to Silver Nanoparticles *320*
18.5   Conclusion *323*
             References *324*

19   Life Cycle of Nanoparticles in the Environment *333*
Jean-Yves Bottero, Mark R. Wiesner, Jérôme Labille, Melanie Auffan, Vladimir Vidal, and Catherine Santaella
19.1   Introduction *333*
19.2   Transport and Bioaccumulation by Plants *334*
19.3   Indirect Agricultural Application of NMs through Biowastes *336*
19.4   Transformations of NPs in Soils after Application *339*
19.4.1   Direct Application *339*
19.4.2   Indirect Applications from Biosludges *340*
19.5   Conclusion *342*
             Acknowledgments *343*
             References *343*

Part Five Governance of Nanotechnology and Societal Dimensions *347*

20 **The Politics of Governance: Nanotechnology and the Transformations of Science Policy** *349*
Brice Laurent
20.1 An Issue of Governance *349*
20.2 Operationalizing the Governance of Nanotechnology *352*
20.2.1 ELSI and ELSA Projects *352*
20.2.2 Voluntary Codes *353*
20.2.3 Public Engagement *355*
20.3 The Constitutional Project of Governance *356*
20.3.1 The Politics of Responsible Research and Innovation *356*
20.3.2 How to Think Critically about Governance *358*
References *360*

21 **Potential Economic Impact of Engineered Nanomaterials in Agriculture and the Food Sector** *363*
Elke Walz, Volker Gräf, and Ralf Greiner
21.1 Introduction *363*
21.2 Potential and Possible Applications of Nanomaterials in the Food Sector and Agriculture *364*
21.3 Nanotechnology: Market Research and Forecasts *366*
21.3.1 Methodology of Market Research and Forecasts *366*
21.3.2 Market Forecasts of Nanotechnology in Food and Agriculture: Publicly Available Data *367*
21.4 Critical Considerations and Remarks Concerning Market Reports and Forecasts *367*
21.5 Obstacles Regarding Commercialization of Nanotechnologies in Food and Agriculture *370*
21.6 Conclusion *372*
References *372*

22 **Conclusions** *377*
Monique A.V. Axelos and Marcel Van de Voorde

Index *381*

Foreword

As defined by the European Commission in 2005,[1] "Nanosciences and nanotechnologies are new approaches to research and development that concern the study of phenomena and manipulation of materials at atomic, molecular and macromolecular scales, where properties differ significantly from those at a larger scale."

Over the last three to four decades, these approaches have emerged as a very fast expanding research area at the confluence of physics and chemistry, with a wide range of interactions with other scientific fields, such as life sciences, optics, electronics, and a broad spectrum of potential applications in many areas, such as information and communication technologies, manufacturing of materials, transportation, instruments, energy, medicine and healthcare, security, food, agriculture, water, and the environment.

However, the wide breadth and sharp growth of this key emerging scientific and technological area have resulted in debates about the very definition of nanotechnology and nanomaterials. Ten years ago, the difficulty to reach a consensual definition of nanoscience and nanotechnology was indeed well illustrated by the inaugural issue of *Nature Nanotechnology*.[2] Nevertheless, across the variety of perspectives reflected by the 13 recognized scientists, industrialists, and stakeholders who contributed to the "Feature" section of this issue, a few common points appear: (i) the size of the studied objects, typically 1–100 nm in one or the other of their dimensions; (ii) the very close links between science and technology, that is, between understanding and modeling the specific and often unexpected physical, chemical, or biological properties of nano-objects and the capacity to manipulate individual atoms and molecules and to manufacture nanostructures, nanodevices, and nanosystems; (iii) the broad spectrum and the magnitude of the potential applications and the subsequent need to consider and assess the diversity of impacts of nanotechnology.

1) EU Commission (2005) *Nanosciences and Nanotechnologies: An Action Plan for Europe 2005–2009*. Communication from the Commission to the Council, the European Parliament, and the Economic and Social Committee, 12 pp.
2) T. Feith et al. 2005. Feature: Nano·tech·nolo·gy n. *Nat. Nanotechnol.*, **1**, 8–10.

Since 2006, the number of programmes, projects, and publications in nanoscience and nanotechnology has continued to increase at a very fast pace. For example, the number of papers that fall under the category *Nanoscience & Nanotechnology* of the *Core Collection* of the *Web of Science* has nearly quadrupled between 2005 (9,445 papers) and 2015 (37,751 papers). During the same period, the number of these publications that also refer to agriculture, agronomy, food, and veterinary sciences has even grown faster, by more than 10×. However, it should be noted (i) that the latter number remains fairly low (about 2–3%), compared to the total number of publications in nanoscience and nanotechnology or to the total number of papers in food, agriculture, and veterinary sciences and technologies; and (ii) that the rise of publications in nanotechnology applied to the agrifood sector started in the late 1990s, about 10 years later than in most other sectors.

The potential applications of nanotechnology in the agriculture and food sectors are manifold: from the development of sensors for monitoring the environment to the treatment of wastewater and the remediation of contaminated soils; from increasing crop yield (e.g., nanopesticides or nanofertilizers) to biosecurity (e.g., sensors for detecting pathogens along the whole food chain from the farm to fork); from cellulose-derived nanoparticles and new biomaterials to functional packaging; and from food processing to the delivery of specific food additives and ingredients. Some applications have existed for the last several years, but most of them are still under development, and it remains to be seen which will have a real economic impact. Moreover, natural nanoparticles have been existing for ever and natural nanomaterials are part of conventional food and conventional food processing.

Nanoscience and nanotechnology thus generate plenty of new opportunities for the agrifood sector and more widely for bioeconomy. Simultaneously they raise concerns about their potential impacts on environment and human health. These concerns are especially critical for the agrifood sector, because of the strong environmental footprint of agriculture and because food, along with water and air, is one of the major sources of exposition of humans to their environment. There is thus a strong need to develop nanotoxicology and nanoecotoxicology as new research areas, for example, by investigating the uptake and translocation of nanomaterials by the gastrointestinal tract. As for other technologies that have the potential to generate disruptive innovations, it is also worth assessing and monitoring the impacts of nanotechnology on the economic and social organization of the agrifood sector.

As underlined by the Joint Ethics Committee of Inra and Cirad in 2012,[3] agricultural and food scientists should not only contribute to understanding and predicting the specific properties of agricultural and food products related to their nanoscale structure, to exploring the potential applications of nanotechnology in the bioeconomy, including the assessment of their environmental,

---

3) Comité d'éthique Comité consultatif commun d'éthique pour la recherche agronomique (2012) *Avis n°4 sur les nanosciences et les nanotechnologies*, Inra & Cirad, Paris, 33 pp.

health, social, and economic impacts, and to informing the agencies in charge of the regulation of new agrifood nanomaterials and nanoproducts. They should also more broadly inform the society and interact with the citizens. This book is therefore most welcome, as it brings together various sources of expertise on the different aspects related to the application of nanoscience and nanotechnology in the agrifood sector.

Chairman of French AllEnvi Alliance *François Houllier*
and former President and CEO of INRA
Montpellier, August 21, 2016

Introduction

Due to the growing world population and increasingly varying climate change, leading to lower yields and increasing harvest losses, feeding global population has become an international major issue. The food resources from the field to fork need to be used wisely, with minimum waste and maximum nutritional efficiency.

For this purpose nanotechnologies can play an important role. It is envisaged that the convergence between nanotechnology, plant science and agriculture will lead to revolutionary developments and advances in the next decades to improve food security and sustainability through, for examples, the re-engineering of crops at cellular level, the precision agriculture leading to water and nutrient control for more sustainable farming, the identification systems for tracking plants from origin to consumption or through the precise and the controlled release of fertilizers and pesticides, etc. In the domain of food technology, nano-biosensors will contribute to the identification of harmful molecules such as toxins or pesticides and to quick identification of spoilage processes in food. The development of nanoscience-based food with improved nutritional and palatable benefits will allow to increase food nutritional efficiency and the addition of nanoscale materials for food packaging will extend shelf life and retain quality, both contributing to waste reduction.

As for all new technologies, their application offers great potential but raise ethical questions, and when food is concerned, issues on food safety, risk and benefits, and consumer mistrust become the key ones.

This book provides detailed coverage on the state of the art and the importance of nanoscience in agriculture and food and highlights the perspectives of a science-based nanotechnology in these domains in the future. Through concrete examples, it points out the major role of nanotechnology in the improvement of food supply and in studies and applications ranging from agricultural processes and productivity to nutritional improved foodstuffs, including packaging materials for more effective storage and for secure tracking from source to the consumer to reduce spoilage. It details means to ensure safety for human and for the environment to address the current evolution of the European science policy.

The book will be of interest to students of agriculture and food sciences, physics, chemistry, and biosciences, as well as those working or planning to work in the restaurant and hospitality sectors. It will be of value to food scientists, policy makers, agrochemists and industrialists, and those with a role in consumer bodies, associations, and government agencies. Nanofoods has a global dimension, of particular importance for Europe, Japan, and the United States, but also becoming important for highly populated countries such as China, India, and South America. "Nanofoods" is becoming a hot topic and this book provides clarity and confidence.

Monique Axelos
Marcel Van de Voorde

**Part One
Basic Elements of Nanofunctional Agriculture
and Food Science**

1
Nanotechnologies for Agriculture and Foods: Past and Future

Cecilia Bartolucci

National Research Council of Italy, Foresight Group, Department of Chemical Sciences and Technology of Materials, Institute of Crystallography, Via Salaria km 29.300, Rome 00015, Italy

Nanomaterials and nanoparticles are not an invention of the twentieth century. Examples of nanostructured materials can be found throughout the fourth to the seventeenth century. Important examples are vividly colored stained glass windows in European cathedrals obtained through the use of gold nanoparticles; silver or copper nanoparticles used in the Islamic world to give luster to their ceramics; and finally carbon nanotubes and cementite nanowires present in the famous Damascus saber blades. These materials, showing unusual characteristics, were generally produced empirically by talented craftsmen, often through the use of high temperature.

The intentional manipulation at atomic level or molecular scale to manufacture nanoparticles or nanostructured materials, however, requires the understanding and the control of matter at dimensions between 1 and 100 nm, approximately, and was possible only after the advent of high-powered microscopes, in particular the scanning tunneling microscope by Gerd Binning and Heinrich Rohrer in 1981, which for many marked the birth of nanotechnologies. From that moment, tools were developed that allowed imaging, measuring, modeling, and manipulating matter at nanoscale to achieve altered characteristics that could differ greatly from those on the macroscale. One should talk about nanotechnologies only if the correlation between the nanostructure of the novel materials and the resulting highly unique properties is recognized and deliberately applied. This criterion excludes naturally occurring nanoparticles and hence naturally formed biomolecules and material particles, and separates these from the particles resulting from nanotechnological applications. It is also clear from the above description of nanotechnologies that these encompass a whole group of different technologies and involve many different disciplines. Soon, several countries recognized the applicability of nanotechnologies in several different sectors such as medicine, biotechnology, electronics, materials science, energy, and more. In 2000, the US National Nanotechnology Initiative

Nanotechnology in Agriculture and Food Science, First Edition. Edited by Monique A.V. Axelos and Marcel Van de Voorde.
© 2017 Wiley-VCH Verlag GmbH & Co. KGaA. Published 2017 by Wiley-VCH Verlag GmbH & Co. KGaA.

(NNI) was created to support this highly interdisciplinary technological development, while in 2009, the European Commission recognized nanotechnology as one of the six key enabling technologies [1]. Several developing countries such as India, Brazil, South Africa, Thailand, the Philippines, Chile, Argentina, and Mexico invested millions in pursuing nanotechnologies during the first decade of the twenty-first century, while in 2005, the number of nanotechnology patent applications from China ranked third, behind the United States and Japan [2].

While consumer products making use of nanotechnologies and engineered nanomaterials began appearing on the marketplace in everyday products such as cosmetics, clothing, sporting goods, and computer processors, the applications in the agriculture and food sector lagged behind. The main reason for this different development is probably due to different levels of risk/benefit factors attributed to distinct applications. In fact, while the benefits due to the use of nanotechnologies in medicine are, despite possible risks, recognized as being very important by most stakeholders, including consumers, the applications of engineered nanomaterials or nanoparticles in, or around food cause alarm. In 2004, Britain's Royal Society and the Royal Academy of Engineering published a report [3] in which they illustrated not only the opportunities provided by nanotechnologies but also the necessity for an open debate and the need to address uncertainties about the health and environmental effects of nanoparticles. They also recommended the evaluation of nanospecific regulations. In 2011, the European Commission published a "Recommendation on the definition of nanomaterials," which uses size as the only defining property of the material (i.e., size range 1–100 nm) [4]. Regulations on food information followed soon afterwards [5], requiring indication of nanomaterials in the list of ingredients. Specifically, the ingredients to be labeled "XX (nano)" are "engineered nanomaterials," further characterized as "any intentionally produced material," with size on the order of 100 nm, or above, retaining "properties that are characteristics of the nanoscale." These characteristics are related to the large specific surface area, and/or physicochemical properties different from those of the nonnanoform of the same material [5]. The discrepancies between the recommendation and the regulation underline the regulatory uncertainties regarding nanolabelling, uncertainties which still exist also outside the European Union.

In the past years, there have been great national and international efforts in developing risk assessment and risk management approaches that propose and implement strategies to identify potential hazards. Today, the need for a differentiated debate involving all actors is becoming increasingly necessary. For years, the word "nano" has been used as an advertising tool by both supporters and detractors of nanotechnologies. The former used it to underline unprecedented, possibly all-resolving characteristics; the latter as an overall warning sign. In particular, in the public perception, anything "nano" applied to agriculture and food runs counter-current to the trends on "organic," "natural," and "environment-friendly." A study conducted in 2012 [6] showed that while there was an increased effort in addressing the complexity of the "nano issue" among the experts community (both scientists, policy makers, and regulatory bodies), the results were communicated insufficiently, the processes were less transparent, and the industries remained, or became silent. As a consequence, the knowledge among consumers regarding nanotechnologies and the benefits of their applications decreased, while uncertainty and expectation of the risks to health and environment increased significantly. It has been shown [7] that communication of scientific uncertainty for a given risk will give rise to a disproportionate increase in the seriousness of risk perception. The type of risk is less important; the uncertainty itself and the trust in the source of information are critical to risk acceptability.

In their first axiom about communication, Paul Watzlawick *et al.* [8] said: "One cannot not communicate." Even if communication is being avoided, this is a form of communication and leaves room to the development of one's own frames and patterns, and often nourishes mistrust in those who fail to communicate. The success of technological innovation, particularly in a field as close to the consumer, both literally as well as emotionally, as food, is tightly linked to consumer acceptance. Therefore, it is mandatory to reinforce communication and transparency, as well as every form of knowledge acquisition and education. Inventions need to provide real benefits and they become innovations only if they can be adopted effectively by users or other parties to improve what they are doing [8]. The needs of society are the most important drivers for responsible development and innovation. Whenever addressing technological advances, technologies should never be the starting point. The key question should be how their applications can benefit a broad community and which societal needs they can address.

Considering the projected increase in the world's population in the next decades, some of the greatest challenges to mankind will be to sustainably and equitably provide better living conditions, to deliver vital goods and services, and to support human health and well-being. Few studies consider the interaction between all these challenges. However, in the future, it will be imperative to address them in a concerted way and design strategies that will support a more holistic approach. In particular, there is a need to provide global food security.

Food is a necessity for all, making each of us a stakeholder in this important sector. It is such a critical need that the implications connected to food security are enormous, and extend from physical and mental health and well-being to development, economy, migration, and conflict. While the demand for food may increase 70% by 2050, the production of food worldwide has a high impact on natural resources on which it is fundamentally dependent. In a very recent report on "Food Systems and Natural Resources" [9] the UNEP provides evidence of unsustainable and/or inefficient practices used globally by current food systems. According to this report, 33% of the world's soil is moderately to highly degraded; at least 20% of the world's aquifers are overexploited; 60% of global terrestrial biodiversity loss is related to food production; over 80% of the input of minerals does not reach consumers' plates, implying very large nutrient losses to the environment. It is clear that one of the greatest challenges of our time is to address both food security and sustainability. If we want to ensure food security while maintaining healthy ecosystems, we also need to consider climate change, weather variability, and possible increase in the number of extreme events, habitat loss, constraints in available water and energy resources, competition for arable land and urbanization, as well as the use of fertilizers and other inputs, which constitute huge challenges on the resilience of the food system. Furthermore, a changing population, not just a growing one, also poses a challenge to meeting the growing global demand for food and nutrition, thus further complicating the system. Higher average incomes, urbanization, a more aged and more educated population are all factors that will contribute to increased food consumption and dietary changes, with a greater proportion of resource-intensive food such as meat and dairy products [10]. In fact, according to the World Resource Institute one-third of the expected growth on food demand will be attributed to the increased purchasing power [11].

Meeting food security requires addressing availability, access, and utilization over time. The World Food Summit of 1996 defined food security as existing, "when all people at all times have access to sufficient, safe, nutritious food to maintain a healthy and active life." Currently, the global food production, which has almost tripled during the past 50 years [12], is enough to feed the entire world population (2700 kcal/person/day produced vs 1800 to 2200 kcal/person/day required) as estimated by the World Health Organization. Yet, more than 800 million people face hunger daily, and over 2 billion still suffer from vitamin and mineral deficiencies, in particular iron, vitamin A, iodine, followed by zinc, folate, and calcium [13]. An estimated 162 million children experience stunted growth, reflecting chronic undernutrition during the early stages of life. This phenomenon, which occurs predominantly during the first 2 years of life, causes mental and physical growth failures. Simultaneously, 42 million children under 5 years of age are overweight, and two-thirds of these children reside in low- and middle-income countries. Globally, more than 2 billion people are overweight or obese [14], conditions that are linked to an increase in chronic diseases such as diabetes, cardiovascular disease, and cancer.

Clearly, there is something seriously wrong with our current food systems. Recent Foresight studies [9,15–17] agree that there is an urgent need to address critically the failures of the present agriculture and food sector and that substantial changes throughout the whole system will be required. The concept of a food system approach is recurrent among experts involved in the great challenge of a sustainable food production, able to ensure food security. There is also consensus in the demand to acknowledge that without mitigation of climate change and maintaining biodiversity and ecosystems services, there is no chance to achieve sustainability.

Rather more complicated is the evaluation of supply and demand projected to 2050. While the numbers regarding population growth recurrently point to an increase from nearly 7.2 billion today to 8 billion by 2030, and more than 9 billion by 2050 under a medium growth scenario [18], numbers regarding the quantification of potential demand and necessary supply are often divergent, even though all point to the need to increase production. These variations are due to applying different measures and premises and considering different drivers of demand and supply such as population growth, income growth, socioeconomic development, climate change, and bioenergy expansion. In a recent study [19], it was shown that while results depend largely on the chosen scenarios, variations in food demand are more sensitive to socioeconomic assumptions than other factors such as climate change. The most frequently found number, however, is a required increase in food production of 70%, based on a paper written in 2009 by FAO [20]. Since then, there have been papers reporting the required increase in the production of specific agricultural products, ranging from 45% for cereals to 89% for oil crops [21]. Murray, from the Institute for Health Metrics and Evaluation, went even further and showed, while examining current global diets and human requirements, that the current production of certain food items is higher than what is required for a healthy diet [22]. In particular, the production of whole grains and fish is currently 50% higher, while the production of red meat is 568% higher. Hence, according to Murray, an increase in food production is currently necessary only for certain items, for example, vegetables by 11%, seeds and nuts by 58%, fruits by 34%. This aspect is very important, because it directly links supply to nutrition and health and introduces the factor quality where usually only the factor quantity is taken into consideration.

Quantity is, in fact, the leading aspect in most considerations about food security, and increase in crop yields is one of the main targets. Agricultural productivity is usually evaluated using the standard definition of yield, which is in tonnes per hectare (or similar units). Cassidy *et al.* in their paper proposed to calculate agricultural productivity by determining the actual food delivery, expressing yield in calories of human-consumable product per hectare, or people nourished per hectare [23]. This is one way to stress food availability, considering that crops are allocated to different uses besides just food. Currently, 36% of the calories produced by crops are being used for animal feed, and eventually only 12% of these feed calories contribute to animal product calories [23]. To further complicate the picture, human-edible crops are used to produce biofuel, for example, in 2010 United States and Brazil combinedly used 6% of global crop production (by mass) for this purpose [24]. Cassidy *et al.* argued that by growing crops exclusively for human consumption, global calories availability could be increased by as much as 70%, enough to feed additional 4 billion people.

The focus on food quality can be pushed even further by considering the real nutritional value of food in addition to merely calculating the calories. The scientific world is becoming increasingly aware of the link existing between health and diet, and of the importance of a nutritious, diversified diet. The providing of sufficient calories does not protect from malnutrition, which is often caused by micronutrients deficiencies. Further proof of a diet rich in "empty calories" is the constant increase in obesity rates in poor communities and underdeveloped countries [25]. The food quality, defined as the nutritional value of the food, is essential in providing a healthy diet.

While there is a greater awareness of the complexity of the food system(s) and the necessity to address all processes starting even before the production of raw materials and running through the whole food chain to food loss and food waste, there is still a need to bridge the gap between theory and implementation in this crucial sector. No single technology should be advertised as a panacea. However, nanotechnologies can have a disruptive impact at every step of the food chain, provided that other technologies, such as biotechnology, system biology, and information and communication technology, converge toward its development and application. It is imperative however to analyze the opportunities offered by the introduction of such technologies with a forward looking approach, and to define a medium and a long-term vision, in order to elaborate coherent research strategies. Four main areas of application for nanotechnologies in the agricultural and food sector had been originally identified, namely, agriculture, food processing, food packaging, and supplements; but the most exciting innovation in nanoscience should be investigated at the intersection of these areas. This kind of research requires a highly interdisciplinary system approach and encourages the transfer of knowledge from one sector to another. Nanotechnologies are *per se* inter- and transdisciplinary and therefore best suited for this endeavor.

The new food system will have to be driven by the necessity to create a balance at many different levels. It will be mandatory to find a balance between demand and supply, quality and quantity, and the needs of the developed and the developing world, just to mention a few. The spread and implementation of existing knowledge and technologies can already contribute to addressing these challenges, however, investments in research and development will be essential.

A key role for addressing food demand is often assigned to input intensification. According to preliminary results from the GFWS platform [16], a continued increase in crop yield productivity by at least 0.5% per year should be sufficient to meet food requirements of a crop-based food supply by 2050. This goal can be reached only with an increased use of fertilizers and water, which would put an unsustainable stress on our planet [26]. In fact, the production of nitrogen fertilizers is not only highly energy-intensive but also contributes considerably to greenhouse gas emissions. Fertilizer technology is 100 years old. Still, the fertilizer – nitrogen – use efficiency by crops is not more than 30–50%. The remainder is lost via volatilization, denitrification, leaching, and stabilization into soil organic matter. It will therefore be necessary to find new ways to deliver the nitrogen, essential to food production. In the past years, there has been a remarkable development in nanoagrochemicals, which include nanofertilizers and nanopesticides, and there have been increasing incentives in the scientific community to develop nanoproducts that are more efficient and less harmful to the environment compared to conventional agrochemicals. Nanotechnology could support the development of new products offering benefits such as increased efficiency, durability, and reduction in the amount needed. While researchers were originally interested in inorganic nanoagrochemicals, organic-based nanomaterials such as nanodelivery systems used in medical applications are now being investigated intensively. Equally interesting are nano-enabled formulations, for example, emulsions or microcapsules showing a well-defined nanopore network. Eventually, these new particles and formulations should allow the introduction of an essential functionality: the synchronization between crop demand and release of required inputs.

Multifunctional nanomaterials could provide this intelligent feature of synchronizing demand and response, a characteristic which is also of great advantage in addressing other problems, for example, water supply. Agriculture, including irrigation, livestock and aquaculture, is responsible for 70% of water withdrawal [27]. It also contributes to the pollution of groundwater through the use of pesticides, fertilizers, and other chemicals. Nanotechnological applications could offer new and affordable solutions for the remediation and purification of water: ligodynamic metallic nanoparticles, nanoporous fibers, and nanoporous foams are being developed to be used in microbial disinfection; nanocomposite membranes offer a low energy alternative for desalination; functionalized ligand-based nanocoatings will soon be available for the removal of heavy metals. In the future nanodevices, delivery systems and nanocapsules could play an important role in the controlled release of water in response to different signals, and, linked through a network of nanosensors, they could eventually support the diffusion of precision agriculture, which combines accurate data collection with a controlled response.

A wireless monitoring system developed through nanosensors is also a tool that can be used to address environmental stresses and crop conditions, allowing for responses that are optimized for the needs of specific plants, soil, and climate conditions. The result would be a more tailored and on-demand supply of inputs and a more controlled decision-making process, which could greatly contribute to a sustainable use of resources.

Also starting at the field level is the improvement of the quality of the food that we consume. Within the food chain, the protection or the introduction of nutrients should start as early as possible. Taking the whole system into consideration, all possible points of improvement should be identified. Technologies that until now have been used mainly to enhance productivity in terms of quantity should be used and developed to enhance quality. The quality of soil can be improved through the application of intelligent nanoagrochemicals, which would avoid temporal overdose and reduce the amount of input needed, minimizing impact on environment and reducing waste. The quality of the crops could be enhanced by using nutrient delivery systems, allowing for a targeted uptake from roots and leaves.

Nutrient fortification, for example, through micro- and naonoencapsulation is yet another way to enhance the quality of the raw materials, protecting the targeted compounds and increasing their bioavailability. Processing technologies using nanodelivery systems or nanoemulsions could intervene at a later stage in the production of food products. In fact, smaller particle size confers improved bioavailability of bioactive agents, while nanoemulsions offer a preferred means of fortifying aqueous products with functional ingredients. Nanotechnologies not only provide the means to add active ingredients to food, they also allow the production of foods with reduced fat or salt content or the creation of new foods with novel textures, flavors, and tastes. While all these attributes are part of food itself, it is important to keep in mind that in the creation of new nanoproducts there should be a real benefit, a real added value. It is the responsibility of the scientific community and regulatory bodies to show the benefits and the safety of products for the consumer. Validation procedures and safety tests will need to be introduced at each step of the chain, in particular when new technologies or new materials will be used. There is still a need to acquire a basic knowledge of food structures on the micro- and nanoscale, and of the existing link between raw material, food processing, and food structures. One should also consider a reverse approach "from fork to farm," starting with the analysis of food absorption, particularly at gut level, and back to the structure dynamics, checking the efficiency and safety of the proposed solution.

Dietary requirements are varied and in order to have adequate nutrition, we will need to be able to monitor changes in metabolism, and to evaluate nutrient needs in a dynamic way that takes into consideration the complexity of the whole system. The tools used to achieve that should eventually reach the consumer and deliver the necessary information to allow knowledge-based decisions. Here too intelligent, responsive nanosensors could play an important role and support the development of a preventive, personalized nutrition in combination with a preventive and personalized medicine. The stress lies on prevention and it is mandatory that scientific-based knowledge and new communication strategies support a change in attitude that also recognizes a scale of action. Acknowledging the current longer life expectancy, it is important to realize that investing in healthy nutrition today will result in a better quality of life tomorrow.

Highly nutritious, healthy food should be strictly connected to safe food. Along the whole food chain there are points of intervention where nanotechnologies can not only help in identifying contaminated or spoiled food, they can also provide tools to prevent contamination and spoilage. Nano- and biosensors, connected or not to a remote sensing system, can monitor soil conditions (moisture, soil fertility, nutrients, etc.) as well as pathogens, insects, and weeds. They can provide information about when and how much pesticide or herbicide needs to be administered, eventually triggering an *in situ* response only when and where those substances are really needed, thereby avoiding overuse and unnecessary exposure of nontarget organisms. Applications of nanotechnology-enabled gene sequencing could also contribute to the effective identification and utilization of plant trait resources, improving their capability to react against environmental stresses and diseases.

Nanomaterials can also play a fundamental role in maximizing food safety both during processing and during packaging. Coatings for food production machinery (e.g., biofilm formation), nanostructured sieves, filters and membranes (e.g., enabling cold sterilization), nanostructured as well as nanoscale adsorbents and catalysts are only few of the applications providing benefits in food processing. In particular, nanofood contact materials can add novel self-cleaning and antiseptic properties, useful in the production of safe food. In addition to surface biocides, nanotechnology in plastics and bioplastics in packaging can provide improvement in barrier properties and greater protection and preservation of food, and facilitate new and more efficient active functionalities. One property necessary for promoting food safety is traceability, which allows, for example, the removal of all tainted products from the market and the system during a recall process. It also ensures authenticity, adding value to the product. One could envisage, in the future, the placing of nanodevices not only on the packaging, but embedded inside the food, or even inside the raw materials allowing consumers to trace back the origin of all ingredients and providing information about the processes used to produce the products.

Food packaging applications currently form the largest share of nano-enabled products in the food sector on the market. They provide an opportunity "to do things right" at all levels, scientific, regulatory, and social, as well as at the economic and market level. Consumers are more open to accepting nanotechnologies applied outside the food, for example, packaging, rather than in the food, for example, nanoemulsions or nanoencapsulation. It is the responsibility of all stakeholders to implement methods to measure exposure and toxicity, and to develop risk-benefit assessment procedures, including impact on humans and environment. Developed countries should also adapt the innovation to the specific requirements of the food market in developing countries.

In fact, the needs and practices in developed and developing countries are quite different and often divergent. Potentially, successful technological innovation should be adaptable to different realities. A meaningful example is given by food waste, which plays a fundamental role in sustainability. It has been estimated that about 30% of all food grown worldwide is lost or wasted. Loss occurs mainly at the farm end in developing countries, while waste is produced at the fork end in developed countries. In the Foresight report of the British Government the importance of reducing food loss and waste is quantified in this statement: "Halving the total amount of food waste by 2050 is considered to be a realistic target . . . If the current global estimate of 30% waste is assumed, then halving the total could reduce the food required by 2050 by an amount approximately equal to 25% of today's production." [15] There is waste, and consequently potential for improvement at every stage, adding value to waste by enabling its usage. Through the new acquisition of knowledge, the application of innovative technologies, and especially through a better-developed system approach, we should be able to substantially minimize waste. In high-income communities, the introduction of sensor technologies described for quality testing and traceability should persuade consumers to rely on more specific information rather than the "best before" label, responsible for a great amount of wasted food. Waste is generally associated with quantity not quality. However, emphasizing nutritious, quality food is an essential component in the reduction of waste. Innovation and education should be combined and used to induce a permanent cultural change in behavior. This might take generations, but would have a great impact both on society's health and on sustainability. What is essential in developing, low-income countries is to reduce post-harvest loss. Technologies, including nanotechnologies, developed for precision agriculture apt to avoid spoilage and waste of inputs can contribute to reaching this goal. Furthermore, food-contact nanomaterials for more efficient, safer processing and storage, and all technologies that support agricultural practices more adaptable to environmental and climate changes, could contribute to reducing food loss.

Developing countries face larger barriers regarding the applications of nanotechnologies than the developed countries for several reasons, including lack of funding and human capacity. Prioritization is hence particularly important and funding programs should focus on those applications that could provide maximum benefit-risk ratio for the poor [28]. Nanotechnology has the advantage that it often does not require technological expertise to be adopted. The final users, who eventually determine the acceptability of any new technology, need to know it exists and what its purpose is. However, the implementation of safety regulations could put an additional strain to poorer societies. Efforts should be put into communicating and knowledge sharing between developed and developing countries, in order to avoid a "nano divide," shifting the focus of nanotechnologies applications even further away from the necessities of the poor.

One of the pillars of food security is availability. Producing more food, particularly more nutritious food, minimizing waste, and changing dietary habits are measures that all need to be tackled at the same time. Still, it would not be enough to ensure food security unless one ensures that food reaches everyone. In recent decades, and in many countries, food production has evolved into an ever more centralized model. While this has generated notable advances in productivity, enabling us to produce enough calories globally, it has failed to distribute adequately the food produced and hence to meet the nutritional needs of our societies. There is now a need to decentralize food production and give greater priority to rural development. A decentralized production, able to use local resources would also be more easily adaptable to specific requirements posed by environment, health, and diverse economic, cultural, and social challenges. This should allow for greater availability and affordability of different nutrient sources, which supports both the concept of a food supply tailored to specific needs and an on-demand production. Ideally, food should be produced where needed and in the quantity needed. In the food system a responsible, evidence-driven adoption of nanotechnologies integrated with other converging technologies can greatly contribute to a distributed and networked food production supply. However, to increase public acceptance, it is imperative to evaluate carefully the use of new technologies, particularly of nanotechnologies, and to assess the risk of new nanomaterials. The assessment of food value chain sustainability should integrate natural, social, and political sciences and also consider "nontraditional" sustainability dimensions such as health and ethics.

It is urgent that we address food safety and sustainability problems. In doing so, we have a chance to review how we produce and consume food and radically change our approach. A more holistic view will be necessary, requiring a highly inter- and transdisciplinary food system approach. Knowledge sharing and knowledge transfer are prerequisites for such a change, and communication among different sectors will be necessary in order to integrate innovative technologies into social change. This is what this book is about.

References

1 Commission of the European Communities (2009) Preparing for our future: developing a common strategy for key enabling technologies in the EU. COM/30.09.2009/512 final, Brussels.
2 Salamanca-Buentello, F., Persad, D.L., Court, E.B., Martin, D.K., Daar, A.S., and Singer, P.A. (2005) *PLoS Med.*, **2** (5), e97.
3 The Royal Society and the Royal Academy of Science (2004) Nanoscience and nanotechnologies: opportunities and uncertainties. Available at http://www.nanotech.org.uk/finalReport.htm (accessed June 10, 2016).
4 European Commission (2011) Commission recommendation of 18 October 2011 on the definition of nanomaterial. *Official J. Eur. Union*, **L275**, 38.
5 European Commission (2011) Regulation (EU) No. 1169/2011 of the European Parliament and of the Council of 25 October 2011. *Official J. Eur. Union*, **L304**, 18.
6 Grobe, A., Rissanen, M., Funda, P., De Beer, J., and Jonas, U. (2012). Available at http://www.dialogbasis.de/fileadmin/content_images/Home/Consumerstudy_Nano_20125 Summary_EN.pdf (accessed June 10, 2016).
7 Miles, S. and Frewer, L.J. (2003) Public perception of scientific uncertainty in relation to food hazards. *J. Risk Res.*, **6**, 267–283.
8 Watzlawick, P., Beavin Bavelas, J., and Jackson, D.D. (1967) *Pragmatics of Human Communication: A Study of Interactional Patterns, Pathologies, and Paradoxes*, Norton, New York.
9 UNEP (2016) Food Systems and Natural Resources: A Report of the Working Group on Food Systems of the International Resource Panel. Westhoek, H., Ingram J.,Van Berkum, S., Özay, L., Hajer, M., United Nations Environment Programme .
10 Sonnino, A. (2015) Meeting the growing global demand for food and nutrition: current situation and outlook, in *World Food Production, Facing Growing Needs and Limited Resources*, Vita e Pensiero/Ricerche, Milano, Italy, pp. 55–71.
11 WRI (2013) Creating a sustainable food future: interim findings, World Resources Institute, Washington DC.
12 FAO (2016) Database collection of the Food and Agriculture Organization of the United Nations, faostat3.fao.org/home/E (accessed June 5, 2016).
13 FAO (2013) The State of Food and Agriculture: food systems for better nutrition, Food and Agriculture Organization of the United Nations, Rome.
14 WHO (2014) WHO Facts Sheet No. 311, reviewed May 2014, World Health Organization, Geneva.
15 Foresight (2011) The Future of Food and Farming. A Foresight Report. Available at http://www.gov.uk/government/publications/future-of-food-and-farming (accessed June 11, 2016).
16 Quentin Grafton, R., Williams, J., and Jiang, Q. (2015) Food and water gaps to 2050: preliminary results from the global food and water system (GFWS) platform. *Food Secur.*, **7** (2), 209–220.
17 Bartolucci, C. (2016) Report "Diversified Adaptable Food", Science and Technology Foresight, National Research Council of Italy. Available at www.foresight.cnr.it (accessed June 15, 2016).

18 United Nations, Department of Economic and Social Affairs, Population Division (2013) World Population Prospects: The 2012 Revision, Volume II, Demographic Profiles.

19 Valin, H., Sands, R.D., van der Mensbrugghe, D., Nelson, G.C., Ahammad, H., Blanc, E. *et al.* (2014) The future of food demand: understanding differences in global economic models. *Agr. Econ.*, **45**, 51–67.

20 FAO/Food and Agriculture Organization of the United Nations (2009) How to Feed the World in 2050. Available at http://www.fao.org/fileadmin/templates/wsfs/docs/expert_paper/How_to_Feed_the_World_in_2050.pdf (accessed June 14, 2016).

21 Alexandratos, N. and Bruinsma, J. (2012) World agriculture towards 2030/2050: the 2012 Revision. ESA Working Paper No. 12-03, June 2012, Agricultural Development Economics Division, Food and Agriculture Organization of the United Nations. Available at http://www.fao.org/economics/esa (accessed June 14, 2016).

22 Murray, C.J.L. (2014) Institute for Health Metrics and Evaluation. Available at www.healthdata.org (accessed June 5, 2016).

23 Cassidy, E.S., West, P.C., Gerber, J.S., and Foley, J.A. (2013) Redefining agricultural yields: from tonnes to people nourished per hectare. *Environ Res. Lett.*, **8**, 034015.

24 Food and Agricultural Policy Research Institute (FAPRI) (2011) World Biofuels: FAPRI-ISU 2011 Agricultural Outlook. Available at http://www.fapri.iastate.edu/outlook/2011/ (accessed June 15, 2016).

25 WHO/World Health Organization (2016) Available at http://www.who.int/mediacentre/factsheets/fs311/en/ (accessed June 15, 2016).

26 Rockström, J., Falkenmark, M., Allan, T., Folke, C., Gordon, L., Jägerskog, A. *et al.* (2014) The unfolding water drama in the anthropocene: towards a reslience-based perspective on water for global sustainability. *Ecohydrology*, **7**, 1249–1261.

27 FAO/Food and Agriculture Organization of the United Nations (2016) Available at http://www.fao.org/nr/water/aquastat/water_use/index.stm (accessed on June 15, 2016).

28 IFPRI/International Food Policy Research Institute Gruère, G., Narrod, C., and Abbott, L. (2011) Agriculture, Food and Water Nanotechnologies for the Poor: Opportunities and Constraints. IFPRI Policy Brief 19.

2
Nanoscience: Relevance for Agriculture and the Food Sector

Shahin Roohinejad and Ralf Greiner

Max Rubner-Institut, Federal Research Institute of Nutrition and Food, Department of Food Technology and Bioprocess Engineering, Haid-und-Neu-Straße 9, Karlsruhe 76131, Germany

2.1
Introduction

The practice of agriculture is the process of producing food and feed by the cultivation of different crops and breeding and raising of livestock. Agriculture (directly or indirectly) provides food for humans and is the backbone of most developing countries' economies and a key driver for their growth and progress. The population of the world is estimated to be 8 billion people by 2025 that shows the importance of agricultural productivity for global food security to feed a rapidly growing world population. Changes in climate, limitations of energy-resource, and fast-growing global population may put severe pressure on food and water resources [1].

The Food and Agriculture Organization of the United Nations (FAO) has predicted that the annual production of meat should increase to 455 million tons by 2050 to respond to the food needs of a growing population [2]. This predicted enhancing demand for meat puts further pressure on agricultural land, since farmers are required to grow crops to produce more animal feed. This has caused agriculture to become increasingly important as a source of income and food in a world of diminishing resources and enhancing population [1]. Thus, due to the increasing world population, it is essential to apply modern technologies such as nanotechnology in agricultural and food sectors.

New nanomaterials are now being developed for different applications and quickly entering all industries around the world. The increase in innovation of nanoscience in these fields is mainly attributed to the high investments made by government and private sectors in R&D [3]. For instance, the feasibility of using nanotechnology to revolutionize in health care systems, textile materials and products, biotechnology, cognitive sciences, information and communication technologies, energy, as well as food and agriculture sections has been publicized over the last years.

Nanotechnology in Agriculture and Food Science, First Edition. Edited by Monique A.V. Axelos and Marcel Van de Voorde.
© 2017 Wiley-VCH Verlag GmbH & Co. KGaA. Published 2017 by Wiley-VCH Verlag GmbH & Co. KGaA.

The application of nanotechnology in the agricultural and food industries was first reported by the US Department of Agriculture (USDA) roadmap published in September 2003. The prediction is that nanotechnology will transform the entire food industry, changing the way food is produced, processed, packaged, transported, and consumed. Since that time, several studies have shown the application of nanotechnology to monitor plant growth, detect plant and animal diseases, slow release of pesticides and developing diagnostic tools, development of functional food systems, produce interactive edible nano wrappers, targeted release of chemicals, packaging, extensive nano surveillance, and reduce waste for "sustainable intensification and increase global food production" [1,4]. Generally, the application of nanotechnology in the agricultural and food sectors has opened up new avenues to understand how physicochemical characteristics of nanosized substances can modify the structure, texture, and quality of foodstuffs and agricultural products [5].

Convergence of nanoscience with other disciplines is also providing additional innovations that expected to make a major influence on production, processing, storage, transportation, health, and safety of foodstuffs. For instance, nanotechnology integration with biotechnology and information technology (IT) has opened new opportunities for the development of nanobiosensors that are used for detection of pathogens and contaminants in food systems. Moreover, such technologies integration has lead to the development of an electronic tongue that can be used to describe the taste attributes of foodstuffs [5]. In this chapter, the fundamentals of nanoscience, applications of nanotechnology in the agriculture and food sectors, and the possible implications of such developments in relation to consumer safety are highlighted.

2.2
Fundamental of Nanoscience

Nanotechnologies involve the study and use of materials at very small scale, generally between 1 and 100 nm. Some materials can display different properties at this ultra-small scale compared to those have larger scale. Generally, nanomaterials and nanoparticles can have any of the following forms: nanoparticles, nanotubes, fullerenes, nanofibers, nanowhiskers, and nanosheets [6]. A nanoparticle is known as a separate entity that has three dimensions on the order of 100 nm or less [7]. There is no scientific explanation to support this specific upper limit. However, in order to avoid a multitude of conflicting definitions of "nanomaterial," on October 18, 2011, the European Commission (EC) has published a recommendation for a definition [8]. According to the recommendation, a nanomaterial means a "natural, incidental, or manufactured material containing particles, in an unbound state or as an aggregate or as an agglomerate and where, for 50% or more of the particles in the number size distribution, one or more external dimensions is in the size range of 1–100 nm." However, this recommendation is so far not used in legislation. According to the regulation (EU)

No 1169/2011 on the Provision of Food Information to Consumers "all ingredients present in the form of engineered nanomaterials must be indicated in the list of ingredients with the word "nano" in brackets" [9]. An "engineered nanomaterial" means any intentionally produced material that has one or more dimensions on the order of 100 nm or less or that is composed of discrete functional parts, either internally or at the surface, many of which have one or more dimensions on the order of 100 nm or less, including structures, agglomerates or aggregates, which may have a size above the order of 100 nm but retain properties that are characteristic of the nanoscale.

"Top-down" and "bottom-up" are two popular approaches for fabricating nanomaterials. Fabrication of nanomaterials using top-down approach is achieved by means of a physical processing of the food and agricultural materials, such as grinding and milling [10]. For instance, dry-milling method can be applied to achieve wheat flour of fine size with higher water-binding capacity [11]. Schematic representation of the bottom-up and top-down approaches to make nanoparticles is shown in Figure 2.1. Shibata [12] reported that the antioxidant effect of green tea can be improved by top-down size reduction method. Reducing the powder size of green tea to 1000 nm by dry milling resulted in an increase in the oxygen-eliminating enzyme activity. Other techniques such as homogenization, which is widely used in dairy science to reduce the size of fat

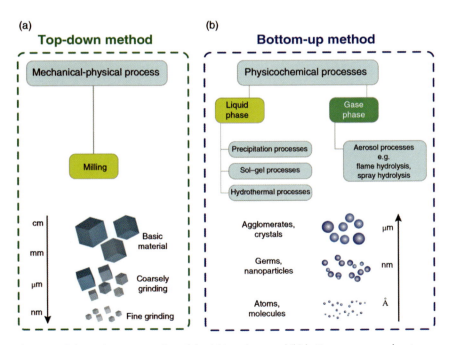

Figure 2.1 Schematic representation of the (a) top-down and (b) bottom-up approaches to make nanoparticles. (Adapted from Royal Society and Royal Academy of Engineering (2004) [13].)

globules, lasers and vaporization followed by cooling, are also considered as top-down size reduction methods [14].

Bottom-up manufacturing, which is based on atomic or molecular manipulation, is the alternative fabrication method of nanomaterials. This approach is normally applied to fabricate more complex molecular structures based on self-organization of biological components. In other words, these methods are used to arrange the molecules step by step to design the particles so that they have particular properties [6]. Approaches of bottom-up manufacturing include crystallization, chemical synthesis, layer-by-layer deposition, solvent extraction/evaporation, self-assembly, microbial synthesis, positional assembly techniques, and biomass reactions [14,15]. For instance, molecular self-assembly involves the application of supramolecular chemistry to allow the molecules to self-assemble into a particular configuration. The formation of casein micelles or starch, the folding of globular proteins, and aggregation of the proteins are examples of self-assembly structures that cause stable entities [10]. Nanometer scale self-organization can be obtained by providing a balance between the different non-covalent forces [16]. Compared with the top-down methods, bottom-up approaches are able to fabricate devices in parallel and are much cheaper, nanomaterials prepared by these methods could be overwhelmed as the size and complexity of the desired assembly enhances.

2.3
Applications of Nanotechnology in the Agriculture Sector

Applications of nanotechnology in materials science and biomass conversion technologies have revolutionized the agricultural industry to provide different foods, feeds, fibers, and fuels. Figure 2.2 shows the diagrammatic representation of nanotechnology applications in modern agriculture. For several years, the research and development on the applications of agricultural nanotechnology have been ongoing, attempting to provide solutions to challenges facing agriculture and environment such as sustainability, improving varieties, and increasing productivity. The increase in the number of scientific publications and patents has shown the importance of agricultural nanotechnology, particularly for management of diseases and crop protection [17,18]. Nanomaterials in agriculture have been suggested to decrease the amount of sprayed chemicals by smart delivery of active compounds, to minimize the nutrient losses in fertilization and to enhance the yields through optimizing the water and nutrient management [18]. Application of nanotechnology-developed devices is also under investigation in other agricultural fields such as plant breeding and genetic transformation [19].

However, despite the above mentioned potential benefits of nanotechnology, its applications in the agricultural sector have not been widely applied in the market compared with other industrial sectors. In other words, the application of nanoscience in this sector has been mostly claimed by the academic

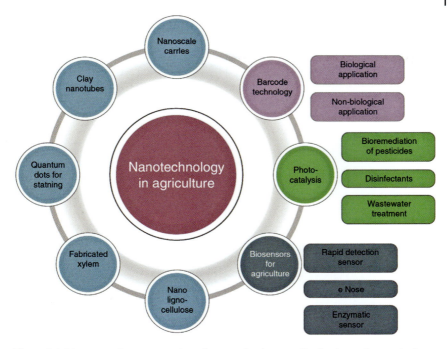

Figure 2.2 Diagrammatic representation of nanotechnology application in modern agriculture. (Adapted from Dasgupta *et al.* (2015). From [20] with permission © 2015 Elsevier Ltd.)

researches or small enterprises, whereas big industries have the main patent ownership. Although the trends of patent applications, especially from agrochemical companies, are constantly growing, no new products prepared by nanotechnology have reached the agricultural sector market. In plant-based agriculture, nanotechnology can be used for the precise control of manufacturing at the nanometer scale and a number of new possibilities in enhancing the precision farming practices are possible.

2.3.1
Delivery of Agriculture Chemicals

Different nanoscale delivery systems such as encapsulation and entrapment, polymers and dendrimers, surface ionic and weak bond attachments, can be applied to store, protect, deliver, and release the nutrients, fertilizers, pesticides, herbicides, and plant growth regulators in plant production processes [21]. These carriers in nanoscale have "self-regulation," which means that the medication on the needed amount only be delivered into plant tissue. One of the benefits of nanoscale delivery systems is its improved stability of the payloads to the environment degradation and increasing the effectiveness alongside with decreasing the amount applied. This reduction assists in addressing agricultural chemicals run-off and reduce the environmental consequence. In other words, controlled release mechanisms through nanoscale delivery systems avoid temporal overdose, reduce the use of high amount of agricultural chemicals, and minimize input and waste. Currently, different studies are being conducted to make novel delivery systems that can respond to environmental changes. The main aim is to tailor the products in such a way that they release their ingredients in a slow/targeted manner in response to the various signals such as pH, moisture, temperature, ultrasound, magnetic fields, and so on.

Application of nanoencapsulation and controlled release methods have revolutionized the use of pesticides and herbicides in agricultural section. These technologies are used for efficient transition of the chemicals to the plants/insects cells [15]. Effective penetration of chemicals through cuticles and tissues, which results a gentle and constant release of the active compounds, can be carried out by nanocapsules. Compared to traditional pesticides, particular nanoencapsulated pesticides have been designed to only kill the targeted insects, thus decreasing the effective dose. Although conventional pesticides are washed away in the rain, nanoencapsulated pesticides are adsorbed on the plant surface, which facilitate their prolonged release that takes longer [15].

Applications of nanosilver and titanium dioxide nanoparticles in management of plant diseases have been suggested [22]. It was previously reported that silver nanoparticles (at 100 mg/kg) could inhibit the growth of mycelia and conidial germination on cucumber and pumpkin against powdery mildew, the most damaging foliar diseases of cucurbits [23]. In another study, ethanolic suspension of hydrophobic alumina silicate nanoparticles was used for treatment of *Bombyx mori* leaves with grasserie disease and resulted a significant reduction in the viral load [24]. Combined with a smart delivery system, pesticides and herbicides can be used only when necessary, resulting in better production of plants and less injury to labours working in agricultural section.

The production of nanocapsules in smaller size will make them more potent. Numerous formulations that contain nanocapsules within the 100–250 nm size range have been fabricated, which are able to effectively dissolve in water more than existing ones, thus enhancing their activity [25]. Some companies fabricate suspensions of nanoscale particles (e.g., nanoemulsions, nanosuspensions), which can be either water- or oil-soluble and contain same suspensions of pesticidal or herbicidal nanoparticles in the range of 200–400 nm. These products can be easily used in different medias such as gels, creams, or liquids and have several applications for preventative measures or treatment/preservation of harvested products. Moreover, nanoencapsulation can be used for delivery of DNA and other chemicals into plant tissues to protect the plants against insect pests [19]. Diffusion, dissolution, biodegradation, and osmotic pressure with particular pH are the mechanisms of releasing encapsulated ingredients [26,27].

The use of chemical fertilizers plays an important role in improving crop production. Nanoencapsulation techniques can be applied to increase the efficiency of fertilizers and overcome the chronic problem of eutrophication. Nanoencapsulation of fertilizers can be performed in three different ways. The nutrients can be encapsulated inside nanoporous materials, coated with thin polymer film, or delivered as particles or emulsions of nanoscale dimensions [28]. Application of nanofertilizers can decrease the nitrogen loss because of the leaching, emissions, and long-term incorporation by soil microorganisms. Controlled or the slow release of the fertilizers may also improve soil by reducing the toxicity of fertilizers usage [29].

2.3.2 Nanosensors/Nanobiosensors

Nanobiosensors can be effectively used in agriculture sector for sensing a wide range of materials including fertilizers, herbicides, pesticides, insecticides, pathogens, moisture, and soil pH, and their controlled use could improve the sustainability of agriculture and the crop productivity [28]. Application of smart sensors can help the farmers to enhance their agricultural productivity by providing a better fertilization management, decreasing of inputs, and better time and environment management. The agricultural natural resources such as water, nutrients, and chemicals could be applied efficiently using nanosensors. Moreover, nanosensors can be used to detect the presence of plant pathogens (e.g., viruses) and the level of soil nutrients [30,31]. Nano-smart dust and gas sensors could be used to quickly evaluate the levels of environmental pollutions [32]. To monitor the quality of agricultural products, nanobarcodes and nanoprocessing can also be used [33].

Recently, the development of a nanobiosensor based on an atomic force microscopy (AFM) tip functionalized with the acetolactate synthase enzyme in the detection of enzyme-inhibiting herbicides was investigated [34]. It was reported that the development of sensors/biosensors based on specific interactions was a superb alternative to provide greater sensitivity and selectivity, making AFM more effective in detecting herbicide metsulfuron-methyl through the acquisition of force curves. Otles and Yalcin [35] reported the application of nanobiosensors for quantification and rapid detection of bacteria and viruses to increase food safety.

In another study, metal (gold, palladium, platinum)/DNA/single-walled carbon nanotube (SWCNT) hybrid nanostructure-based gas sensor arrays were manufactured by means of inkjet printing of metal ion chelated DNA/SWNTs on microfabricated electrodes, followed by electroless deposition to decrease metal ions to metal [36]. The results on the sensitivity and selectivity of the gas sensors were reported by analyzing their response to different gases (e.g., H_2, H_2S, NH_3, and NO_2) at room temperature and showed the increase in the sensitivity and selectivity to certain analytes by functionalizing with various metal nanoparticles such as Pd/DNA/SWCNTs for H_2 and H_2S. The combined responses provided a unique pattern or signature for each analyte by which the system identified and quantified an individual gas [37].

Recently, a one-step and label-free optical biosensor for determination of aflatoxin B1 was investigated [38]. Gold nanorods were used as a sensing platform and revealed high stability under high ionic strength conditions without any stabilizing agent addition. The proposed nanobiosensor was proved to be sensitive, selective, and simple, and provided a viable alternative for rapid screening of toxins in agriculture products and foods. The whole experiment from taking samples to final analysis was carried out within 45 min. Moreover, the feasibility of using this approach for aflatoxin B1 detection was proved in artificially contaminated peanut samples.

In the field of sensor research, application of bionanotechnology has potential to make significant contributions and to change the approach sensors are designed, fabricated, and implemented. The development of a biosensor based on the bi-immobilization of laccase and tyrosinase phenoloxidase enzymes for determination of toxic compounds and smart biosensors for determination of mycotoxines were previously studied [38,39]. A good compatibility between membranes and enzymes without altering the conformation of the enzyme molecule was reported by biosensor design and binding was carried out outside the enzyme active centers.

2.3.3
Diagnosis and Control of Plant Diseases

Diagnosis of diseases is a complicated process due to the extremely low concentrations of biochemicals and the presence of very low amount of detectable virus, fungal, or bacterial infections in plants. Application of nanoparticles has been suggested as an effective, ecofriendly, and cost effective method for the control of pathogens and consequently management of plant diseases [40]. Nanoforms of carbon, silver, silica and aluminosilicates are the examples of nanoparticles which can be used for this purpose. Silver can affect various biochemical processes in the microorganisms including the alterations in routine functions and plasma membrane [41]. Nanosilver is the most investigated and utilized nanoparticle in biosystem. Application of nanosilver particles as antimicrobial agents for controlling different plant pathogens in agricultural sector has become more common and made their production economical. Compared to the unencapsulated (bulk) silver, nanosilver has high antimicrobial effect, due to the high surface area and high fraction of surface atoms. The expression of ATP production associated proteins can also be prevented by silver nanoparticles [42].

2.3.4
Waste Reduction and Production of High-Value Added Products

The use of agricultural wastes for the generation of energy and electricity is an attractive and promising option for the environment. Among the different solutions, nanotechnology approaches could be used to convert agricultural wastes into useful products. Previous studies evaluated the application of nanotechnology in transesterification, gasification, pyrolysis, hydrogenation, and reforming of biomass-derived compounds [43]. Application of cellulose-based nanocrystals has been of high interest as potential nanoreinforcing filler into bionanocomposite for biomedical and other industrial applications [44]. The metabolism of microorganism metabolism could be stimulated by nanomaterials. In this situation, the efficiency of the lipid extraction could be improved by the use of nanomaterials and even accomplish it without damaging the microalgae cells. It has been reported that calcium and magnesium oxide nanoparticles can be used as biocatalyst carriers or as heterogeneous catalysts in oil transesterification into biodiesel [45]. Nanotechnologies are among the most suitable conversion technologies for the biofuels for the future. Currently, most of the attempts in second-generation conversion to liquid biofuels are relying on biomass cellulosics to ethanol and biodiesel to meet rising transportation fuel demands. The future of biofuels will depend on the development and diffusion of novel technologies such as nanotechnology with an appropriate and market-friendly regulatory environment.

2.4
Applications of Nanotechnology in the Food Sector

The food and beverage sector is a high finance global industry and many food companies have been conducting research to improve production efficiency, food safety, and nutritional properties. The effect of nanoscience in the food industry has become more demanding over the last few years and brought various advantageous into the food sector, while it could also have risks for consumers like the other emerging technologies (Figure 2.3) [46]. Many of the world's

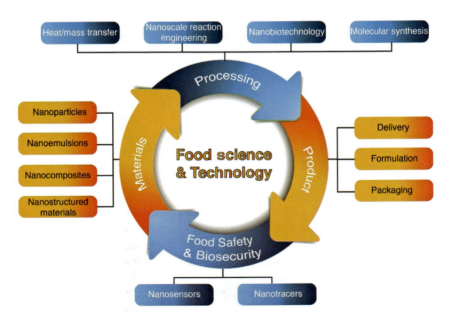

Figure 2.3 Nanotechnology applications in food science and technology. (Adapted from Ravichandran (2010). From [47] with permission © 2010 Taylor & Francis.)

largest food companies are constantly looking for ways to improve the efficiency, safety, quality, and nutritional properties of food products. Nanotechnology in food sector can be used to develop promising nanoprocesses, fabricate eco-friendly processes and intelligent nanopackaging, manufacture products with required texture and tastes, produce low-calorie food and beverage products with the aim of changing the lifestyles into healthy ones [5,46]. Advances in other areas (e.g., electronics, computing, data storage, communication) and the increasing use of integrated devices may indirectly influence the food industry in relation to food safety, authenticity, and waste reduction [6].

Regarding the risks of using nanotechnology in food industry, most of the nanomaterials entering the gut via oral administration are absorbed through intestine cells and designed in a way that they do not allow other large or foreign particles to pass through them. However, the nanosized compounds are able to cross these barriers, and there is a potential risk in bringing up gastric diseases, thus they need to be evaluated through *in vivo* studies [46]. In the following sections, the application of nanotechnology in food processing, development of food packaging materials, and fabrication of nanosensors for the detection of microbial contamination is highlighted.

2.4.1
Delivery of Active Compounds

Recently, several studies have evaluated the use of nanoencapsulation of valuable compounds (e.g., flavours, vitamins, minerals, antimicrobials, drugs, colorants, antioxidants, probiotic microorganisms, and micronutrients) in the food sector [46,48]. A large number of delivery systems (e.g., emulsions, biopolymer matrices, simple solutions, and association colloids) have been reported to maintain active ingredients at appropriate levels for long-time periods. However, compared to the conventional encapsulation methods, nanoparticles have better properties for encapsulation and higher efficiency to release active ingredients. These nanoparticles can be used to encapsulate functional ingredients and released in response to particular environmental triggers (e.g., particle dissolution or porosity can be achieved by the change in solution conditions) [49].

The efficiency of delivery systems can be enhanced by dendrimer, a unique class of polymer-coated particles. Dendrimers can be used as sensors, catalysts, and for drug delivery due to their nontoxicity, nonimmunogenecity, and biodegradability [50]. Also, cochleates are small-sized stable delivery systems that can be used to encapsulate different poor water soluble, peptides, proteins, and large hydrophilic bioactive molecules [51].

Most of the nanoparticles applied in food sector traditionally belonged to the group of colloids such as micelles, emulsions, mono-, and bilayers. Micelles are spherical particles, which are formed by the self-aggregation of the surfactant in aqueous solution, and have the ability to encapsulate different nonpolar molecules (e.g., lipids, flavorants, antimicrobials, antioxidants, and vitamins). Colloids are widely used for encapsulation and delivery of polar, nonpolar, or amphiphilic functional compounds.

Nanoemulsions are produced using high-pressure homogenizers. This emulsion-based delivery system can be applied to encapsulate value-added compounds and reduce their chemical degradation [52]. Microemulsions are thermodynamically stable, isotropic, and transparent emulsions with droplet size smaller than 100 nm. The application of microemulsions as a food delivery system has attracted a considerable degree of interest during the last few decades. More emphasis has been placed on edible microemulsions and their applications in foods and beverages [53]. Liposomes are spherical lipid bilayers (50–1000 nm) that serve as convenient delivery vehicles for bioactive compounds. Liposomes can be used to encapsulate enzymes and vitamins to enhance the speed of cheese ripening and the nutritional quality of dairy products, respectively [49]. These studies revealed that the use of liposomes can enhance the bioactivity of nutrients against degradation in food.

2.4.2
Food Packaging

One of the most promising benefits of using nanotechnology in the food sector is fabrication of novel food packaging materials. Although the application of nanotechnology in food industry struggles with public concerns over safety, development of the novel nanoparticles for using in food packaging is moving full-speed ahead. Using nanotechnologies to improve packaging materials is very costly and will not be introduced before optimization of the methods. Thus, food packaging is a main focus of food industry-related nanotechnology R&D. Currently, different companies are producing packaging materials based on nanotechnology to extend the shelf life of foods and drinks by improving the food safety [6]. The application of nanoparticles can improve the mechanical and heat resistance properties of foodstuffs packaging and consequently enhance shelf life, by affecting the permeability of gas or water vapor. For instance, although polymers are not naturally impermeable to gases or water vapor, polymer silicate nanocomposites have reported to improve the physical properties (e.g., gas barrier, the mechanical stability, the heat resistance, etc.) of food packaging [49].

Current trends in packaging of food products are based on the incorporation of nanomaterials into the plastic polymers in order to develop novel food packaging materials. The examples are: (1) polymer nanocomposites with improved flexibility, durability, temperature/moisture stability, and gas-barrier properties, (2) active packaging based on polymers incorporated nanomaterials with antimicrobial/antioxidant properties, (3) active nanocoatings for using in hygienic food contact materials, and lipophilic nanocoatings for self-cleaning surfaces, and (4) nanosensors and nanobiosensors for smart packaging for detecting food pathogens and improving food safety [54]. Examples of these novel materials are plastic polymers with nanoclay, nanosilver and nanozinc oxide, nanotitanium dioxide, nanotitanium nitride, and nanosilica.

It should be noted that there may be some risks to the consumer in case of migrating particulate nanomaterials from food contact materials into the food. So far only few studies have fully evaluated the results of this type of exposure and the lack of such data limit the assessment of risks caused by the consumption of foods contacted with nano packaging [6]. For instance, the European Food Safety Authority (EFSA) Scientific Panel on FCMs, Enzymes, Flavorings and Processing Aids (CEF) concluded that titanium nitride nanoparticles at a level of 20 mg/kg in PET bottles did not migrate and therefore were not a toxicological risk for food [55].

2.4.3
Other Applications

There are a number of other applications of nanoscience in the food sector including water filtration using nanoporous nanomaterials, removal of unpleasant tastes, and quality control of foods with sensors (e.g., time–temperature integrators, gases detection, O_2 sensors, detection of pathogens) that could offer innovative solutions to the food and related sectors [20,54]. For instance, nanomaterials such as zero valent iron can be used in water decontamination. Chaudhry *et al.* [56] studied the impact of titanium dioxide nanocoatings for photocatalytic sterilization of surfaces and water, nano(bio)sensors for food safety, and nanobarcodes for food authenticity. Application of nanotechnologies for water filtration and desalination may bring a lot of benefits to developing countries in terms of using safe potable water and reduce the infrastructural costs. Multianalyte detection of bacteria and biological markers of disease and food contaminants can be carried out using nano(bio)sensors [54]. The potential advantages of using these nanosensors include microbial and chemical safety of food products to protect consumer health.

Indicators of time–temperature integrators (TTIs) can be used to monitor, record and indicate foods safety. According to Singh [57], the TTIs can be divided into three basic categories including: abuse indicators, partial temperature history indicators, and full temperature history indicators. This technique is normally used in the foods stored at nonoptimal conditions. For example, application of TTI shows whether a food product that was supposed to be frozen properly has been inadequately exposed to higher temperatures.

Gas sensors can be used to detect the food spoilage caused by microorganisms in order to translate the chemical interactions between particles. Gases are usually detected by nanosensors based on metal oxides. Moreover, nanosensors fabricated based on conducting polymers that can quantify and/or identify pathogens based on their gas emissions could be used in the food sector. The change in resistance of the nanosensors forms a pattern related to the gas under examination [58].

Novel nontoxic and irreversible O_2 nanosensors can also be used to assure complete removal of O_2 from oxygen-free food packages. A UV-activated colorimetric O_2 indicator was previously designed using TiO_2 nanoparticles to photosensitize the reduction of methylene blue by triethanolamine in a polymer encapsulation medium [59]. Upon UV irradiation, the nanosensor bleached and remained colorless until it was exposed to oxygen. The rate of change in the color of sensor was reported to be proportional to the level of O_2 exposure [60].

2.5 Challenges of Using Nanotechnology in Agriculture and Food Sectors

Although the use of nanotechnology brings significant benefits to the agri-food sector, there are some health and safety issues that must be considered. Research developments on the potential benefits of using nanotechnology in the food and agriculture sectors have been published at an increasing rate, so there is an urgent requirement to analysis the risk. The risk of using this technology is mainly because of the small particle size of nanoparticles and their large surface area that cause easy dispersion, might cross anatomical barriers, reach more distal regions of the body, and display potential toxicity [20]. In the agriculture sector, handling of nano fertilizers and pesticides, which can be easily dispersed into the soil, water, or atmosphere, may increase the health risk of farmers. Also, the potential risks and advantages of using nanosilver as an antibacterial agent in health care products are being considered globally [1].

In the food sector, potential risks to human health may occurr by the migration of particulate nanomaterials from food contact materials into food. This depends on the level of the toxicity of the used nanomaterials as well as migration and the consumption rates of the particular food [6]. Application of novel materials may cause potential risks in the food sector and risk assessments must be highly accurate and reliable to identify and quantify these risks. Thus, detailed life cycle analysis, uptake of nanoparticles by plants, biodistribution, entry in the food chain, and so on need to be investigated before these tools are widely used as products in agriculture and food sectors [20].

Over 10 years ago, the European Commission has started to investigate the regulatory challenges caused by the application of nanotechnologies in the agri-food sector [61]. The following EU regulation already refer to size: regulation (EC) No 1333/2008 on food additives [62], regulation (EU) No 10/2011 on plastic materials and articles intended to come into contact with food [63], regulation (EC) No 450/2009 on active and intelligent materials and articles intended to come into contact with food [64], regulation (EU) No 1169/2011 on the provision of food information to consumers [9] and regulation (EU) No 2283/2015 on novel foods and novel food ingredients [65]. Recently, a new biocidal product regulation (EU 528/2012) was adopted in the European Union (EU). The regulation specifically needs evaluation and approval of active nanomaterial biocidal ingredients. In order to use the full potential of nanotechnology in food and agriculture sectors, the advantageous of using this technology for different purposes over existing technologies must be conveyed to the public.

2.6
Conclusions

Application of nanotechnology has become one of the fastest growing branches of modern science and play an important role in both agriculture and food production systems. Generally, two different processes are applied to fabricate nanomaterials including "bottom-up" processes (e.g., self-assembly) that create nanoscale materials from atoms and molecules, and "top-down" processes (e.g., milling) that create nanoscale materials from their macroscale counterparts. Compared to the macroscale materials, nanoscale form that have macroscale counterparts normally show different or enhanced properties. The main focus of nanoparticles application in the agriculture sector has been on the water quality management, delivery of agriculture chemicals, diagnosis of plant diseases, quality control with nanosensors, and so on which have brought more profit to farmers. Moreover, it is undeniable that nanotechnology has many beneficial applications in the food sector especially in food processing, food packaging, delivery of active ingredients and quality control with nanosensors, water filtration, removal of unpleasant tastes, and so on. However, despite the aforementioned benefits and a steady increase in the number of research on this topic, the current level of nanotechnology applications in agriculture and food sectors is still new and the new developments in most countries is still at basic research and development level. Also, there is a lack of scientific data among research communities for various regulatory agencies to identify hazards, assess risks, and provide risk management guidelines. The market for nanotechnology-derived products in the agrofood sector is predicted to grow rapidly in the coming years, thus it is required to increase the knowledge and awareness of nanotechnology applications.

Acknowledgment

Shahin Roohinejad would like to acknowledge the Alexander von Humboldt Foundation for his postdoctoral research fellowship award.

References

1 Sekhon, B.S. (2014) Nanotechnology in agri-food production: an overview. *Nanotechnol. Sci. Appl.*, **7**, 31–53.
2 Bogdan, A.T. *et al.* (2010) Prospects of agrifood green power in 2050 and forecasting for 2100 with sustenable solutions based on ecobioeconomics new paradigm. *Bull. Univ. Agric. Sci. Vet. Cluj-Napoca. Anim. Sci. Biotechnol.*, **67** (1–2), 1–18.
3 Norman, S. and Hongda, C. (2013) IB IN DEPTH special section on nanobiotechnology, Part 2. *Ind. Biotechnol.*, **9**, 17–18.
4 Misra, A.N., Misra, M., and Singh, R. (2013) Nanotechnology in agriculture and food industry. *Int. J. Pure Appl. Sci. Technol.*, **16** (2), 1.

5 Chaudhry, Q. et al. (2008) Applications and implications of nanotechnologies for the food sector. *Food Addit. Contam.*, **25** (3), 241–258.

6 Cushen, M. et al. (2012) Nanotechnologies in the food industry – Recent developments, risks and regulation. *Trends Food Sci. Tech.*, **24** (1), 30–46.

7 Som, C. et al. (2010) The importance of life cycle concepts for the development of safe nanoproducts. *Toxicology*, **269** (2), 160–169.

8 European Commission (2011) Commission recommendation of 18 October 2011 on the definition of nanomaterial. *Official J. Eur. Union*, **275**, 38.

9 European Commission (2011) Regulation (EU) No 1169/2011 of the European parliament and of the council of 25 October 2011 on the provision of food information to consumers, amending regulations (EC) No 1924/2006 and (EC) No 1925/2006 of the European parliament and of the council, and repealing commission directive 87/250/EEC, council directive 90/496/EEC, commission directive 1999/10/EC, directive 2000/13/EC of the European parliament and of the council, commission directives 2002/67/EC and 2008/5/EC and commission regulation (EC) No 608/2004. *Official J. Eur. Union*, **304**, 18.

10 Sozer, N. and Kokini, J.L. (2009) Nanotechnology and its applications in the food sector. *Trends Biotechnol.*, **27** (2), 82–89.

11 Degant, O. and Schwechten, D. (2002) Wheat flour with increased water binding capacity and process and equipment for its manufacture. German Patent DE10107885A1.

12 Shibata, T. (2002) Method for producing green tea in microfine powder. Google Patents.

13 Royal Society and Royal Academy of Engineering (2004) Nanoscience and nanotechnology: opportunities and uncertainties, in *Policy Document 19/04*, Royal Society, London.

14 Brody, A.L. et al. (2008) Scientific status summary. *J. Food Sci.*, **73** (8), R107–R116.

15 Scrinis, G. and Lyons, K. (2007) The emerging nano-corporate paradigm: nanotechnology and the transformation of nature, food and agri-food systems. *Int. J. Sociol. Agric. Food*, **15** (2), 22–44.

16 Dickinson, E. (2003) *Food Colloids, Biopolymers and Materials*, vol. **284**, Royal Society of Chemistry.

17 Sastry, K., Rashmi, H., and Rao, N. (2010) Nanotechnology patents as R&D indicators for disease management strategies in agriculture. *J. Intellec. Prop. Rights*, **15**, 197–205.

18 Gogos, A., Knauer, K., and Bucheli, T.D. (2012) Nanomaterials in plant protection and fertilization: current state, foreseen applications, and research priorities. *J. Agric. Food Chem.*, **60** (39), 9781–9792.

19 Torney, F. et al. (2007) Mesoporous silica nanoparticles deliver DNA and chemicals into plants. *Nat. Nanotechnol.*, **2** (5), 295–300.

20 Dasgupta, N. et al. (2015) Nanotechnology in agro-food: from field to plate. *Food Res. Intern.*, **69**, 381–400.

21 Chen, H. and Yada, R. (2011) Nanotechnologies in agriculture: New tools for sustainable development. *Trends Food Sci. Tech.*, **22** (11), 585–594.

22 Soni, N. and Prakash, S. (2012) Efficacy of fungus mediated silver and gold nanoparticles against *Aedes aegypti* larvae. *Parasitol. Res.*, **110** (1), 175–184.

23 Lamsal, K. et al. (2011) Inhibition effects of silver nanoparticles against powdery mildews on cucumber and pumpkin. *Mycobiology*, **39** (1), 26–32.

24 Goswami, A. et al. (2010) Novel applications of solid and liquid formulations of nanoparticles against insect pests and pathogens. *Thin Solid Films*, **519** (3), 1252–1257.

25 Sharon, M., Choudhary, A.K., and Kumar, R. (2010) Nanotechnology in agricultural diseases and food safety. *J. Phytol.*, **2** (4), 83–92.

26 Ding, W. and Shah, N.P. (2009) Effect of various encapsulating materials on the stability of probiotic bacteria. *J. Food Sci.*, **74** (2), M100–M107.

27 Vidhyalakshmi, R., Bhakyaraj, R., and Subhasree, R. (2009) Encapsulation "the future of probiotics"–a review. *Adv. Biol. Res.*, **3** (3–4), 96–103.

28 Rai, V., Acharya, S., and Dey, N. (2012) Implications of nanobiosensors in agriculture. *J. Biomater. Nanobiotechnol.*, **3** (2A), 315.

29 Prasad, R., Jain, V., and Varma, A. (2010) Role of nanomaterials in symbiotic fungus growth enhancement. *Curr. Sci.*, **99** (9), 1189–1191.

30 Jones, P.B. (2006) A nanotech revolution in agriculture and the food industry. Information Systems for Biotechnology. Available at http://www.isb.vt.edu/articles/jun0605.htm.

31 Brock, D.A. *et al.* (2011) Primitive agriculture in a social amoeba. *Nature*, **469** (7330), 393–396.

32 Mousavi, S.R. and Rezaei, M. (2011) Nanotechnology in agriculture and food production. *J. Appl. Environ. Biol. Sci.*, **1** (10), 414–419.

33 Li, Y., Cu, Y.T.H., and Luo, D. (2005) Multiplexed detection of pathogen DNA with DNA-based fluorescence nanobarcodes. *Nat. Biotechnol.*, **23** (7), 885–889.

34 Da Silva, A.C. *et al.* (2013) Nanobiosensors based on chemically modified AFM probes: a useful tool for metsulfuron-methyl detection. *Biosensors*, **13** (2), 1477–1489.

35 Otles, S. and Yalcin, B. (2010) Nano-biosensors as new tool for detection of food quality and safety. *LogForum*, **6** (4), 67–70.

36 Su, H.C. *et al.* (2013) Metal nanoparticles and DNA co-functionalized single-walled carbon nanotube gas sensors. *Nanotechnology*, **24** (50), 505502.

37 Xu, X. *et al.* (2013) A simple and rapid optical biosensor for detection of aflatoxin B1 based on competitive dispersion of gold nanorods. *Biosens. Bioelectron.*, **47**, 361–367.

38 ElKaoutit, M. *et al.* (2007) Dual laccase-tyrosinase based Sonogel-Carbon biosensor for monitoring polyphenols in beers. *J. Agric. Food Chem.*, **55** (20), 8011–8018

39 Grabchev, I., Betcheva, R., and Yotova, L. (2007) Photophysical and biological properties of fluorescent PAMAM dendrimer. Biotechnical Fictionalization of Renewable Polymeric Materials, Graz, Austria, 12–14.

40 Al-Samarrai, A. (2012) Nanoparticles as alternative to pesticides in management plant diseases. *Int. J. Sci. Res. Publ.*, **2** (4), 1–4.

41 Pal, S., Tak, Y.K., and Song, J.M. (2007) Does the antibacterial activity of silver nanoparticles depend on the shape of the nanoparticle? A study of the gram-negative bacterium Escherichia coli. *Appl. Environ. Microbiol.*, **73** (6), 1712–1720.

42 Yamanaka, M., Hara, K., and Kudo, J. (2005) Bactericidal actions of a silver ion solution on Escherichia coli, studied by energy-filtering transmission electron microscopy and proteomic analysis. *Appl. Environ. Microbiol.*, **71** (11), 7589–7593.

43 Ramsurn, H. and Gupta, R.B. (2013) Nanotechnology in solar and biofuels. *ACS Sustain. Chem. Eng.*, **1** (7), 779–797.

44 Kumar, A. *et al.* (2014) Characterization of cellulose nanocrystals produced by acid-hydrolysis from sugarcane bagasse as agro-waste. *J. Mater. Phys. Chem.*, **2** (1), 1–8.

45 Zhang, X. *et al.* (2013) Biodiesel production from heterotrophic microalgae through transesterification and nanotechnology application in the production. *Renew. Sustain. Energy Rev.*, **26**, 216–223.

46 Katouzian, I. and Jafari, S.M. (2016) Nano-encapsulation as a promising approach for targeted delivery and controlled release of vitamins. *Trends Food Sci. Tech.*, **53**, 34–48.

47 Ravichandran, R. (2010) Nanotechnology applications in food and food processing: innovative green approaches, opportunities and uncertainties for global market. *Int. J. Green Nanotechnol. Phys. Chem.*, **1** (2), P72–P96.

48 Singh, H. (2016) Nanotechnology applications in functional foods; opportunities and challenges. *Prev. Nutr. Food Sci.*, **21** (1), 1–8.

49 Rashidi, L. and Khosravi-Darani, K. (2011) The applications of nanotechnology in food industry. *Crit. Rev. Food Sci. Nutr.*, **51** (8), 723–730.

50 Khosravi-Darani, K. *et al.* (2007) The role of high-resolution imaging in the evaluation of nanosystems for bioactive encapsulation and targeted nanotherapy. *Micron*, **38** (8), 804–818.

51 Gould-Fogerite, S., Mannino, R., and Margolis, D. (2003) Cochleate delivery vehicles: applications to gene therapy. *Drug Deliv. Technol.*, **3** (2), 40–47.

52 McClements, D. and Decker, E. (2000) Lipid oxidation in oil-in-water emulsions: impact of molecular environment on chemical reactions in heterogeneous food systems. *J. Food Sci.*, **65** (8), 1270–1282.

53 Roohinejad, S. *et al.* (2015) Formulation of oil-in-water β-carotene microemulsions: effect of oil type and fatty acid chain length. *Food Chem.*, **174** (0), 270–278.

54 Chaudhry, Q. and Castle, L. (2011) Food applications of nanotechnologies: an overview of opportunities and challenges for developing countries. *Trends Food Sci. Technol.*, **22** (11), 595–603.

55 National Nanotechnology Initiative (2009) *National Nanotechnology Initiative: Research and Development Leading to a Revolution in Technology and Industry, Supplement to President's FY 2010 Budget*, Office of Science and Technology Policy, Washington, DC, p. 40.

56 Chaudhry, Q., Castle, L., and Watkins, R. (2010) *Nanotechnologies in Food*, Royal Society of Chemistry.

57 Singh, R. (2000) Scientific principles of shelf-life evaluation, in *Shelf Life Evaluation of Foods*, Springer, **2**, pp 3–17.

58 Arshak, K. *et al.* (2007) Characterisation of polymer nanocomposite sensors for quantification of bacterial cultures. *Sens. Actuators B Chem.*, **126** (1), 226–231.

59 Lee, S.-K., Sheridan, M., and Mills, A. (2005) Novel UV-activated colorimetric oxygen indicator. *Chem. Mater.*, **17** (10), 2744–2751.

60 Gutiérrez-Tauste, D. *et al.* (2007) Characterization of methylene blue/TiO2 hybrid thin films prepared by the liquid phase deposition (LPD) method: application for fabrication of light-activated colorimetric oxygen indicators. *J. Photochem. Photobiol. A: Chem.*, **187** (1), 45–52.

61 Salvi, L. (2015) The EU's soft reaction to nanotechnology regulation in the food sector. *Eur. Food Feed Law Rev.*, **10**, 186–193.

62 European Community (2008) 1333 European Community Regulation (EC) No. 1333/2008 of the European Parliament and of the Council of 16 December 2008 on food additives. *Official J. Eur. Union*, **L354**, 16–33

63 European Commission (2011) Regulation (EC) No. 10/2011 of 14 January 2011 on plastic materials and articles intended to come into contact with food. *Official J. Eur. Union*, **L12**, 1–89.

64 European Commission (2009 Commission regulation (EC) No 450/2009 of 29 May 2009 on active and intelligent materials and articles intended to come into contact with food. *Official J. Eur. Union*, **L135**, 3–11.

65 European Commission (2015) Regulation (EU) 2015/2283 of the European parliament and of the council of 25 November 2015 on novel foods, amending regulation (EU) No 1169/2011 of the European parliament and of the council and repealing regulation (EC) No 258/97 of the European parliament and of the council and commission regulation (EC) No 1852/2001. *Official J. Eur. Union*, **L327**, 1.

3
Naturally Occurring Nanostructures in Food

Saïd Bouhallab,[1] Christelle Lopez,[1] and Monique A.V. Axelos[2]

[1]*INRA, UMR1253 Science et Technologie du Lait et de l'Oeuf, 35042 Rennes, France*
[2]*INRA, UR1268 Biopolymères Interactions Assemblages, 44316 Nantes, France*

3.1
Introduction

Nanostructures are the natural building blocks of a lot of multiscale organization in plant and animal tissues. These nanostructures made with proteins, lipids, or polysaccharides, alone or mixed, even in association with inorganic molecules, play a major role in bestowing biological materials their unique properties, such as mechanical properties, for example, in spider silk or nacre, energy storage properties like oleosome in plant, or transport properties like high-density lipoprotein in human. These nanostructures result from the bottom-up assembly of molecules via hydrophobic or electrostatic interactions or covalent links depending on the chemical nature of the molecules involved and on environmental conditions. In food there are a lot of different naturally occurring particles that fall in the category of nanoparticles because of their size in the range of some nanometers like casein micelles or fat globules in milk. Very often only a fraction is really at a nanometric size (<100 nm) because of the large size distribution of these natural structures. In food, nanoparticles come directly from plant or animal raw products used and may also be generated during processing due to the thermomechanical conditions and physicochemical environments. Since 10 years the determination of structural changes occuring during processing and the knowledge of the interaction mechanisms have allowed to better understand the preservation, or alteration, of the native nanoparticles and the creation of new ones. Nanotechnology provides new ways of controlling and structuring food with greater functionality and value. Nanotechnolo
gy in food science is now becoming an active research field especially with the development of functional foods. In this chapter we focus on protein-based and lipid-based native nanostructures.

Nanotechnology in Agriculture and Food Science, First Edition. Edited by Monique A.V. Axelos and Marcel Van de Voorde.
© 2017 Wiley-VCH Verlag GmbH & Co. KGaA. Published 2017 by Wiley-VCH Verlag GmbH & Co. KGaA.

3.2
Protein-based Nanostructures

Proteins themselves are nanoparticles that are able to self-assemble into, ordered or amorphous, larger structures. The shape and size of the final architecture result from a coexistence of long-range repulsive forces and short-range attractive forces between the proteins building blocks. By changing the balance between these various forces a large diversity of structures can be obtained. Either covalent or noncovalent interactions are generally involved in these assemblies. Depending on the strength and/or number of interactions per proteins, structures may occur of varying lifetimes. The thermodynamic equilibrium of these assemblies is highly dependent on several factors: protein composition and concentration, solvent conditions like pH, ionic strength, nature of ions, and physical conditions such as temperature, local pressure, and so on. The bottom-up method of molecular self-assembly that exists in nature is a powerful technique to construct complex nanostructures. Throughout a variety of intra- (folding) or inter-molecular noncovalent interactions such as hydrophobic effect, hydrogen bonding, electrostatic interaction, and/or metal binding, proteins can fold and assemble spontaneously into structurally well-defined and functional nano-scaled structures (see Ref. [1] for more details). The nature explores the dynamic and adaptive natures of noncovalent, reversible interactions to construct protein nanostructures with a diversity of conformations according to the environmental stimuli. As such, natural foods contain nanoparticles with size ranging from one to few hundred nm [2,3]. This is illustrated in this chapter by proteins from foods as widely consumed as milk and eggs.

Proteins are complex natural macromolecules. Proteins are distinguished from each other by their covalent-linked backbone of amino acids, depending upon the sequence of amino acids they form β sheet or α helix or nonstructured coiled domains, these domains interact each other through noncovalent interactions to lead to the tertiary structure (globular form, fibrous form, intrinsically unfolded conformation) and to give proteins their biological functions. Beside these three structural levels, some proteins have a quaternary structure that results from the association of two or more identical or different amino acid chains (protein subunits). In solutions, the overall protein conformation fluctuates between a large number of conformations. The native protein structure is defined as the structure of lowest energy or structure of highest probability. Proteins are considered as zwitterions because they contain both positive and negative charges in a specific proportion. Hence, proteins are weak polyelectrolytes whose charge sign and density show a strong dependence on the physicochemical conditions. Away from their isoelectric point (pI), where the net charge is zero, proteins in solution are stable because the protein molecules carry charges of the same sign and repel each other. Protein sizes, especially globular ones, are generally on the order of few nm, thus constituting an example of naturally occurring nanoparticles. In contrast to intrinsically unfolded proteins, numerous globular proteins are also stable close to their pI at low or medium ionic strength due to the presence of residual charged patches on the protein surface that counterbalance short-range attractive interactions. In some conditions (elevated temperatures, presence of denaturants, etc.), the native protein unfolds (lose of its native conformation) or denatures. Each protein (except intrinsically unfolded proteins) has a denaturation temperature that is dependent on the solvent conditions (pH, ionic strength, dielectric constant). Upon denaturation, the hydrodynamic size, the flexibility, and also the reactivity of the proteins increase because of the exposure of reactive groups to the surface of the proteins. This denaturation may lead to protein aggregation and to some extent to a loss of solubility.

This chapter is focused on the description of the structure of several food proteins as they exist naturally in food. In some cases we will also address the spontaneously formed new nanostructures that can be formed under native conditions following the interaction of individual proteins with naturally coexisting bioactive molecules. For artificial self-assembled protein materials, for example, following processing, the reader should refer to Chapter 4 of this book.

Most of the food proteins obtained from animal or plant origins are globular structures, with some exceptions such as the caseins, present in milk, which display complex colloidal structures called casein micelles and myofibrillar proteins that display a fibrous structure with a structuring role in meat tissue. Proteins presented below are principally from milk (β-lactoglobulin (β-Lg), α-lactalbumin (α-La), bovine serum albumin (BSA), lactoferrin, caseins) or egg white (ovalbumin, lysozyme, ovotransferrin) because of their large occurrence in foods. Food proteins exist mainly in the monomeric form, but are able to self-assemble into oligomers or aggregates in some specific conditions. These conditions are also addressed.

3.2.1
Examples of Protein Nanostructures Present in Foods

3.2.1.1 β-Lactoglobulin
Bovine β-lactoglobulin (β-Lg) is a globular protein of 162 amino acids with a molar mass of 18.3 kDa and an isoelectric point of about 5.2 [4]. Native β-Lg monomer is about 3.6 nm in length and its molecular structure is well established. Basically, β-Lg has 10–15% α-helix, 43% β-sheet, and 47% unordered structures, including β-turn (Figure 3.1a). Its structure contains nine β-sheets (in yellow) that are organized to form a calyx and a C-terminal α-helix, as determined by X-ray crystallography. Two disulfide bonds play an important role in the sensibility of β-Lg toward denaturation. Bovine β-Lg also contains one free sulfhydryl group, which is buried within the protein structure and consequently is nonreactive under physiological conditions but plays an important role in stabilizing the protein tertiary structure. In the pH range of 5.5–7.5, and at room temperature, native β-Lg exists as a stable noncovalent dimer but its oligomerisation sate is dependent on the medium conditions leading to fractal aggregates

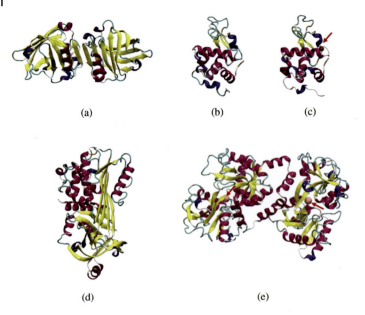

Figure 3.1 Tridimensional structures of five globular proteins representing the diversity of naturally occurring proteins nanostructures present in food. (a) Dimer of β-lactoglobulin (PDB code 1BEB), acidic protein of 2x 18.3 kDa. (b) Lysozyme (PDB code 2VB1), basic protein of 14.3 kDa. (c) α-Lactalbumin (PDB code 1F6S), acidic protein of 14.3 kDa with calcium binding site (red arrow). (d) Ovalbumin (PDB code 1OVA), acidic glycoprotein of 45 kDa. (e) Lactoferrin (PDB code 1BLF), basic glycoprotein of 80 kDa with two distinct domains containing one iron binding site each (red arrow). Secondary structure: α-helix (purple); β-sheet (yellow); β-turn (blue).

which fractality depends on the ionic strength. Screening the protein charges by NaCl addition leads to an increase in size and protein density of aggregates. Under other pH values, for example, below pH 5.5, β-Lg self-assemble into higher ordered oligomers (tetramers) or dissociate into monomers. At pH 2 fibrillar structures may be formed. Until now, the exact biological function of β-Lg is not really known. However, it belongs to the lipocalin superfamily, sharing the common β-barrel calyx structural feature as an ideal binding site for hydrophobic ligands. Lipocalins are generally transporters of small hydrophobic molecules. Consequently, the role of β-Lg in the protection and transport of small bioactive ligand is a potential biological function.

3.2.1.2 Serum Albumin

Serum albumin is a well-known, abundant, and multifunctional protein [5]. Bovine Serum Albumin (BSA) is a helical protein having a polypeptide chain of 582 amino acids, a molar mass of 66.4 kD and a pI of 4.9. The tridimensional structure of BSA monomeric nanoparticle is an ellipsoid with a radius of about 3.5 nm. BSA contains 1 sulfhydryl group and 17 disulfide bonds close to each other along the polypeptide chain, which strongly stabilize the structure of the protein. For this reason, BSA is extensively used as standard protein for calibration purposes. BSA is also extensively used as a model to study hydrophobic ligand interactions, because it exhibits several binding sites on its surface. The biological function has been associated with its property and ability to bind fatty acid and lipid. Until today, BSA is the most plentiful protein in blood plasma; each protein molecule can carry seven fatty acid molecules. From *in vitro* studies, it was suggested that the protein play a role in mediating lipid oxidation.

3.2.1.3 α-Lactalbumin and Lysozyme

Lysozyme (Lys) from hen egg white (Figure 3.1b) and α-lactalbumin (α-La) from milk (Figure 3.1c), exhibit an extraordinary close homology in sequence and tertiary structures but with several biochemical, biophysical, and biological differences [6]. Among the 123 amino acid residues of α-La, 54 are identical to corresponding residues in lysozyme and a further 23 residues are structurally similar. α-La consists of 123 amino acids and has a molecular weight of 14.2 kDa and a radius of 2.01 nm. It is an acidic protein with an isoelectric point between 4.2 and 4.5. α-La consists of 26% α-helix, 14% β-sheet, and 60% unordered structure. α-La contains eight cysteine residues, all engaged in disulfide bonds that stabilize the tertiary structure of the protein. In addition, α-La binds specifically one calcium ion per molecule with high affinity ($K_a = 108\,\text{M}^{-1}$) in a highly negatively charged pocket containing four aspartate residues. Decreasing the pH leads to the formation of calcium-free apo form of α-La, which is highly heat-sensitive. The denaturation temperature of apo α-La is around 30 °C whereas the native holo α-La is stable up to around 60 °C. Calcium pocket of α-La can also bind other divalent such as Zn^{2+}, Mg^{2+}, Mn^{2+}. α-La performs an important function in mammary secretory cells as one of the two components of lactose synthase, which catalyzes the final step in lactose biosynthesis in the lactating mammary gland. Other studies showed that some conformational forms of α-La can induce apoptosis in tumor cells suggesting that this protein possesses many important biological functions.

Lysozyme from egg white (C-type family without a specific metal binding site) is a globular protein of 129 amino acids (14.3 kDa) and a radius of about 2.1 nm. In contrast to α-La, Lys is rich in lysine and arginine residues leading to a strong basic character (pI ≈ 10.7). Like α-La, Lys is a protein consisting of two domains (domains α and β) linked together by a long helix–loop–helix (residues 87–114). The secondary structure of the protein is formed by 39% α-helix gathered mainly in the domain α and 11% β-sheet involved in a three-strand antiparallel β-sheet that constitutes, with some helices, the β-domain. Lys has eight cysteine residues, all of which are involved in intramolecular disulfide bonds making the protein compact and stable. It was noticed that under specific conditions of protein concentration, temperature, and ionic strength, Lys forms transient nanoclusters of 10–20 nm. These nanoclusters result from short-range attractions, leading to surface energy reduction upon cluster formation, and long-range repulsions that increase the Coulomb energy of the clusters and thus limits their growth.

3.2.1.4 Ovalbumin and Avidin

Ovalbumin (Ova) is the major protein in egg white and was one of the first proteins to be isolated in a pure form [7]. Ovalbumin is a phosphoglycoprotein of 385 amino acids and a molar mass of about 45 kDa and a radius of about 3.05 nm. Its amino acid sequence contains about 50% hydrophobic residues and about 33% charged residues, mostly acidic, giving the protein a pI of 4.5. The sequence includes six cysteine residues, two of which are involved in a disulfide bond, and the N-terminal residue of the protein is an acetylated glycine. Ovalbumin has one glycosylation site and two phosphorylation sites. Ovalbumin contains 32% β-sheets and 30% α-helix as determined by X-ray crystallography (Figure 3.1d). Some evidences indicate that Ovalbumin molecules self-assemble depending on protein concentration and pH. Ovalbumin is monomeric at concentration lower than about 0.1% and forms oligomers (dimer, trimer, and tetramer) at higher protein concentrations; this association behavior is favored when pH decreases from a neutral value to the protein pI. Although Ovalbumin belongs to serpin superfamily, its exact biology is not well defined. It was suggested that ovalbumin plays a role in the development of chick embryo.

Avidin is found in egg white of reptiles, amphibians, and birds. Chicken avidin is known to be an essential protein for extensive types of biotechnological application [8]. Avidin molecule is a noncovalent homotetrameric glycoprotein with a pI of 9.5 and a radius of about 3.4 nm. Each monomer contains 128 amino acid residues, giving the protein an overall molecular weight of about 57 kDa. Ten percent of the overall molecular weight is attributed to carbohydrate moieties. Although each monomer shares interactions with the three others, avidin is usually regarded as a dimer of dimers. The dimers are constituted of monomers highly stabilized by a multitude of polar and hydrophobic interactions, and the interaction between the two dimers was reported to be weaker and mainly stabilized by hydrophobic interactions. About the biological function, it has been postulated that Avidin may act, in the oviduct, as a bacterial growth-inhibitor, by binding biotin helpful for bacterial growth. Also, it was reported that functional avidin is found only in raw egg, as the biotin affinity for the protein is destroyed by cooking.

3.2.1.5 Transferrins

Transferrins are a group of glycoproteins with well-known iron binding properties and are present in different biological secretions [9]. In their monomeric state, transferrins have a radius of about 3.6 nm. Among the known four types of transferrins, two types are present in food matrices: (i) lactoferrin (Lf) is found in milk and other mammalian secretions; it binds iron more tightly over a larger pH range than other transferrin proteins and has a diverse range of biological activities including innate defense; (ii) ovotransferrin (Ovo) is an acidic glycoprotein that constitutes about 12–13% of avian egg white, providing an antimicrobial defense mechanism to the avian egg.

Lactoferrin is an iron-binding glycoprotein composed of 689 amino acids and has a molar mass of about 80 kDa. It has approximately 41% α-helix and 24% β-sheets and the polypeptide chain is folded in two homologous lobes (N- and

C-lobes) connected to each other by three turns of an α-helix (residues 334–344) (Figure 3.1e). Each lobe is composed of two domains (N1, N2 and C1, C2) forming a cleft in between; the inter-domain cleft forms a binding site for one iron or another metallic ion associated with a carbonate anion. Lf tertiary structure is stabilized by 17 disulfide bonds. Compared to the holo form, which has a very stable conformation, the metal-free apo form of Lf is much more flexible and more susceptible to denaturation. Lf is a basic protein with a pI of 8.6–8.9 and consequently carries a positive net charge at neutral pH, which gives it an ability to interact specifically with acidic proteins (see below). The charge distribution on Lf surface is uneven with some highly positive patches on the N-lobe and the inter-lobe region. Additional negative charges on the Lf surface are also carried by sialic moieties, which increase the hydration volume on the protein surface and stabilize Lf. Lf exhibits ionic strength-dependent aggregation behavior. These studies reported an Lf monomer–aggregate equilibrium that is sensitive to the variation of ionic strength. At low ionic strength (and neutral pH), Lf molecules are mainly monomeric (radius = 3.6 nm) and positively charged. When ionic strength increased, Lf forms neutral or negatively charged nanosized aggregates of about 10–20 nm.

Ovotransferrin is a single glycopeptide chain containing 686 amino acids, with a molecular weight of 78 kDa, a radius of ≈ 3.5 nm, and an isoelectric point of around 6.0. It contains 15 disulfide bonds without free sulfhydryl groups. Like other members of the transferrin family, the chain is folded into two globular lobes, N and C, with two subdomains per lobe linked by an alpha helix of nine amino acid residues that can be released by protease digestion. The N and C lobes associate natively through noncovalent interactions and it has been reported that isolated N and C lobes can reassociate spontaneously in solution. As for Lf, each lobe of Ovo has the capability to reversibly bind one Fe^{3+} ion associated with one bicarbonate anion but with different iron-binding affinities, with $K_a \approx 1.5 \times 10^{18}$ for the C-terminal lobe and 1.5×10^{14} for the N-terminal lobe.

In their natural forms, Lf and Ovo exhibit different biological activities including antimicrobial, antiviral, antifungal, and immunomodulatory activities. Some of these activities are linked to their iron-binding ability rendering iron unavailable for microbial growth. Lf was also shown to be involved in the regulation of cell growth.

3.2.1.6 Osteopontin and Lactoperoxydase

Osteopontin (OPN) is a multifunctional protein present in most tissues and body fluids. OPN is a secreted phosphoprotein, present at concentration approximately 10-times higher in human milk than in bovine milk. Human OPN is a 36 kDa, highly negatively charged (pI ≈ 3.6), predominantly unstructured glycoprotein [10]. OPN can go through posttranslational modifications, in particular phosphorylation, which increase its apparent molecular weight to about 44 kDa. Given its intrinsic properties, in particular the high negative charge density, OPN has been hypothesized to act as a transporter of the immune-modulating and antimicrobial proteins to their site of action.

Lactoperoxidase is another basic protein present in milk that consists of a single polypeptide chain with a molecular weight of approximately 78 kDa, a radius of 3.6 nm, and a pI around 8. Lactoperoxydase contains eight internal disulfide bridges and about 10% of carbohydrate. It plays an important role in protecting the lactating mammary gland and the intestinal tract of the newborn against pathogenic microorganisms.

3.2.2
Formation of Natural Nanostructure Subsequently to Molecular Interaction/Complexation

Proteins can form slightly higher ordered nanocomplexes through their interactions between themselves and with other small molecules present in the biological fluid. Many recent studies try to generate artificially induced nanostructures with lipids, polysaccharides, proteins, or mixture of them as vehicles for nutriments or bioactive substances such as, minerals, vitamins, anti-oxidants, or fatty acids. Such complexes already exist in foods. As mentioned above, some globular proteins, for example, lactoferrin and α-lactalbumin possess specific binding sites for mineral, iron, and calcium, suggesting a specific physiological role. Similarly, β-lactoglobulin and BSA are able to complex and transport hydrophobic substances through binding in specifically designed pockets with affinities ranging from 10^5 to $10^8 \, M^{-1}$ (Figure 3.2, Table 3.1). Since the discovery of β-lg–vitamin A complex [11] and further extensive studies, β-lg is considered today as a natural nanovehicle forming nanocomplexes with different molecules, in particular vitamins and fatty acids [12]. More recently, we reported on the ability of positively charged Lf to form spontaneously nanocomplexes with negatively charged folic acid (vitamin B9) [13]. One Lf molecule bound more than 10 molecules of folic acid to form complexes that self-associate into nanocomplexes with finite size of around 16 nm. We can then assume that some proteins

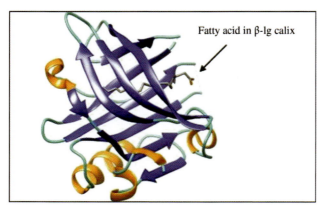

Figure 3.2 A schematic view of the main-chain fold of bovine β-lactoglobulin in interaction with linoleic acid inserted in its central cavity. RCSB PDB code 4DQ4 [14].

Table 3.1 Examples of the size on naturally formed nanoparticles between food proteins and small ligands.

Protein	Ligand	Size of formed nanoparticles (nm)	References
β-lactoglobulin and BSA	Retinol	<5	[11]
β-lactoglobulin	Folic acid	<10	[15]
Lactoferrin	Folic acid	16–18	[13]
β-lactoglobulin	Lipids (oleate or linoleate)	up to 170	[12]

present in foods or in food ingredients act as naturally occurring vehicle for nutritional and bioactive substances. Interestingly, the final size of formed nanocomplexes depends on the protein and the considered ligand (see Table 3.1 for examples) but the biological functions of the nanocomplex according to their size are still speculative. As summarized in the review by Le Maux et al. [12], putative roles would be: (i) increase of ligand solubility and absorption; (ii) modification of the kinetics of the enzymatic hydrolysis of the protein; (iii) protection of sensitive molecules against oxidation and other stresses; and (iv) modification of the bioaccessibility of the ligands.

Nanocomplexes can also naturally occur between different proteins. In milk, the negatively charged Osteopontin is highly phosphorylated enabling it to bind calcium ions, increasing consequently its solubility. Osteopontin can also form bioactive natural complexes of 5–10 nm with positively charged proteins like lactoferrin, lactoperoxidase, and immunoglobulin [16]. Lactoferrin and osteopontin interact with each other through a complex mechanism involving multiple cationic lactoferrin molecules binding to a single anionic molecule of osteopontin. Two classes of distinct lactoferrin binding sites were identified, with a highest affinity constant $K_a \approx 10^6 \, M^{-1}$ for the first interacting sites. Ca^{2+} binding to osteopontin or Fe^{3+} binding to lactoferrin, had little effect on the overall formation of natural nanocomplexes between the two proteins. Considering that the regions of electrostatic complementarity between osteopontin and lactoferrin mediate the numerous biological functions of each protein, it was suggested that osteopontin may act as a carrier for lactoferrin in milk, and modulate the potent antimicrobial and immunestimulatory activities of the lactoferrin protein. Hence, the interaction between osteopontin and lactoferrin to form nanocomplexes could be biologically important, behind the specific biological role of individual proteins.

3.2.3
Special Case: Casein Micelles

Due to its size, fascinating structure, and dynamics, casein micelle present in milk of all mammalians, constitutes the most cited supramolecular structures as example of naturally occurring protein nanostructure in the food we consume. Also, given its mixed organic–inorganic composition, it represents a natural vehicle and delivery system of bioactives and nutriments including amino acids, calcium, and phosphorus for neonates. In parallel, due to its unique structure, casein micelles have received much attention in many fields such as medicine, food, and cosmetics.

3.2.3.1 Casein Micelle Composition

Casein micelles are made of four distinct caseins called α_{s1}, α_{s2}, β, and κ in proportion of 3:1:3:1 in bovine milk, in mixture with about 8% in mass of phosphate and calcium ions [17]. Caseins are acidic proteins of about 20–24 kDa (from 169 to 209 amino acids): α_{s1}-casein (23.6 kDa, pI $\approx$ 4.95), α_{s2}-casein (25.3 kDa, pI $\approx$ 5.25), β-casein (24 kDa, pI $\approx$ 5.15), and κ-casein (19 kDa, pI $\approx$ 5.60). The four casein molecules can represent up to 80–85% of total proteins in bovine milk (25 g/l). They are a group of flexible, intrinsically unstructured proteins sharing some common features such as the presence of ester-bound phosphate (organic phosphate) in their structure and high number of charged (glutamic acid, lysine) and uncharged (leucine, isoleucine, proline) residues, but low number of sulfur-containing amino acids [17,18]. In addition, caseins especially β- and κ-caseins consist of well-separated hydrophobic and hydrophilic domains. There is a big microheterogeneity regarding their degree of phosphorylation and glycosylation. β-casein and the two α_s-caseins are extensively phosphorylated, whereas κ-casein has only one or two ester-bound phosphates but it is glycosylated. Given their high phosphorylation degree, α_s- and β-casein interact strongly with calcium ions inside the natural casein micelle. Caseins are then considered as rheomorphic proteins as their structure is highly sensitive to environmental variations as illustrated in Figure 3.3 [19]. Any small change in the environmental conditions (pH, temperature, ions) immediately alters the subtitle equilibrium between micelle components and the dispersing phase [19].

3.2.3.2 Casein Micelle Structure

In milk, casein molecules are organized into colloidal particles referred to as casein micelles. The mean native casein micellar size varies between species in the range of 140–300 nm. Casein micelle from human milk is among the smallest ones with a mean size of around 140 nm. The four caseins are self-assembled to form micelles that are stabilized by calcium phosphate nanoclusters. α_s- and β-caseins are mainly located in the interior of the micelle and are bound to calcium phosphate nanoclusters by their phosphoserine domains. The surface of the micelle is covered with negatively charged κ-casein at physiological pH ensuring steric and electrostatic stabilization of the nanostructure. However, while the exact composition of casein micelle is now well known, the exact location of each component (proteins, minerals) inside the structure is still under debate. This is because casein micelles have a very wide size distribution and a great instability. Any experimental action modifies the casein environment and consequently the native structure (Figure 3.3; [19]). Several models describing

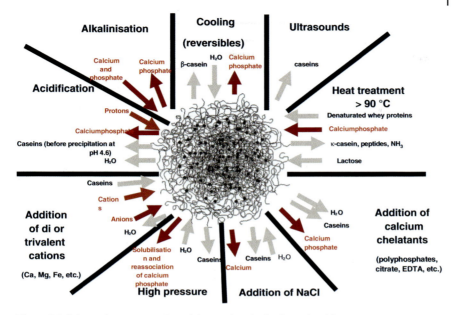

Figure 3.3 Schematic representation of the casein micelle dynamic with component exchanges as a function of different environment [19].

the casein micelles exist: (i) the submicelle model, which describes the micelle as made of closely packed submicelles of <15 nm in diameter linked together by calcium phosphate nanoclusters (Figure 3.4a); (ii) the homogeneous model, not shown, in which the internal structure is made of a loose and uniform casein matrix with randomly distributed mineral nanoclusters. The calcium phosphate

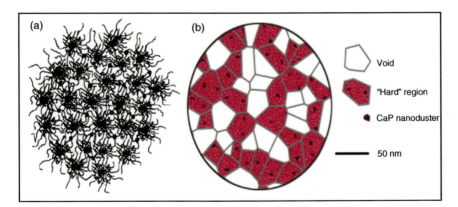

Figure 3.4 Two models of casein micelle structure. (a) The Holt model showing a matrix in which calcium phosphate nanoclusters (2 nm) are attached to the phosphoseryl groups of the caseins [21]. (b) The sponge-like model recently proposed by Bouchoux et al. [22].

are dispersed as very small nanoclusters ($\approx$2 nm radius) on which the phosphate groups of the caseins are attached [20]; (iii) the core–shell model, not shown, equivalent to a homogeneous model but in which the mineral nanoclusters are preferentially located near the micelle surface [21]. Recent studies report new additional information on the internal structure of casein micelles using a combination of osmotic stress and SAXS experiments [22]. The authors proposed a sponge-like, highly hydrated structural model of casein micelle with three structural levels (Figure 3.4b): the first level consists of the calcium phosphate nanoclusters that serve as anchors for the casein molecules forming the second level (10- to 40-nm hard regions), the third level or casein micelle itself, with an average size of 100 nm, is formed by a continuous and porous material of connected hard regions. The common point of all proposed models is that the glycosylated forms of κ-casein are located at the surface of casein micelles, conferring the above mentioned stability due to electrostatic repulsion and steric hindrance.

Casein micelle constitutes a natural functional nanoparticle that is consumed daily and serves for delivery purposes. By its intrinsic supramolecular structure, the casein micelles stabilize and increase the bioavailability of calcium and phosphate ion, two vital micronutrients. Today, the use of nanosized vehicles for the protection and controlled release of nutrients and bioactive food ingredients is a growing area of interest in the scientific community. Casein micelle is a natural nanovehicle ready for use for this purpose.

3.3
Lipid-Based Nanostructures

3.3.1
Lipid Nanodroplets

The lipid droplets are spherical particles constituted by a hydrophobic core rich in triacylglycerols (TAG, esters of fatty acids and glycerol) covered by surface-active molecules such as phospholipids and proteins. These TAG-rich droplets are involved in the storage and delivery of energy. Examples of the natural nanostructures formed by lipids in foods are the lipoproteins found in egg yolk, the oleosomes in plant seeds, and the small fat globules in milk (Figure 3.5). The lipoproteins found in egg yolk are spherical particles (15–60 nm in diameter) with a core containing TAG and cholesterol esters surrounded by a monolayer of phospholipids and apoproteins [23]. Hen egg yolk is an efficient ingredient in many food products as it combines functional properties (emulsifying, coagulating, and gelling properties), organoleptic, and nutritional properties. The oleosomes in plant seeds are the natural oil droplets (0.1–10 μm in diameter) carrying structures composed by a core of TAG covered by a monolayer of phospholipids and proteins. These nanostructures are submitted to degradation during seed maturation and germination, or during oil extraction performed by industry [24].

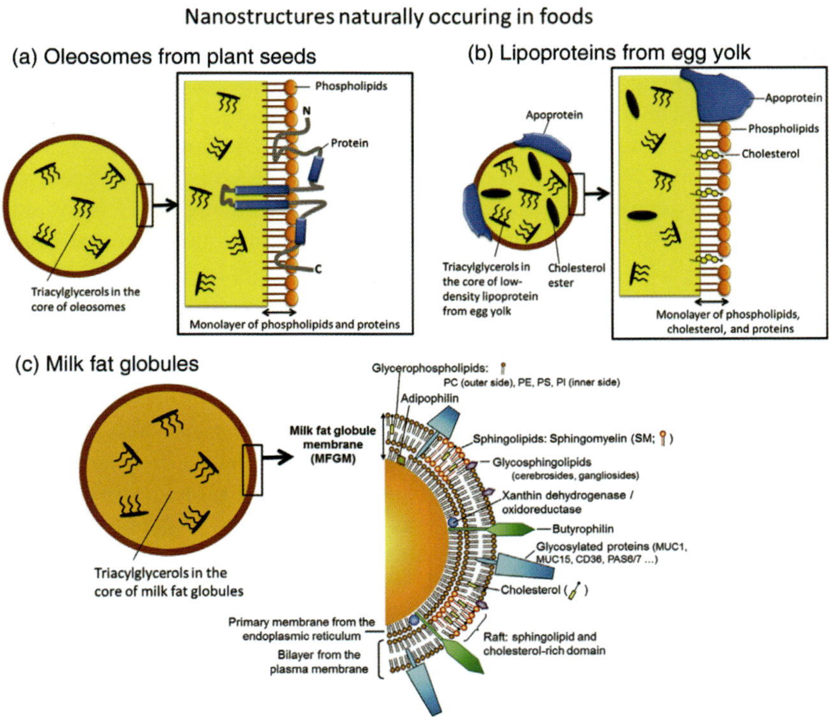

3.3 Lipid-Based Nanostructures

Figure 3.5 Schematic representation of natural nanostructures formed by lipids in food.

3.3.2
Special Case: Milk Fat Globules

Milk lipids are secreted by the mammary epithelial cells in the form of lipid droplets called the milk fat globules [25]. Milk fat globules have a core of TAG covered by a biological membrane composed by a trilayer of phospholipids and proteins (Figure 3.5c). These biological assemblies are composed by lipids with different chemical properties (apolar lipids, polar lipids) and characterized by a multiscale organization of these lipids that can be affected by external factors such as temperature.

Apolar lipids (triacylglycerols, TAG: esters of fatty acids and glycerol; 98% of milk lipids) are located in the core of milk fat globules. The fatty acid composition depends on mammal species and is also highly sensitive to the diet. Milk TAG molecules are rich in saturated fatty acids characterized by a high melting point. As a consequence, milk TAG molecules are partially crystallized below physiological temperature (i.e., for $T < 37\,°C$). Storage conditions (room temperature; fridge: $4-7\,°C$) induce milk TAG crystallization [26]. In the solid state, TAG molecules associate to form lamellar structures (thickness = 4–7.5 nm) that associate to form crystals and network of crystals.

Polar lipids (glycerophospholipids: phosphatidylethanolamine PE, phosphatidylserine PS, phosphatidylcholine PC, and phosphatidylinositol PI, and sphingolipids, mainly sphingomyelin) together with cholesterol and proteins are located in the biological membrane surrounding milk fat globules. This membrane is organized as a trilayer of polar lipids resulting from the mechanisms of milk fat globule secretion. Recently, a lateral organization of polar lipids has been revealed in the outer bilayer of the milk fat globule membrane [27]. Polar lipids with a high phase transition temperature segregate to form domains that could have similarities with lipid rafts present in the cell membranes.

The first function of milk fat globules is to bring about 50% of the energy and bioactive molecules for an optimal growth of all mammal newborns. The size of milk fat globules ranges from about 50 nm to 10 µm (mean diameter = 4 µm). This size distribution raises questions about the potential different biological functions of milk fat globules according to their size. The smallest lipid droplets secreted in milk are natural nanostructures that could have specific biological functions that are currently unknown. However, milk fat globules as naturally secreted are scarcely consumed by infants and adults since the dairy industry uses homogenization, that is, a mechanical treatment able to increase the physical stability of milk fat globules for a longer duration of storage of milk. Homogenization of milk leads to a decrease in the size of milk fat globules (0.15–1 µm) and to the disruption of the biological membrane with adsorption of milk proteins [28]. Natural milk fat globules and processed milk lipid droplets are structurally different, which could have metabolic consequences [29]. Milk fat globules are also involved in the manufacture of many dairy products such as cream and cheese.

Other complex lipid nanostructures have been found in milk, they correspond to lactosomes and exosomes that contain mRNA and miRNA [30–32]. Their biological functions are currently not elucidated.

3.4
Concluding Remarks and Future Prospects

As shown, there are a large number of nanostructures in the food we eat. These structures act naturally as essential carriers for a lot of lipophilic components like ω3 fatty acids, vitamins, or aroma and also hydrophilic elements like minerals. Increasing our knowledge on: (i) the complex mechanisms of proteins or lipids self-assembly that occurs naturally or during processing and cooking and (ii) on the noncovalent interactions between proteins or lipids with bioactive molecules, open the way for the design of tailor-made nanostructures that could carry, protect, and deliver, in a control way, molecules with a high nutritional and health importance (Chapter 4 and [33,34]). The naturally occurring nanostructures coming from milk and eggs constitute a large part of the building blocks already used to design new nanostructures. Recently, there is an increasing interest in using alternative protein and lipid sources (mainly plant-derived sources) in order to meet the constantly growing demand for more sustainable sources and to explore new potentialities. All these mimetic nanostructures are very promising, nevertheless the digestive and metabolic fates of these processed nanostructures are still poorly known and require further studies to assess their real benefits and evaluate the exact risk [35].

References

1 Wang, A., Shi, W., Huang, J., and Yan, Y. (2016) Adaptive soft molecular self-assemblies. *Soft Matter*, **12**, 337–357.

2 Morris, V.J. (2010) Natural food nanostructures, in nanotechnology in food, in *RSC Nanoscience & Nanotechnology 14* (eds Q. Chaudhry, L. Castle, and R. Watkins), Royal Society of Chemistry, Cambridge, UK, pp. 50–68.

3 Rogers, M.A. (2015) Naturally occurring nanoparticles in food. *Curr. Opin. Food Sci.*, **7**, 14–19.

4 Brownlow, S., Morais Cabral, J.H., Cooper, R., Flower, D.R., Yewdall, S.J., Polikarpov, I., North, A.C., and Sawyer, L. (1997) Bovine beta-lactoglobulin at 1.8Å resolution – still an enigmatic lipocalin. *Theochem. J. Mol. Struct.*, **5** (4), 481–495.

5 Majorek, K.A., Porebski, P.J., Dayal, A., Zimmerman, M.D., Jablonska, K., Stewart, A.J., Chruszcz, M., and Minor, W. (2012) Structural and immunologic characterization of bovine, horse, and rabbit serum albumins. *Mol. Immunol.*, **52**, 174–182.

6 Permyakov, E.A. and Berliner, L.J. (2000) α-Lactalbumin: structure and function. *FEBS Lett.*, **473**, 269–274.

7 Huntington, J.A. and Stein, P.E. (2001) Structure and properties of ovalbumin. *J. Chromatogr. B*, **756**, 189–198.

8 Livnah, O., Bayer, E.A., Wilchek, M., and Sussman, J.L. (1993) Three-dimensional structures of avidin and the avidin-biotin complex. *Proc. Nat. Acad. Sci. USA*, **90**, 5076–5080.

9 Wu, J. and Acero-Lopez, A. (2012) Ovotransferrin: structure, bioactivities, and preparation. *Food Res. Intern.*, **46**, 480–487.

10 Christensen, B. and Sørensen, E.S. (2016) Structure, function and nutritional potential of milk osteopontin. *Int. Dairy J.*, **57**, 1–6.

11 Futterman, S. and Hiller, J. (1972) The enhancement of fluorescence and the decreased susceptibility of enzymatic oxidation of retinol complexed with bovine serum albumin, lactoglobulin and the retinol-binding protein of human plasma. *J. Biol. Chem.*, **247**, 5168–5172.

12 Le Maux, S., Bouhallab, S., Giblin, L., Brodkorb, A., and Croguennec, T. (2014) Bovine ß-lactoglobulin/fatty acid complexes: binding, structural, and biological properties. *Dairy Sci. Technol.*, **94**, 409–426.

13 Tavares, G.M., Croguennec, T., Lê, S., Lerideau, O., Hamon, P., Carvalho, A.F., and Bouhallab, S. (2015) Binding of folic acid induces specific self-aggregation of lactoferrin: thermodynamic characterization. *Langmuir*, **31**, 12481–12488.

14 Loch, J., Polit, A., Bonarek, P., Ries, D., Kurpiewska, K., Dziedzicka-Wasylewska, M., and Lewinski, K. (2012) Bovine beta-lactoglobulin complex with linoleic acid. Available at www.cgl.ucsf.edu/chimera.

15 Pérez, O.E., David-Birman, T., Kesselman, E., Levi-Tal, S., and Lesmes, U. (2014) Milk protein–vitamin interactions: Formation of beta-lactoglobulin/folic acid nano-complexes and their impact on *in vitro* gastro-duodenal proteolysis. *Food Hydrocolloid.*, **38**, 40–47.

16 Azuma, N., Maeta, A., Fukuchi, K., and Kanno, C. (2006) A rapid method for purifying osteopontin from bovine milk and interaction between osteopontin and other milk proteins. *Int. Dairy J.*, **16**, 370–378.

17 Holt, C. and Sawyer, L. (1993) Caseins as rheomorphic proteins: interpretation of

primary and secondary structures of the αs1-,β- and k-caseins. *J. Chem. Soc. Faraday Trans.*, **89**, 2683–2692.

18 Fox, P.F. and Brodkorb, A. (2008) The casein micelle: Historical aspects, current concepts and significance. *Int. Dairy J.*, **18**, 677–684.

19 Broyard, C. and Gaucheron, F. (2015) Modifications of structures and functions of caseins: a scientific and technological challenge. *Dairy Sci. Technol.*, **95**, 831–862.

20 Mezzenga, R. and Fischer, P. (2013) The self-assembly, aggregation and phase transitions of food protein systems in one, two and three dimensions. *Rep. Prog. Phys.*, **76**, 046601.

21 Holt, C. (1992) Structure and stability of bovine casein micelles. *Adv. Protein Chem.*, **43**, 63–151.

22 Bouchoux, A., Gésan-Guiziou, G., Pérez, J., and Cabane, B. (2010) How to squeeze a sponge: casein micelles under osmotic stress, a SAXS study. *Biophys. J.*, **99**, 3754–3762.

23 Anton, M. (2013) Egg yolk: structures, functionalities and processes. *J. Sci. Food Agric.*, **93**, 2871–2880.

24 Huang, A.H.C. *et al.* (2012) Synthesis and degradation of lipid bodies in the *Scutella* of maize, in *The Metabolism, Structure, and Function of Plant Lipids* (eds Paul K. Stumpf, J. Brian Mudd, and W. David Nes), Springer, 239–246.

25 Lopez, C. (2011) Milk fat globules enveloped by their biological membrane: unique colloidal assemblies with a specific composition and structure. *Curr. Opin. Colloid Interface. Sci.*, **16**, 391–404.

26 Lopez, C., Bourgaux, C., Lesieur, P., and Ollivon, M. (2007) Coupling of time-resolved synchrotron X-ray diffraction and DSC to elucidate the crystallisation properties and polymorphism of triglycerides in milk fat globules. *Lait*, **87**, 459–480.

27 Lopez, C., Madec, M.-N., and Jimenez-Flores, R. (2010) Presence of lipid rafts in the bovine milk fat globule membrane revealed by the lateral segregation of phospholipids and heterogeneous distribution of glycoproteins. *Food Chem.*, **120**, 22–33.

28 Lopez, C., Cauty, C., and Guyomarc'h, F. (2015) Organization of lipids in milks, infant milk formulas and various dairy products: role of technological processes and potential impacts. *Dairy Sci. Technol.*, **95**, 863–893.

29 Michalski, M.C. (2007) On the supposed influence of milk homogenization on the risk of CVD, diabetes and allergy. *Br. J. Nutr.*, **97**, 598–610.

30 Argov, N., Lemay, D.G. *et al.* (2008) Milk fat globule structure and function: nanoscience comes to milk production. *Trends Food Sci. Technol.*, **19**, 617–623.

31 Argov-Argaman, N., Smilowitz, J.T. *et al.* (2010) Lactosomes: structural and compositional classification of unique nanometer-sized protein lipid particles of human milk. *J. Agric. Food Chem.*, **58**, 11234–11242.

32 Izumi, H., Tsuda, M. *et al.* (2015) Bovine milk exosomes contain microRNA and mRNA and are taken up by human macrophages. *J. Dairy Sci.*, **98**, 2920–2933.

33 Singh, H. (2006) The milk fat globule membrane – a biophysical system for food applications. *Curr. Opin. Colloid Interface Sci.*, **11**, 154–163.

34 Sağlam, D., Venema, P., van der Linden, E., and de Vries, R. (2014) Design, properties, and applications of protein micro- and nanoparticles. *Curr. Opin. Colloid Interface Sci.*, **19**, 428–437.

35 Suran, M. (2014) A little hard to swallow? *EMBO Rep.*, **15** (6), 638–641.

4
Artificial Nanostructures in Food

Jared K. Raynes,[1] Sally L. Gras,[2] John A. Carver,[3] and Juliet A. Gerrard[4,5,6]

[1]*CSIRO, Agriculture & Food, 671 Sneydes Road, Werribee VIC 3030, Australia*
[2]*The ARC Dairy Innovation Hub and Bio21 Molecular Science and Biotechnology Institute, The University of Melbourne, Department of Chemical and Biomolecular Engineering, 30 Flemington Rd, Parkville VIC 3052, Australia*
[3]*Australian National University, Research School of Chemistry, Acton ACT 2601, Australia*
[4]*University of Auckland, School of Biological Sciences and School of Chemical Sciences, Private Bag 92019, Auckland 1010, New Zealand*
[5]*Massey University, Riddet Institute, University Ave, Palmerston North 4474, New Zealand*
[6]*Victoria University, MacDiarmid Institute for Advanced Materials and Nanotechnology, 6 Kelburn Parade, Kelburn, Wellington 6140, New Zealand*

4.1
Introduction

Artificial organic nanostructure research has blossomed in all areas of the agri-food industry over the last decade. This research has provided potential routes to the manufacture of particles and materials that can be fine-tuned to give precise properties and functions at a very small scale [1]. Potential benefits of utilizing artificial organic nanostructures in the agri-food industry include improved nutrition (e.g., targeted delivery of nutrients), improved food quality (e.g., increased shelf life or fat and sugar replacement), safer food (e.g., active-microbe detecting materials), or more sustainable food production (e.g., efficient delivery of pesticides or micronutrients) [2]. For the benefits to be realized and products accepted, regulators need to ensure the safety of all new nanotechnology and consumers must be properly informed about the advances and benefits of nanotechnology in the agri-food industry.

Nature produces all manner of nanostructures such as proteins, DNA, polysaccharides, lipids, viruses, and bacteria. Researchers also harness architectural cues from nature as the basis for the design of artificially produced nanostructures [3]. The description "artificial" in this chapter is applied to any organic material that has been intentionally altered to produce a structure on the nanoscale. Humans have been consuming such organic nanostructures for thousands

---

Nanotechnology in Agriculture and Food Science, First Edition. Edited by Monique A.V. Axelos and Marcel Van de Voorde.
© 2017 Wiley-VCH Verlag GmbH & Co. KGaA. Published 2017 by Wiley-VCH Verlag GmbH & Co. KGaA.

of years. The intentional production of nanostructures for utilization in foods, however, is a relatively new concept. Cooking of food is a classic example of altering native nanostructures where the structure of proteins is altered. For example, when an egg is cooked, ovalbumin, the major protein in the egg white, denatures to form an aggregated, cross-linked protein gel [4]. The process of cooking is considered by many to be a natural process because the intention is not to produce nanostructures with unique properties, nevertheless such structural changes occur at the nanoscale.

During the manufacture of food products, foods can be subjected to a variety of external forces and environments that can modify or change their structure. Examples during processing include heating, shear (as occurs during mixing, pumping, filtration), high pressure, pulsed electric fields, sonication, and drying [1]. In addition, traditional food preservation techniques such as fermentation, acidification, and salting transform food structures. With most of these processes, their use is not intended to produce artificial nanostructures but to sterilize and increase the shelf life of the food product. A relatively new food production technology that harnesses multiple external forces to create new artificial food structures is high moisture extrusion cooking (HMEC). This process simultaneously introduces shear, heat, and pressure to convert protein, lipids, and polysaccharides into novel food structures such as meat analogues [5]. The fibrous meat-like protein structures that form are distinct from the globular structures of these proteins that exist under normal conditions. This technology provides a good example of differentiation between natural and artificial structures, because HMEC purposefully produces artificial food structures. There are many different types of meat analogues currently on the market and public perception and acceptance of these artificially produced foods is positive, with the main barrier to eating them being their low sensory appeal, not their method of production [6].

Artificial organic nanostructures can be produced from the three primary food components: proteins, polysaccharides, and lipids. Nanostructures can be composed of just one component, but are usually a combination of these components. In contrast, hybrid artificial nanostructures incorporate both organic and inorganic components. The two classical approaches to developing inorganic nanostructures, that is, bottom-up or top-down, also apply to the manufacture of artificial organic nanostructures. An example of the bottom-up approach to engineering artificial organic nanostructures is the self-assembly of proteins and peptides into higher order nanostructures such as protein nanotubes [7]. An example of the top-down approach is the formation of lipid nanoparticles by high-pressure homogenization of lipids [8]. There are many different types and methods for producing artificial organic nanostructures dependent upon the functional need for the nanostructure. An overview of the types and functions of protein, polysaccharide, and lipid artificial organic nanostructures, either in current use or with potential use in the agri-food industry is presented in Table 4.1.

Food packaging is the most active area of nanotechnology research that utilizes artificial organic nanostructures for the agri-food industry. Food packaging maintains the quality, safety, and shelf life of food and beverages during

Table 4.1 Overview of selected artificial organic nanostructures utilized in the agri-food industry with examples from the literature (schematic structures are not to scale). (Adapted from Ref. [9]. Published by The Royal Society of Chemistry.)

Artificial organic nanostructure	Method of preparation	Structure	References
Lipid and surfactant based			
Nanoemulsions	High-pressure homogenization, ultrasound-assisted homogenization		[10]
Solid lipid nanoparticles	Hot emulsification of high melting lipids		[11]
Lipid nanocarriers (LNC), nanostructured lipid carriers (NLC)	Hot emulsification of high melting lipids with certain proportion of low melting lipids		[12]
Nanoliposomes/vesicles	Mixture of phospholipids, evaporation of solvent under reduced pressure		[13]
Polysaccharide based			
Molecular complexes: cyclodextrin inclusion complexes, amylose complexes	Solubilization under appropriate conditions		[14]
Nanoparticles	Self-assembly		[15]
Nanocrystals	Comminution of larger materials		[16]
Protein based			
Protein nanotubes, rods, fibres	Self-assembly		[17]
Reassembled milk casein micelles	Self-assembly		[18]
Protein inclusion complexes	Solubilization under appropriate conditions		[19]
Coacervates (electrostatically driven colloids)	Complex coacervate method using oppositely charged polymers		[20]

transportation and storage by controlling the permeation of moisture and gases and offers protection from adulterating flavors and taints. Advances in food packaging utilizing nanotechnology have the potential to offer routes to material manufacture that are not only environmentally friendly but also increase the food safety of packaged food.

In this chapter, an overview is presented of selected types and uses of artificial organic nanostructures derived from proteins, polysaccharides, and lipids with some examples and their potential uses.

4.2
Types and Uses of Artificial Organic Nanostructures Found in Food

4.2.1
Protein Nanostructures

Proteins provide an abundant and highly tuneable source of material for the production of artificial nanostructures. *In vivo* this is evident, as most of the essential components of cells and organs are formed through the assembly of proteins. Enzymes are the workhorses of cells and are generally comprised of proteins. Their unique properties stem from the 20 common amino acids that form their polypeptide chains. The functional side chains of the amino acids provide an array of chemical groups and combinations that give rise to different hydrophobicity, ionization, and polarity. The organization and types of amino acids present in a protein polypeptide chain determine the overall structure and functional properties of the protein. Examples of nanostructures manufactured from proteins include: nanotubes, nanofibrils, nanorods, nanocarriers, nanocomposites, nanomaterials, coacervates, nanocomplexes, and enzymes.

Protein nanotubes, fibrils, and rods (Figure 4.1) fall within a similar group of nanostructures and have potential applications in foods as additives for increasing viscosity, inducing gelation, and for use as an encapsulant. They are grouped together because of their ability to form specific nanostructures via a self-assembly process using as the building blocks either full-length proteins or peptides derived from proteins [21].

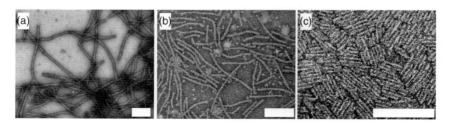

Figure 4.1 Examples of (a) α-lactalbumin protein nanotubes, (b) κ-casein protein fibrils, and (c) fungal hydrophobin rodlets. The scale bars represent 200 nm. (Image (a) is adapted from Ref. [22] with permission from Elsevier, image (b) is from the author, and image (c) is adapted from Ref. [23] with permission from Springer.)

Table 4.2 Amyloid fibril nanostructures formed from proteins found in foods. (Modified from Ref. [1].)

Amyloid forming protein	Source	References
Lysozyme	Egg white	[27]
β-Lactoglobulin (β-Lg)	Milk whey	[28]
Ovalbumin	Egg white	[29]
Hydrophobins	Fungi	[30]
Hydrophobins	Bacteria	[31]
α_{s2}-Casein	Milk	[32]
κ-Casein	Milk	[33]
Bovine serum albumin (BSA)	Red meat	[34]
Glycinin	Soy	[35]
β-Conglycinin	Soy	[35]
7 S Globulin	Kidney bean	[36]

These nanostructures, and in particular amyloid fibril nanostructures, are proposed to be able to be produced from any protein under the appropriate destabilizing conditions [24]. Amyloid fibrils are a protein nanostructure characterized by a repetitive β-sheet structure that runs perpendicular to the fibril axis and can be either natural or artificial in their preparation [25]. There are many types of food-derived proteins that are currently used to produce artificial nanofibrils (Table 4.2). The appeal of protein nanotubes/fibrils stems from their ability to self-assemble under appropriate conditions, their high aspect ratio (length versus width) and great strength, allow them to form entangled networks at low concentrations, which act as efficient gelling agents while preserving a low caloric burden [26].

Amyloid fibril nanostructures have received wide-spread research attention due to their association with over 30 protein misfolding related diseases, for example, Alzheimer's and Parkinson's [25]. The risk of using these nanostructures in food has therefore been questioned [1]. However, recent *in vitro* evidence from Lassé *et al.* [37] indicates that food-derived amyloid fibrils from whey, soy, kidney bean, and egg white are not nontoxic to Caco-2 and Hec-1a human cell lines. As a result, some of the safety questions around the use of these nanostructures in food have been alleviated. However, it is well recognized that alteration in the morphology and functionality of fibrillar nanostructures occurs even with small changes in the fibril-forming conditions [38]. Hence, the toxicity of food-derived amyloid fibrils could be altered if they are made from other food protein sources and if the methodologies to produce the nanostructures are altered.

Currently, there are a number of potential uses for fibrillar nanostructures in food. For example, hydrophobins that are produced by filamentous fungi [30] and some bacteria [31] are being investigated as a protein biosurfactant, foaming agent, and encapsulating agent. Fungal hydrophobins are highly surface-active and form rodlets (amyloid-like nanostructures) at air–water interfaces [30], while bacterial hydrophobins self-assemble at surfaces via a structural rearrangement to form three-stranded helices. One promising application is the use of hydrophobins as a low-calorie surfactant to create air-filled emulsions in ice cream [39] and to reduce the speed of ice cream melt [40]. The many promising applications of hydrophobins are currently limited because they are not authorized for use as a food additive in the European Union due to problems with manufacturing consistently pure preparations [17].

Protein nanoparticles, nanoemulsions, coacervates, and nanocomplexes are another promising source of artificial nanostructures being investigated for use in the food industry. The primary uses of these nanoparticles in food are as vehicles for carrying and delivering functional ingredients including: polyphenols, vitamins, antioxidants, antimicrobials, flavors, colors, and bioactive compounds [41]. Using a carrier can improve the bioavailability of the encapsulated compounds, control their release rates, increase their stability, and thereby increase efficacy of the compounds in humans. Many different proteins are currently being investigated as protein nanoparticle carriers (Table 4.3). These artificial nanostructures are produced using a range of methods including

Table 4.3 Examples of proteins used as nanoparticle carriers for a range of bioactives. (Modified from Ref. [45].)

Nanocarrier	Bioactive	References
Animal proteins		
β-Lg	Folic acid	[46]
β-Lg	Retinol and EGCG	[47]
β-Lg	Curcumin	[48]
β-Lg	Polyphenol extracts of teas, coffee, and cocoa	[49]
β-Lg	Oleic and linoleic acid	[50,51]
β-Lg	Narangin and naringenin	[52]
β-Lg, BSA, α-lactalbumin	Folic acid	[53]
BSA	EGCG	[54]
Casein	Catechin	[55]
Ovalbumin	Caffeine, theophylline, and diprophylline	[56]
Ovalbumin and lysozyme	Tea polyphenol	[57]
Plant proteins		
β-Conglycinin	Vitamin D	[58]
β-Conglycinin	Curcumin	[59]
Canola protein and pea protein isolates	Ketones	[60]
Gliadin and zein	Resveratrol	[61]

high-pressure homogenization, enzymatic cross-linking, self-assembly, and precipitation. The method of production depends on the protein used. For example, casein proteins self-assemble into micelles under physiological conditions, whereas β-lactoglobulin nanoparticles can be produced via precipitation through denaturation by heating. Specific examples of the use of protein nanoparticle carriers include to: (i) increase the solubility of hydrophobic polyphenols such as curcumin [42], (ii) increase the bioavailability of compounds like vitamin D3 [43], and (iii) increase the stability of polyphenols against external factors such as light, heat, processing, and storage, for the controlled release of compounds [44].

Protein nanomaterials represent an exciting opportunity for the production of novel packaging as proteins are biodegradable, can be eaten, and have good mechanical properties. Through the use of enzymes, protein materials can also act as antimicrobial agents (e.g., glucose oxidase produces H_2O_2) or oxygen scavengers. Many different proteins are currently being investigated for producing nanomaterials; they include whey protein isolate, soy, gelatin, zein, gluten, amaranth, and canola seed protein. Among all of these protein sources, gelatin is currently the most extensively studied for alternative food packaging as it produces a strong and transparent film [62].

Protein nanomaterials that are being produced for food packaging, need to be nontoxic, stable, of high strength, optically transparent, have good barrier properties, and derived from a cheap source, such as a byproduct stream from another food processing activity [63]. Recently, Oymaci and Altinkaya [64] produced a whey protein isolate (WPI) film incorporating nanoparticles made from zein (a protein from corn meal). These films had improved mechanical and water barrier properties compared to the WPI-only film, showing their potential as a new food packaging material.

4.2.2
Polysaccharide Nanostructures

Polysaccharides are polymers of carbohydrates made of monosaccharides linked together through glycosidic bonds. The two functional roles of polysaccharides in biology are to provide structural support (e.g., cellulose in plants or chitin in arthropods) or act as a source of energy (e.g., glycogen in animals and fungi and starch in plants). Polysaccharides get their functionality from their monosaccharide building blocks, the types of linkage joining them, isomeric forms, modifications (e.g., methylation), branching, and periodicity of the monomers [65]. Polysaccharides are commonly classified as homopolysaccharides (contains a single sugar backbone, for example, amylose) or heteropolysaccharides (contains more than one type of sugar, for example, pectin).

Polysaccharide nanostructures have received widespread attention for their use in the agri-food industry, particularly in the fields of controlled release and packaging, due to their almost limitless availability. For example, cellulose and chitin, derived from plants and crustaceans/insects, respectively, are the first and second most abundant biopolymers on earth [66]. Polysaccharides also have a long history of safe use in food applications, are biocompatible, nontoxic, and degrade in the environment. Some polysaccharides also possess the ability to self-assemble in response to a change in environmental conditions, which can be desirable for maintaining structural integrity or releasing carrier payloads. For example, chitosan, a derivative of chitin, self-assembles to form a gel via a pH change from acidic to neutral/basic conditions, whereas alginate self-assembles to form a gel upon the addition of Ca^{2+} [67]. Polysaccharides can also be modified to induce self-assembly by the addition of molecules, such as esters, to give the polysaccharide amphiphilicity. Many different types of polysaccharides are used to manufacture artificial organic nanoparticles for use in the agri-food industry (Table 4.4).

Table 4.4 Examples of food-grade polysaccharides used for manufacturing nanoparticles. (Modified from Ref. [45].)

Name	Source	Main structure	Major monomer/s
Alginate	Algae	Linear	D-Manuronic and guluronic acids
Amylose (from starch)	Plants (e.g., potato tubers)	Linear	α-1,4-β-D-Glucose
Amylopectin (from starch)	Plants (e.g., potato tubers)	Branched	α-1,4-β-D-Glucose and α-1,6-β-D-glucose
Arabic gum	Acacia spp.	Branched	3,6-β-D-Galactopyranose and other sugars
Carrageenan	Algae	Linear/helical	Sulfated D-galactose and 3.6 anhydrogalactose
Cellulose	Bacterial (*Glucoacetobacter* spp.)	Fibrils	β-D-Glucose
Cyclodextrins	Bacterial (*Bacillus* spp.)	Cyclic	Cyclic α-1,4 amyloses (C6–C8)
Chitosan	Crustaceans/invertebrates	Linear	2-Amino-2-desoxy-β-D-glucose
Guar gum	Guar beans	Short branched	β-1,4-Mannose and 1,6-galactose
Inulin	Plants or bacteria	Linear	β(2,1)-D-Fructose
Kefiran	Bacteria	Branched	Glucose and kefirose
Pectin	Plant cell walls	Branched/coiled	Methoxylated galacturonic acids
Pululan	Fungal (*Aureobasidium pullulans*)	Linear	α-1,4 Maltotriose
Xanthan	Bacterial (*Xanthomonas campestris*)	Linear/helical (at high MW)	Glucose, mannose, and glucuronic acid (2:2:1)

There are some challenges associated with the use of polysaccharides as a material for nanostructure manufacture. One major limitation is that there are only a restricted number of conformations and chemistries available. This can limit their use when specific properties are required. Heterogeneity is another challenge, as natural biosynthesis produces polysaccharides with differing molecular weights, branching, and chemical modifications [68]. To overcome some of these limitations, chemical modifications can be introduced, for example, cellulose which is usually insoluble can be made water-soluble by a variety of chemical modifications [69]. A more novel approach is to manipulate exocellular polysaccharides by bacteria through genetic engineering that produces novel polysaccharides [70].

There are many different routes to produce polysaccharide nanoparticles including covalent cross-linking, ionic cross-linking, polyelectrolyte complexation and self-assembly (Figure 4.2) [71]. Complexation of molecules with polysaccharides is a well-known strategy to increase their solubility, oral bioavailability and to provide protection from temperature, oxygen, and photodegradation. For example, curcumin is a natural hydrophobic polyphenol with potent anti-inflammatory properties and many other medical applications, but its use is currently constrained by poor solubility, photodegradability, chemical instability, rapid metabolism, and a short half-life, which can be overcome by polysaccharide complexation [72]. This is a good example of the potential of polysaccharide nanostructures, where complexation of a valuable but inherently unstable molecule can be dramatically increased.

Cyclodextrins, commonly regarded as the smallest nanoparticles known [74], are a class of cyclic oligosaccharides of α-(1,4) linked glucopyranose units that are popular for encapsulation. Because of their cyclic nature, cyclodextrins possess a hydrophilic exterior and a hydrophobic internal cavity. As a result of this cavity, cyclodextrins are able to form inclusion complexes with hydrophobic molecules. For example, Mangolim *et al.* [75] developed a curcumin-β-cyclodextrin inclusion complex that exhibited higher sunlight, pH, storage, and heating stability compared to unmodified curcumin. They also developed a proof-of-concept food, ice cream, which contained the curcumin-β-cyclodextrin inclusion complex and obtained good sensorial acceptance by tasting panellists, providing a potential new product with health benefits.

Covalently cross-linked polysaccharide nanoparticles (Figure 4.2a) can be prepared using various cross-linkers, including succinic acid, malic acid, tartaric acid, and citric acid in combination with a condensation agent such as carbodiimide [76]. Recently, Li *et al.* [77] produced cross-linked chitosan nanoparticles by cross-linking with vanillin and demonstrated efficient encapsulation of a model hydrophilic drug, 5-fluorouracil. These nanoparticles show promise as a vehicle for the controlled delivery of anticancer drugs to the colon, which could be incorporated into food.

The use of ionic cross-linking (Figure 4.2b) has several advantages compared to covalent cross-linking, including mild and simple conditions for preparation. For example, polycationic and polyanionic polysaccharides can be cross-linked

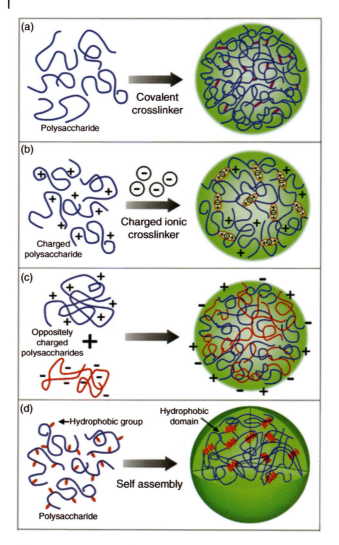

Figure 4.2 Common mechanisms for polysaccharide-based nanoparticle preparation. (a) Covalent cross-linking. (b) Ionic cross-linking. (c) Polyelectrolyte complexation. (d) Self-assembly of hydrophobically modified polysaccharides. (Adapted from Ref. [73] with permission from The Royal Society of Chemistry.)

using low MW polyanions and polycations, respectively. Tripolyphosphate (TPP) is the most commonly used ionic cross-linker used to produce chitosan nanoparticles [78]. This method has been used by Grillo *et al.* [79] to produce chitosan/TPP nanoparticles loaded with paraquat herbicide for controlled release to limit the toxicity in nontarget organisms.

Polyelectrolyte complexation (Figure 4.2c) works in a similar way to ionic cross-linking, whereby polyelectrolyte polysaccharides form complexes with oppositely charged polymers by electrostatic interactions. Again, chitosan, being the only naturally available polycationic polysaccharide, has been used to form complexes with all manner of negatively charged polymers including polysaccharides, peptides, and the polyacrylic acid family [71]. A potential application of such polyelectrolyte nanoparticles is in the delivery and controlled release of proteins, for example, to the colon. Yang *et al.* [80] showed the potential of this approach by producing chitosan–calcium–gellan gum beads that encapsulate albumin as a model for new functional ingredients.

Self-assembling polysaccharide nanoparticles (Figure 4.2d) can be prepared by modifying hydrophilic polysaccharides with hydrophobic moieties to create amphiphilic polysaccharides. When these modified polysaccharides come into contact with an aqueous environment, the amphiphilic nature of the polysaccharides causes them to self-assemble into micelles or micelle-like aggregates of varying size, dependent on the polysaccharide and the added hydrophobic moiety [81]. To illustrate the morphology of these nanoparticles, an example of decanoate modified β-cyclodextrin (β-CDd) viewed by electron microscopy is shown in Figure 4.3. The self-assembly process produces highly repetitive structures with either concentric layers or polygonal particles with a crystal-like core surrounded by a shell of concentric layers. These micelles have been recognized as promising drug and functional food carriers due to their ability to encapsulate hydrophobic molecules.

For food related applications, Raja *et al.* [83] developed self-assembling oleate alginate ester nanoparticles and demonstrated their applicability for encapsulation using curcumin. The self-assembly of these nanoparticles is due to the introduction of hydrophobic methyl oleate to hydrophilic alginate. The encapsulated curcumin showed a significant improvement in aqueous solubility and stability, as well as a sustained and reproducible release profile. These

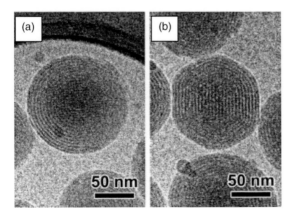

Figure 4.3 Cryo-EM images of β-CDd nanoparticles embedded in vitreous ice: (a) a particle entirely made of concentric layers; (b) a polygonal particle with a crystal-like core surrounded by a shell of concentric layers. (Adapted with permission from Ref. [82]. Copyright (2006) American Chemical Society.)

nanoparticles show promise for the delivery of curcumin and other sensitive hydrophobic molecules.

Polysaccharides are naturally made up of crystalline and amorphous regions. To produce polysaccharide nanocrystals, the amorphous regions are dissolved by acid hydrolysis leaving the nanocrystals. The predominant polysaccharide used to produce nanocrystals is cellulose, although chitin and starch are also used. The morphology and dimensions of the nanocrystals strongly depend on the source and extraction method of the nanocrystals but for cellulose nanocrystals they are typically between 5 and 70 nm wide and from 100 nm to several microns in length [84]. The high aspect ratio and strength (calculated to be potentially stronger than Kevlar) of cellulose nanocrystals means that one of their most promising uses is as a reinforcing agent in composite materials [85]. For example, Cheng *et al.* [86] produced a novel biodegradable guar gum/nanocrystalline cellulose film (GG/NCC), which had increased strength, increased light transmittance, and decreased air permeability when nanocrystalline cellulose was incorporated. The GG/NCC film showed the same preservative properties as the control film and therefore has potential as a novel biodegradable food packaging material.

4.2.3
Lipid Nanostructures

There are many types of lipophilic biomolecules that are recognized as having health benefits in addition to their nutritional role (Table 4.5). However, these bioactive molecules are usually highly hydrophobic and often have poor bioavailability and stability. Their incorporation into lipid nanostructures can therefore greatly improve the utility and potential application of these biomolecules.

Table 4.5 Examples of lipophilic ingredients that have potential health benefits but suffer from poor solubility, stability, or bioavailability and could benefit from encapsulation in lipid nanostructures. (Adapted from Ref. [87].)

Ingredient	Examples	Potential health benefits
Antioxidants	Tocopherols, polyphenols (e.g., curcumin, catechin), resveratrol, coenzyme Q10	Coronary heart disease, urinary tract disease, cancer
Carotenoids	β-Carotene, astaxanthin, zeaxanthin, lycopene, lutein	Coronary heart disease, cancer, macular degeneration, cataracts
Fatty acids	Conjugated linoleic acid, butyric acid, ω-3 fatty acids	Coronary heart disease, bone health, immune response disorders (e.g., diabetes, inflammatory bowel diseases), obesity, mental and visual acuity, cancer, stroke prevention
Phytosterols	β-Sitosterol, campesterol, stigmasterol	Coronary heart disease

The types of lipid nanostructures that can be used for encapsulating biomolecules are nanoliposomes, nanoemulsions, solid lipid nanoparticles (SLNs), and nanostructured lipid carriers (NLCs). The choice of lipid nanostructure is dependent upon the biomolecule that is being encapsulated, the chosen lipid encapsulant (liquid or solid) and the end use of the nanostructure. For example, a lipid nanostructure for delivering biomolecules to the colon will need different stability and breakdown characteristics compared to a biomolecule carrier for the intestine.

Nanoliposomes are spherical particles that encapsulate an aqueous medium and have one or more bilayer membranes composed of polar lipids that feature a lipophilic and hydrophilic group on the same molecule [8]. Upon interaction with the aqueous medium, the polar lipids self-assemble to form the nanoliposomes, encapsulating both hydrophilic and lipophilic molecules either in the trapped aqueous phase or in the lipid section, respectively. A good example of the use of liposomes in food is from Marsanasco et al. [88], in which the authors developed nanoliposomes made of soy phosphatidylcholine containing ω-3 and ω-6 fatty acids that also encapsulated vitamin C and vitamin E for incorporation into orange juice. The nanoliposomes were found to exhibit good stability particularly for thermolabile vitamin C. Sensory evaluation indicated no difference in the acceptability of the orange juice containing liposomes over the control, showing that such liposomes can potentially be used to encapsulate valuable bioactives in food without any detriment to consumer acceptance.

Unlike nanoliposomes, nanoemulsions are made of two immiscible fluids, usually defined as the oil phase and aqueous phases, one being dispersed in the other. Nanoemulsions usually range in size between 10 and 100 nm, are thermodynamically unstable and therefore suffer the drawback of requiring surfactants and/or cosurfactants to stabilize the emulsion [89]. They also usually require a high input of mechanical energy such as a high-pressure homogenization step, or the use of a microfluidizer or sonication to produce the dispersed phases [90]. Nanoemulsions have advantages over traditional-sized emulsions, because their size can be smaller than the wavelength of light and so they only weakly scatter light. This assists with their incorporation into foods because the appearance of the food is not altered. ω-3 Fatty acids have gained widespread attention for their encapsulation into nanoemulsions as this approach can both protect the fatty acids from oxidation and increase their bioavailability. ω-3 Nanoemulsions have potential uses in foods such as beverages, salad dressings, sauces, dips, and desserts. Research has so far focussed on foods that incorporate ω-3 nanoemulsions into table spreads and yoghurts, however the potential applications are much broader [91,92].

SLNs are similar to nanoemulsions but in this case the nanoemulsions contain lipids with melting points above the body and/or room temperature (i.e., 37 or 25 °C, respectively). Examples of lipids that are solid at room temperature include triacylglycerols such as ticaprin, waxes such as myricyl palmitate, and paraffins above C_{20} [93]. The main advantage of SLNs is that these particles offer a release profile that is both targeted and tuneable by changing the solid lipid component, biodegradability, or biocompatibility of the lipid. These particles are also less expensive to produce than surfactant-based nanoparticles [94]. SLNs have some disadvantages though, including highly purified lipids required for crystallization, poor loading capacity and a high water content (70–99.9%) [93]. A recent example using SLNs is from Pandita *et al.* [95], who produced resveratrol-loaded stearic acid SLNs coated with poloxamer 188. The resveratrol-loaded SLNs gave the resveratrol photostability and an ~eight-fold increase in the bioavailability, illustrating the potential success of this approach.

NLCs are a modified version of a SLN that are considered to be an evolution of the SLN that overcomes the poor loading capacity and susceptibility of SLNs to burst release. NLCs also provide long-term stability than SLNs [93]. Each NLC consists of both a solid lipid and a liquid lipid in the lipid phase, as opposed to only a solid lipid phase of SLNs. For example, Ni *et al.* [96] prepared a quercetin-loaded NLC using hot high pressure homogenization by blending glyceryl monostearate and glycerol monolaurate (solid lipids) with monolaurate caprylic capric triglyceride (a liquid lipid) along with an emulsifier mix (polyglyceryl-10 laurate, polyglycerol-6 monostearate, and sucrose ester-11) to stabilize the particles. The quercetin-containing NLCs had good stability, their antioxidant activity was comparable to free quercetin, they showed sustained release when tested in *in vitro* digestion models and were stable during 60 days of storage in a model beverage.

4.3
Conclusion

The increase in research activity investigating artificial organic nanostructures from proteins and polysaccharides to lipids provides the agri-food industry with a deep toolbox that can be used to assist both food production and product development. In the future the use of nanotechnology in food is likely to increase and the potential benefits for consumers include more efficient food production, environment-friendly alternatives for many current petroleum-based food products (e.g., food packaging) and safer and more nutritious foods.

In terms of the best material for production of artificial organic nanostructures, proteins and polysaccharides have tremendous benefits, including biochemical diversity, the potential for beneficial native biological activity, and high biocompatibility and biodegradability. In comparison, lipids usually require greater refinement, the use of complementary molecules, such as emulsifiers and can be intrinsically less stable. Nevertheless, lipid nanostructures can excel in encapsulating particular lipidic and hydrophobic biomolecules increasing their bioavailability and stability. In conclusion, once safety has been established (refer to the chapter on nanosafety), nature provides a modern nanomaterial scientist with almost limitless options to create new materials and potentially solve problems in our current food production chain.

References

1 Raynes, J.K., Carver, J.A., Gras, S.L., and Gerrard, J.A. (2014) Protein nanostructures in food – should we be worried? *Trends Food Sci. Technol.*, **37** (1), 42–50.

2 Singh, H. (2016) Nanotechnology applications in functional foods; opportunities and challenges. *Prev. Nutr. Food Sci.*, **21** (1), 1–8.

3 Howorka, S. (2011) Rationally engineering natural protein assemblies in nanobiotechnology. *Curr. Opin. Biotechnol.*, **22** (4), 485–491.

4 Brenner, M.P. and Sörensen, P.M. (2015) Biophysics of molecular gastronomy. *Cell*, **161** (1), 5–8.

5 Cheftel, J.C., Kitagawa, M., and Quéguiner, C. (1992) New protein texturization processes by extrusion cooking at high moisture levels. *Food Rev. Int.*, **8** (2), 235–275.

6 Hoek, A.C., Luning, P.A., Weijzen, P. et al. (2011) Replacement of meat by meat substitutes. A survey on person- and product-related factors in consumer acceptance. *Appetite*, **56** (3), 662–673.

7 Raynes, J.K. and Gerrard, J.A. (2013) Amyloid fibrils as bionanomaterials, in *Bionanotechnology: Biological Self-assembly and its Applications*, Caister Academic Press, Norfolk, UK, pp. 85–106.

8 Jesorka, A. and Orwar, O. (2008) Liposomes: technologies and analytical applications. *Annu. Rev. Anal. Chem.*, **1** (1), 801–832.

9 Oehlke, K., Adamiuk, M., Behsnilian, D. et al. (2014) Potential bioavailability enhancement of bioactive compounds using food-grade engineered nanomaterials: a review of the existing evidence. *Food Funct.*, **5** (7), 1341–1359.

10 Ozturk, B., Argin, S., Ozilgen, M., and McClements, D.J. (2015) Nanoemulsion delivery systems for oil-soluble vitamins: influence of carrier oil type on lipid digestion and vitamin D3 bioaccessibility. *Food Chem.*, **187**, 499–506.

11 Salminen, H., Gömmel, C., Leuenberger, B.H., and Weiss, J. (2016) Influence of encapsulated functional lipids on crystal structure and chemical stability in solid lipid nanoparticles: towards bioactive-based design of delivery systems. *Food Chem.*, **190**, 928–937.

12 Dan, N. (2016) Compound release from nanostructured lipid carriers (NLCS). *J. Food Eng.*, **171**, 37–43.

13 Malheiros, P.S., Cuccovia, I.M., and Franco, B.D.G.M. (2016) Inhibition of *Listeria monocytogenes in vitro* and in goat milk by liposomal nanovesicles containing bacteriocins produced by *Lactobacillus sakei subsp.* Sakei 2a. *Food Control*, **63**, 158–164.

14 Aytac, Z., Kusku, S.I., Durgun, E., and Uyar, T. (2016) Quercetin/β-cyclodextrin inclusion complex embedded nanofibres: slow release and high solubility. *Food Chem.*, **197** (Part B), 864–871.

15 Alhaique, F., Matricardi, P., DiMeo, C. et al. (2015) Polysaccharide-based self-assembling nanohydrogels: an overview on 25-years research on pullulan. *J. Drug Deliv. Sci. Technol.*, **30** (Part B), 300–309.

16 Huang, J., Chang, P.R., and Dufresne, A. (2015) Polysaccharide nanocrystals: current status and prospects in material science, in *Polysaccharide-Based Nanocrystals* (eds J. Huang, P.R. Chang, N. Lin, and A. Dufresne), Wiley-VCH Verlag GmbH, pp. 1–14.

17 Khalesi, M., Gebruers, K., and Derdelinckx, G. (2015) Recent advances in fungal hydrophobin towards using in industry. *Protein J.*, **34** (4), 243–255.

18 Haham, M., Ish-Shalom, S., Nodelman, M. et al. (2012) Stability and bioavailability of vitamin D nanoencapsulated in casein micelles. *Food Funct.*, **3** (7), 737–744.

19 Zorilla, R., Liang, L., Remondetto, G., and Subirade, M. (2011) Interaction of epigallocatechin-3-gallate with β-lactoglobulin: molecular characterization and biological implication. *Dairy Sci. Technol.*, **91** (5), 629–644.

20 de Kruif, C.G. (Kees), Pedersen, J., Huppertz, T., and Anema, S.G. (2013) Coacervates of lactotransferrin and β- or κ-casein: structure determined using SAXS. *Langmuir*, **29** (33), 10483–10490.

21 Mandal, D., Shirazi, A.N., and Parang, K. (2014) Self-assembly of peptides to nanostructures. *Org. Biomol. Chem.*, **12** (22), 3544–3561.

22 Graveland-Bikker, J.F. and de Kruif, C.G. (2006) Unique milk protein based nanotubes: food and nanotechnology meet. *Trends Food Sci. Technol.*, **17** (5), 196–203.

23 Morris, V. and Sunde, M. (2013) Formation of amphipathic amyloid monolayers from fungal hydrophobin proteins, in *Protein Nanotechnology* (ed. J.A. Gerrard), Humana Press, pp. 119–129.

24 Dobson, C.M. (1999) Protein misfolding, evolution and disease. *Trends Biochem. Sci.*, **24** (9), 329–332.

25 Chiti, F. and Christopher, M. Dobson (2006) Protein misfolding, functional amyloid, and human disease. *Annu. Rev. Biochem.*, **75**, 333–366.

26 Loveday, S.M., Su, J., Rao, M.A. *et al.* (2011) Effect of calcium on the morphology and functionality of whey protein nanofibrils. *Biomacromolecules*, **12** (10), 3780–3788.

27 Goda, S., Takano, K., Yutani, K. *et al.* (2000) Amyloid protofilament formation of hen egg lysozyme in highly concentrated ethanol solution. *Protein Sci.*, **9** (2), 369–375.

28 Gosal, W.S., Clark, A.H., Pudney, P.D.A., and Ross-Murphy, S.B. (2002) Novel amyloid fibrillar networks derived from a globular protein: β-lactoglobulin. *Langmuir*, **18** (19), 7174–7181.

29 Pearce, F.G., Mackintosh, S.H., and Gerrard, J.A. (2007) Formation of amyloid-like fibrils by ovalbumin and related proteins under conditions relevant to food processing. *J. Agric. Food Chem.*, **55** (2), 318–322.

30 Mackay, J.P., Matthews, J.M., Winefield, R.D. *et al.* (2001) The hydrophobin EAS is largely unstructured in solution and functions by forming amyloid-like structures. *Structure*, **9** (2), 83–91.

31 Hobley, L., Ostrowski, A., Rao, F.V. *et al.* (2013) BslA is a self-assembling bacterial hydrophobin that coats the *Bacillus subtilis* biofilm. *Proc. Natl. Acad. Sci. USA*, **110** (33), 13600–13605.

32 Thorn, D.C., Ecroyd, H., Sunde, M. *et al.* (2008) Amyloid fibril formation by bovine milk α_{s2}-casein occurs under physiological conditions yet is prevented by its natural counterpart, α_{s1}-casein. *Biochemistry*, **47** (12), 3926–3936.

33 Thorn, D.C., Meehan, S., Sunde, M. *et al.* (2005) Amyloid fibril formation by bovine milk κ-casein and its inhibition by the molecular chaperones α_s- and β-casein. *Biochemistry*, **44** (51), 17027–17036.

34 Holm, N.K., Jespersen, S.K., Thomassen, L.V. *et al.* (2007) Aggregation and fibrillation of bovine serum albumin. *Biochim. Biophys. Acta*, **1774** (9), 1128–1138.

35 Tang, C.-H. and Wang, C.-S. (2010) Formation and characterization of amyloid-like fibrils from soy β-conglycinin and glycinin. *J. Agric. Food Chem.*, **58** (20), 11058–11066.

36 Tang, C.-H., Zhang, Y.-H., Wen, Q.-B., and Huang, Q. (2010) Formation of amyloid fibrils from kidney bean 7S globulin (Phaseolin) at pH 2.0. *J. Agric. Food Chem.*, **58** (13), 8061–8068.

37 Lassé, M., Ulluwishewa, D., Healy, J. *et al.* (2016) Evaluation of protease resistance and toxicity of amyloid-like food fibrils from whey, soy, kidney bean, and egg white. *Food Chem.*, **192**, 491–498.

38 Krebs, M.R.H., Devlin, G.L., and Donald, A.M. (2009) Amyloid fibril-like structure underlies the aggregate structure across the pH range for β-lactoglobulin. *Biophys. J.*, **96** (12), 5013–5019.

39 Tchuenbou-Magaia, F.L., Norton, I.T., and Cox, P.W. (2009) Hydrophobins stabilised air-filled emulsions for the food industry. *Food Hydrocoll.*, **23** (7), 1877–1885.

40 Stanley-Wall, N.R. and MacPhee, C.E. (2015) Connecting the dots between bacterial biofilms and ice cream. *Phys. Biol.*, **12** (6), 063001.

41 Zeeb, B., Fischer, L., and Weiss, J. (2014) Stabilization of food dispersions by enzymes. *Food Funct.*, **5** (2), 198–213.

42 Chen, F.-P., Li, B.-S., and Tang, C.-H. (2015) Nanocomplexation of soy protein isolate with curcumin: influence of ultrasonic treatment. *Food Res. Int.*, **75**, 157–165.

43. Diarrassouba, F., Garrait, G., Remondetto, G. et al. (2015) Improved bioavailability of vitamin D3 using a β-lactoglobulin-based coagulum. *Food Chem.*, **172**, 361–367.
44. Munin, A. and Edwards-Lévy, F. (2011) Encapsulation of natural polyphenolic compounds; a review. *Pharmaceutics*, **3** (4), 793–829.
45. Santiago, L.G. and Castro, G.R. (2016) Novel technologies for the encapsulation of bioactive food compounds. *Curr. Opin. Food Sci.*, **7**, 78–85.
46. Pérez, O.E., David-Birman, T., Kesselman, E. et al. (2014) Milk protein–vitamin interactions: formation of beta-lactoglobulin/folic acid nano-complexes and their impact on *in vitro* gastro-duodenal proteolysis. *Food Hydrocoll.*, **38**, 40–47.
47. Keppler, J.K., Sönnichsen, F.D., Lorenzen, P.-C., and Schwarz, K. (2014) Differences in heat stability and ligand binding among β-lactoglobulin genetic variants A, B and C Using ^{1}H NMR and fluorescence quenching. *Biochim. Biophys. Acta*, **1844** (6), 1083–1093.
48. Li, M., Ma, Y., and Ngadi, M.O. (2013) Binding of curcumin to β-lactoglobulin and its effect on antioxidant characteristics of curcumin. *Food Chem.*, **141** (2), 1504–1511.
49. Stojadinovic, M., Radosavljevic, J., Ognjenovic, J. et al. (2013) Binding affinity between dietary polyphenols and β-lactoglobulin negatively correlates with the protein susceptibility to digestion and total antioxidant activity of complexes formed. *Food Chem.*, **136** (3–4), 1263–1271.
50. Le Maux, S., Bouhallab, S., Giblin, L. et al. (2013) Complexes between linoleate and native or aggregated β-lactoglobulin: interaction parameters and *in vitro* cytotoxic effect. *Food Chem.*, **141** (3), 2305–2313.
51. Fang, B., Zhang, M., Tian, M., and Ren, F.Z. (2015) Self-assembled β-lactoglobulin–oleic acid and β-lactoglobulin–linoleic acid complexes with antitumor activities. *J. Dairy Sci.*, **98** (5), 2898–2907.
52. Shpigelman, A., Shoham, Y., Israeli-Lev, G., and Livney, Y.D. (2014) β-lactoglobulin–naringenin complexes: nano-vehicles for the delivery of a hydrophobic nutraceutical. *Food Hydrocoll.*, **40**, 214–224.
53. Liang, L., Zhang, J., Zhou, P., and Subirade, M. (2013) Protective effect of ligand-binding proteins against folic acid loss due to photodecomposition. *Food Chem.*, **141** (2), 754–761.
54. Zheng Li, J.-H.H. (2014) Fabrication of coated bovine serum albumin (BSA)-epigallocatechin gallate (EGCG) nanoparticles and their transport across monolayers of human intestinal epithelial caco-2 cells. *Food Amp Funct.*, **5**, 1278–1285.
55. Haratifar, S. and Corredig, M. (2014) Interactions between tea catechins and casein micelles and their impact on rennetting functionality. *Food Chem.*, **143**, 27–32.
56. Wang, R., Yin, Y., Li, H. et al. (2012) Comparative study of the interactions between ovalbumin and three alkaloids by spectrofluorimetry. *Mol. Biol. Rep.*, **40** (4), 3409–3418.
57. Shen, F., Niu, F., Li, J. et al. (2014) Interactions between tea polyphenol and two kinds of typical egg white proteins—ovalbumin and lysozyme: effect on the gastrointestinal digestion of both proteins *in vitro*. *Food Res. Int.*, **59**, 100–107.
58. Levinson, Y., Israeli-Lev, G., and Livney, Y.D. (2014) Soybean β-conglycinin nanoparticles for delivery of hydrophobic nutraceuticals. *Food Biophys.*, **9** (4), 332–340.
59. David, S., Zagury, Y., and Livney, Y.D. (2014) Soy β-conglycinin–curcumin nanocomplexes for enrichment of clear beverages. *Food Biophys.*, **10** (2), 195–206.
60. Wang, K. and Arntfield, S.D. (2015) Binding of selected volatile flavour mixture to salt-extracted canola and pea proteins and effect of heat treatment on flavour binding. *Food Hydrocoll.*, **43**, 410–417.
61. Joye, I.J., Davidov-Pardo, G., Ludescher, R.D., and McClements, D.J. (2015) Fluorescence quenching study of resveratrol binding to zein and gliadin: towards a more rational approach to resveratrol encapsulation using water-

61 insoluble proteins. *Food Chem.*, **185**, 261–267.
62 Mellinas, C., Valdés, A., Ramos, M. *et al.* (2016) Active edible films: current state and future trends. *J. Appl. Polym. Sci.*, **133** (2), 1–15.
63 Smolander, M. and Chaudhry, Q. (2010) Nanotechnologies in food packaging, in *Nanotechnologies in Food*, Royal Society of Chemistry, pp. 86–101.
64 Oymaci, P. and Altinkaya, S.A. (2016) Improvement of barrier and mechanical properties of whey protein isolate based food packaging films by incorporation of zein nanoparticles as a novel bionanocomposite. *Food Hydrocoll.*, **54** (Part A), 1–9.
65 Kontogiorgos, V. (2014) Polysaccharide nanostructures, in *Edible Nanostructures: A Bottom-up Approach*, Royal Society of Chemistry, UK, pp. 41–68.
66 Synowiecki, J. and Al-Khateeb, N.A. (2003) Production, properties, and some new applications of chitin and its derivatives. *Crit. Rev. Food Sci. Nutr.*, **43** (2), 145–171.
67 Xiong, Y., Qu, X., Liu, C. *et al.* (2015) Polysaccharide-based smart materials, in *Chemoresponsive Materials: Stimulation by Chemical and Biological Signals*, Royal Society of Chemistry, UK.
68 Kajiwara, K. and Takeaki, M. (2004) Progress in structural characterization of functional polysaccharides, in *Polysaccharides: Structural Diversity and Functional Versatility*, 2nd edn, CRC Press, USA, pp. 1–40.
69 Habibi, Y. (2014) Key advances in the chemical modification of nanocelluloses. *Chem. Soc. Rev.*, **43** (5), 1519–1542.
70 Rehm, B.H.A. (2015) Synthetic biology towards the synthesis of custom-made polysaccharides. *Microb. Biotechnol.*, **8** (1), 19–20.
71 Liu, Z., Jiao, Y., Wang, Y. *et al.* (2008) Polysaccharides-based nanoparticles as drug delivery systems. *Adv. Drug Deliv. Rev.*, **60** (15), 1650–1662.
72 Mehanny, M., Hathout, R.M., Geneidi, A.S., and Mansour, S. (2016) Exploring the use of nanocarrier systems to deliver the magical molecule; curcumin and its derivatives. *J. Control. Release*, **225**, 1–30.

73 Mizrahy, S. and Peer, D. (2012) Polysaccharides as building blocks for nanotherapeutics. *Chem Soc Rev*, **41** (7), 2623–2640.
74 Moriyama, H., Saito, Y., and Bagchi, D. (2013) Characterization of cyclodextrin nanoparticles as emulsifiers, in *Bio-Nanotechnology* (eds D. Bagchi, N. Bagchi, H. Moriyama, and F. Shahidi), Blackwell Publishing Ltd., pp. 476–486.
75 Mangolim, C.S., Moriwaki, C., Nogueira, A.C. *et al.* (2014) Curcumin–β-cyclodextrin inclusion complex: stability, solubility, characterisation by FT-IR, FT-Raman, X-Ray diffraction and photoacoustic spectroscopy, and food application. *Food Chem.*, **153**, 361–370.
76 Bodnar, M., Hartmann, J.F., and Borbely, J. (2005) Preparation and characterization of chitosan-based nanoparticles. *Biomacromolecules*, **6** (5), 2521–2527.
77 Li, P.-W., Wang, G., Yang, Z.-M. *et al.* (2016) Development of drug-loaded chitosan–vanillin nanoparticles and its cytotoxicity against HT-29 cells. *Drug Deliv.*, **23** (1), 30–35.
78 Kashyap, P.L., Xiang, X., and Heiden, P. (2015) Chitosan nanoparticle based delivery systems for sustainable agriculture. *Int. J. Biol. Macromol.*, **77**, 36–51.
79 Grillo, R., Pereira, A.E.S., Nishisaka, C.S. *et al.* (2014) Chitosan/tripolyphosphate nanoparticles loaded with paraquat herbicide: an environmentally safer alternative for weed control. *J. Hazard. Mater.*, **278**, 163–171.
80 Yang, F., Xia, S., Tan, C., and Zhang, X. (2013) Preparation and evaluation of chitosan-calcium-gellan gum beads for controlled release of protein. *Eur. Food Res. Technol.*, **237** (4), 467–479.
81 Letchford, K. and Burt, H. (2007) A review of the formation and classification of amphiphilic block copolymer nanoparticulate structures: micelles, nanospheres, nanocapsules and polymersomes. *Eur. J. Pharm. Biopharm.*, **65** (3), 259–269.
82 Choisnard, L., Gèze, A., Putaux, J.-L. *et al.* (2006) Nanoparticles of β-cyclodextrin esters obtained by self-assembling of

biotransesterified β-cyclodextrins. *Biomacromolecules*, **7** (2), 515–520.
83 Raja, M., Liu, C., and Huang, Z. (2015) Nanoparticles based on oleate alginate ester as curcumin delivery system. *Curr. Drug Deliv.*, **12** (5), 613–627.
84 Klemm, D., Kramer, F., Moritz, S. *et al.* (2011) Nanocelluloses: a new family of nature-based materials. *Angew. Chem., Int. Ed.*, **50** (24), 5438–5466.
85 Brinchi, L., Cotana, F., Fortunati, E., and Kenny, J.M. (2013) Production of nanocrystalline cellulose from lignocellulosic biomass: technology and applications. *Carbohydr. Polym.*, **94** (1), 154–169.
86 Cheng, S., Zhang, Y., Cha, R. *et al.* (2015) Water-soluble nanocrystalline cellulose films with highly transparent and oxygen barrier properties. *Nanoscale*, **8** (2), 973–978.
87 McClements, D.J., Decker, E.A., Park, Y., and Weiss, J. (2009) Structural design principles for delivery of bioactive components in nutraceuticals and functional foods. *Crit. Rev. Food Sci. Nutr.*, **49** (6), 577–606.
88 Marsanasco, M., Piotrkowski, B., Calabró, V. *et al.* (2015) Bioactive constituents in liposomes incorporated in orange juice as new functional food: thermal stability, rheological and organoleptic properties. *J. Food Sci. Technol.*, **52** (12), 7828–7838.
89 Sanguansri, L., Oliver, C.M., and Leal-Calderon, F. (2013) Nanoemulsion technology for delivery of nutraceuticals and functional-food ingredients, in *Bio-Nanotechnology*, Blackwell Publishing Ltd., pp. 667–696.
90 McClements, D.J. (2011) Edible nanoemulsions: fabrication, properties, and functional performance. *Soft Matter*, **7** (6), 2297–2316.
91 O' Dwyer, S.P., O' Beirne, D., Ní Eidhin, D. *et al.* (2013) Formation, rheology and susceptibility to lipid oxidation of multiple emulsions (O/W/O) in table spreads containing omega-3 rich oils. *LWT Food Sci. Technol.*, **51** (2), 484–491.
92 Lane, K.E., Li, W., Smith, C., and Derbyshire, E. (2014) The bioavailability of an omega-3-rich algal oil is improved by nanoemulsion technology using yogurt as a food vehicle. *Int. J. Food Sci. Technol.*, **49** (5), 1264–1271.
93 Pardeshi, C., Rajput, P., Belgamwar, V. *et al.* (2012) Solid lipid based nanocarriers: an overview. *Acta Pharm.*, **62** (4), 433.
94 Fathi, M., Mozafari, M.R., and Mohebbi, M. (2012) Nanoencapsulation of food ingredients using lipid based delivery systems. *Trends Food Sci. Technol.*, **23** (1), 13–27.
95 Pandita, D., Kumar, S., Poonia, N., and Lather, V. (2014) Solid lipid nanoparticles enhance oral bioavailability of resveratrol, a natural polyphenol. *Food Res. Int.*, **62**, 1165–1174.
96 Ni, S., Sun, R., Zhao, G., and Xia, Q. (2015) Quercetin loaded nanostructured lipid carrier for food fortification: preparation, characterization and *in vitro* study. *J. Food Process Eng.*, **38** (1), 93–106.

5
Engineered Inorganic Nanoparticles in Food

Marie-Hélène Ropers[1] and Hélène Terrisse[2]

[1]*INRA, UR1268 Biopolymères Interactions Assemblages, Rue de la Géraudière, 44316 Nantes Cedex 3*
[2]*Institut des Matériaux Jean Rouxel (IMN), Université de Nantes, CNRS, 2 rue de la Houssinière, BP 32229, 44322 Nantes Cedex 3, France*

5.1
Introduction

Since the last 20 years, a permanent concern in Western populations deals with nanomaterials. The latter were shown to accumulate in different organs [1] and are suspected of causing cancers. Part of them comes from our food consumption. Nanomaterials provided by food have been classified into three classes: organic, composites, and inorganic [2]. In the same report, the European Food Safety Agency (EFSA) observed that organic nanomaterials (vesicles, nanoemulsions) present a low risk for health as they are probably metabolized as the normal food diet (meal). In contrast, the agency considered that the research effort has to be concentrated on inorganic nanomaterials. Thus, in the first part, we analyze the nature of these nanomaterials and the added value of the nanosize in food products. This list is not exhaustive as the characterization of food nanomaterials is a very recent field. In this respect, the second part describes the different key parameters required to fully characterize the samples and lists the studies carried out to characterize them according to the EU recommendations.

5.2
Engineered Inorganic Materials Containing Nanoparticles

The definition of a nanomaterial according to the European Union [3] is "a natural, incidental or manufactured material containing particles, in an unbound state or as an aggregate or as an agglomerate and where, for 50% or more of the particles in the number size distribution, one or more external dimensions are in the size range 1 nm–100 nm." Moreover, the definition specifies that "in specific cases and where warranted by concerns for the environment, health, safety, or

Nanotechnology in Agriculture and Food Science, First Edition. Edited by Monique A.V. Axelos and Marcel Van de Voorde.
© 2017 Wiley-VCH Verlag GmbH & Co. KGaA. Published 2017 by Wiley-VCH Verlag GmbH & Co. KGaA.

competitiveness the number–size distribution threshold of 50% may be replaced by a threshold between 1 and 50%." The definition is completed by one sentence referring to fullerenes, graphene flakes, and single wall carbon nanotubes. Since they have no application as food ingredients, they are out of scope in this review. The International Organization for Standardization (ISO) gave a slightly different definition to nanomaterials that encompassed nonparticular materials such as proteins or micelles and nanostructured materials with an internal or surface structure ranging between 1 and 100 nm. Here, we do not consider the latter definition but the European one to focus the discussion on materials consisting of particles. The threshold was fixed at 50% for practical reasons but the earlier work of Scientific Committee on Emerging and Newly Identified Health Risks from European Commission, [4] which served as a basis for the European definition, advised to use a threshold value of 0.15%. Considering the work of SCE-NIHR and the toxicity of these products, which is developed elsewhere in this book, we will consider any food grade material with a measurable nanosized fraction. The reader must keep in mind that the definition of nanomaterial is expected to be amended by the European Commission.

Three European countries decided to establish a list of nanomaterials: France since 2013,[1] Denmark since 2015,[2] and Belgium since 2016.[3] Similar databases are expected to be carried out by Sweden and Norway. Even though these registers aim at collecting new knowledge about the occurrence of nanomaterials in products sold in these countries, they are difficult to compare as the extent of declarations is different. The French declaration is on the production, distribution, and importation of substances at nanoscale or nanomaterials, while the other registries restrict the scope of the declaration to some products.

In the French register – giving the best overview of nanomaterial uses, the number of compounds declared for use in the food industry hardly evolved with time. Silica and derivatives (aluminum, calcium, magnesium), and iron hydroxide or oxide continuously appeared in the list. In the early stage (2013), titania and calcium chloride were declared. The next year, calcium chloride did not appear in the list and finally in 2015, titanium dioxide also disappeared from the list and another compound was added (the alumina-silicate attapulgite). The latter is not a food ingredient but rather a drug used as adsorbent to eliminate the bacteria responsible for diarrhea, as mentioned in the review of Willhite et al. [5]. At the European level, a scientific report gathered the list of food additives, food ingredients, novel foods, and supplements containing nanomaterials. This list contains silica, titanium dioxide, zinc oxide, silver, iron oxides, and includes nanoparticles found in fortified foods such as iron and calcium [2]. Engineered inorganic materials identified to be incorporated in food as nanoparticles (irrespective of the information source) are listed in Table 5.1. This list is not exhaustive and is expected to be completed/amended in the future. It was

1) www.r-nano.fr/.
2) http://nanodb.dk/en/.
3) http://www.health.belgium.be/en/environment/chemical-substances/nanomaterials/register.

Table 5.1 List of food additives including a nanosized fraction of nanoparticles.

Chemical name	Chemical formula	Food additive number	Class and function	Sources
Silica and silicates	SiO_2 For silicates, no precise chemical formula due to various compositions according to geographical origin and process.	E 551 Silicon dioxide E 552 Calcium silicate E553a: (i) Magnesium silicate, (ii) magnesium trisilicate E553b Talc E554 Sodium aluminum silicate E555 Potassium aluminum silicate	Mainly as anticaking agent but also as a technical aid in clarification of beverages	French register 2013–2015; [6–14]
Titanium dioxide	TiO_2	E171	Coloring agent (whitening)	French register 2013, 2014; [12,15–22]
Iron oxides and hydroxydes	$FeO \cdot Fe_2O_3$ Fe_2O_3 $FeO(OH) \cdot xH_2O$	(i) E172 (ii) E172 (iii) E172	Coloring agent Black Red Yellow	French register 2013–2015 [23],
Zinc oxide	ZnO		Supplement	EFSA 2014
Silver	Ag	E174	Coloring supplement	[24–27]
Calcium carbonate (less than 1% of nanoparticles)	$CaCO_3$	E170	(1) Anti-caking agent, filling agent (pharmaceuticals), stabilizer in canned fruit (2) White color for surface coating (3) Stabilizer in canned fruit	[28]
Calcium chloride	$CaCl_2$	E509	Thickening agent with polysaccharides	French register 2013
Iron, calcium, and magnesium, selenium	Fe, Ca, Mg, Se		Supplement	[2,26,29]

established according to the French register,[1] the EFSA list [2], and relevant scientific literature, included in Table 5.1.

The materials containing a fraction of primary particles with a size lower than 100 nm belong to three classes of food additives: food coloring, anticaking agents, and food supplements. They include metals or metal oxides where the metallic cation can present various oxidation degrees. Most of them are characterized by their insolubility in water. In contrast to cosmetics, the engineered inorganic particles authorized in foods are usually not coated, except titanium dioxide. This coating is, however, limited to a small percentage (2% for TiO_2 coated with alumina or silica) [30]. In the next sections, we describe the main use of these materials according to European regulations[4] and their main characteristics, in particular the particle size distribution that conducts to list these materials as containing a fraction of nanoparticles.

5.2.1
Silica (SiO_2) and Silicates

Amorphous synthetic silica (SAS) is the most famous nanomaterial used in food. There are various forms of amorphous silica: colloidal silica, precipitated silica, silica gels, and pyrogenic (fumed) silica. Amorphous silica is synthetically produced either by a vapor-phase hydrolysis process, yielding pyrogenic (fumed) silica, or by a wet process, yielding precipitated silica, silica gel, and colloidal silica [30]. Pyrogenic silica is produced in an anhydrous state, whereas the wet process products are obtained as hydrates or contain surface absorbed water. The main functions of synthetic amorphous silica in the food industry (E 551) are anticaking agent (flow aid), carrier, spray drying aid, or milling aid. For these uses, silica is expected to be found in the final products. For other uses, such as refining and clarification, silica is separated from the supernatants and is not present in the consumed products or only in traces. The anticaking properties of amorphous silica are significant to improve the flowability of powders, coffee creamer, confectioner's sugar, flavor, instant beverage, instant soup, spice blend, and table salt.

The different description of SAS can be found in several scientific studies (Table 5.1) and reports from European Commission (see e.g., Ref.[5]). Colloidal silica consists in stabilized dispersion of nonagglomerated, mostly spherical SiO_2 particles, which serve in the food industry, for example, as an aid to clarify wine, beer, fruit juices. Besides, precipitated silica is made up of primary particles in the size range of around 5–100 nm that are aggregated and agglomerated in the final product. Precipitated silica is used as anticaking agent in food powders.[4] In contrast, synthetic silica gels are products of the polymerization process of fine colloidal silica. They have a similar structure as precipitated silica, the difference being that the cross-linked silica particle networks form a nanoporous structure

4) https://webgate.ec.europa.eu/foods_system/.
5) http://ec.europa.eu/health/nanotechnology/docs/swd_2012_288_en.pdf.

that is finer than the porous structure of the aggregated particles in precipitated silica. Silica gels are used in food industry as anticaking agent as well as a carrier for vitamins. Pyrogenic (fumed) silica consists in agglomerated and aggregated primary particles. The agglomerated particles have a size typically between 5 and 100 nm. The aggregates, which are fused and chemically bonded primary particles, typically have a size between 100 and 350 nm. The aggregates in turn form agglomerates typically in the range from 150 nm up to several 100 µm. Pyrogenic or fumed silica is rather used as an aid in the manufacture of decaffeinated coffee and tea, poultry and seafood processing, and oil refining.

The recent literature data confirm the description for pyrogenic and precipitated silica. The as-manufactured samples consist in primary particles with size between 1 and 100 nm, which form aggregates and/or agglomerates with external dimensions higher than 100 nm [31]. In food products, primary particles of silica or agglomerates with external dimensions of 30 nm and larger were detected [31]. In two food grade pyrogenic synthetic amorphous silica samples (with surface area of 300 and 380 m^2/g), the presence of primary nanoparticles of roughly 10 nm was detected along with larger aggregates and agglomerates (50–200 nm) and even much larger aggregates between 300 and 600 nm [8]. Precipitated synthetic amorphous silica is formed of single particles of roughly 50 nm, organized in clusters and aggregates, in prevalence smaller than 600 nm [8]. A broader range of synthetic amorphous silica samples has been recently characterized and clearly shows isolated primary nanoparticles in the size range of 9–26 nm and agglomerates/aggregates ranging in size from below 100 to >500 nm [6,14].

Silica and silicates are permitted in all categories of food, which were authorized to incorporate a food additive and provided that these foods are in a dried form. The maximum level in foods (added individually or in combination with other silicates) is restricted to 2000 mg/kg in processed cereal food and baby foods for infants and young children, to 10 000 mg/kg in dried powdered form for some products such as ripened cheese (sliced or grated cheese, hard and semi-hard cheese), processed cheese, cheese products (sliced or grated cheese, hard and semi-hard cheese), dairy analogs including beverage whiteners, dried powdered foods, in table-top sweeteners in powder form, salt, food supplements, to 20 000 mg/kg in salt substitutes and to 30 000 mg/kg in fats and oil emulsions, as well as in seasoning. Silica and silicates can also be added *at quantum satis* in other food products such as in tablets and coated tablet forms (see the regulations for more details) as well as in table-top sweeteners in tablets, and finally as surface treatment only in some confectionary including breath freshening microsweets, chewing gum, decorations, coatings. For more details, readers are invited to consult the regulations [32]. For all these applications, the acceptable daily intake (ADI) of silica and silicates was not specified. Silica as E551 could be detected in the above cited products like instant coffee or model tomato soup or sweets [7–9,33,34]. The amount detected in processed foods may be higher than the incorporated amount due to the naturally occurring silica in corn and potato plants [34] or the persistence of silica NPs after agrochemical treatments [35].

Regarding the definition of a nanomaterial, the above described forms of silica are effectively nanomaterials. The estimated daily intake of "nanosized" silica is 1.8 mg/kg bw/day for an adult of 70 kg based on products containing E551 [9]. In table salt, silica has been replaced by sodium ferrocyanide (E535) or potassium ferrocyanide (E536). Both of them provide the same properties but they are soluble in water and are expected to be more easily eliminated from the body.

Silicates (E552 calcium silicate, E553a (i) magnesium silicate and (ii) magnesium trisilicate, E553b talc, E554 sodium aluminum silicate, E555 potassium aluminum silicate) are derivatives of silica and are used in the same conditions as silica. Silicates under the code E556 (aluminum calcium silicate), E558 (bentonite), and E559 (kaolin) are no longer used or authorized.

5.2.2
Titania or Titanium Dioxide (TiO$_2$)

Titanium dioxide is a food additive belonging to the Group II of food colors authorized *at quantum satis* in Europe (used in accordance with good manufacturing practice, at a level not higher than is necessary to achieve the intended purpose and provided that the consumer is not misled) and in the limit of 1% in weight in the United States. The food colors of this group, including titanium dioxide are authorized in most food categories such as: (i) dairy products and analogs (flavored fermented milk products and some creams), (ii) cheese and cheese products such as flavored unripened cheese, edible cheese rind, whey cheese, processed cheese, cheese products, and dairy analogs including beverage whiteners, (iii) edible ices, (iv) confectionary (chewing gum, decorations, coatings, and nonfruit-based fillings), (v) surimi and similar products and salmon substitutes, (vi) seasonings and condiments, mustard, soups and broths, sauces, and (vii) food supplements.[4] The whole list and the restrictions are accessible on specialized websites.[4,6] According to these specifications, titanium dioxide was identified in coconut curd, chewing gum, confectionary, coffee creamers and its supplements [1,12,15,18,19]. The highest content of TiO$_2$ was found in chewing gums [19]. The estimated consumption for a US adult was on the order of 1 mg Ti/kg bw/day [19]. These data were recently refined for all food categories, subpopulations, and exposure scenarios [36].

In all these products, titanium dioxide has a whitening effect, due to its high refractive index (>2.5). Although titanium dioxide mainly crystallizes as two structural varieties (anatase and rutile), both of them have whitening abilities, rutile being slightly better than anatase (higher refractive index). These two forms are authorized in foods but until now the characterization of samples in laboratories from the United States and Europe shows that anatase is predominant in food applications [16,18,20,37], presumably because it is less abrasive [38] and requires lower temperatures to be synthesized. Depending on the particle size, titanium dioxide can offer a brightening effect as well. This is achieved for particle size close

6) http://www.fao.org/gsfaonline/additives/index.html.

to 300 nm for which the reflectivity is the highest [33]. The particle size distribution of raw powders of TiO_2 is in fact broader: from 30 to 400 nm [16,19] and 60 to 300 nm in other E171 samples [18]. In the whole set of samples, the nanoparticle size distribution expressed in number is smaller than 50%. The fraction of nanoparticles (<100 nm) was 23% for one sample [16], ranged from 17 to 35% in five different E171 samples [20] and was lower than 10% in other E171 samples [18]. According to the EC definition, E171 should not be labeled as a nanomaterial. However, a recent review pointed out that TiO_2 can be absorbed by the mammalian gastro-intestinal tract and can bioconcentrate, bioaccumulate in the body [39]. In these conditions, this material can no longer be considered as an inactive compound.

5.2.3
Iron Oxides and Hydroxides

Iron oxides and hydroxides belong to the group of food colors authorized *at quantum satis* in numerous categories of foods including dairy products and analogs, cereals and cereal products, soups, spices, bakery wares, salad and protein products, processed fish and fisheries products including molluscs and crustaceans (in only fish paste and crustacean paste, smoked fish, surimi), and beverages, ready-to-eat savories and snacks, seasoning and condiments (in some categories, some restrictions may apply).[4]

The different oxidation degrees of iron lead to different colors. $FeO \cdot Fe_2O_3$ is a black powder of mixed iron oxide (III) and (II), whereas the red and yellow powders are iron oxides (III). The yellow color is obtained in the structure $FeO(OH) \cdot xH_2O$, while the red color results from the fine structure of Fe_2O_3. These coloring agents are not soluble in water [40]. The yellow form was declared as a nanomaterial in the French registry,[1] in agreement with the data of one provider due to the needle shape habit that they possess [2]. No study was found to confirm these data. Industrially, size and shape are crucial for the production of high-quality iron oxide and hydroxide pigments: smaller particle sizes give more intense and greater tinctorial strength [41]. Moreover, the optical properties of the yellow one (the needle-shaped iron oxide $FeO(OH) \cdot xH_2O$) depend also on the length-to-width ratio [42]. Interestingly, there are several processes for the production of iron oxide and hydroxide pigments with controlled mean particle size, particle size distribution, and particle shape [42]. As underlined by EFSA, a more complete characterization step is required for this range of additives [40].

5.2.4
Silver (Ag)

Silver (Ag, E174) added intentionally as food ingredient is authorized in the decoration of chocolates, in liqueurs and in the external coating of confectionary (*at quantum satis*). In this case, it is provided as powder or metal sheets. There are numerous studies about silver nanoparticles in the literature data but there is currently only one relevant to food [27]. The silver layer from the decoration of confectionery pearls was removed and analyzed by two different techniques (TEM and single particle ICP-MS). The found that 20% of the mean total silver concentration in the pearls was released as particles and that 94.70% of the eluted particles had a Feret minimum diameter (distances between two parallel tangents on opposite sides of the image of a particle) smaller than 100 nm [27]. This should lead to label this additive as a nanomaterial according to the EC definition [27]. The authors noticed the contradiction between the measured size of particles and the technological function of silver in the finished product [27]. Since silver is used to make confectionary surfaces shinier and that cannot be achieved by individual nanoparticles of silver, one can assume that these nanoscaled particles are agglomerated. The knowledge about the food additive E174 is very scarce with respect to the EC recommendation [24]. In addition to the particle size, the release of silver ions is a matter of concern due to their toxicity [43,44]. It is worth noting that the amount of silver migrating from food packaging ranges between 5.66 ± 0.02 µg/l and 28.92 ± 0.01 µg/l in the case of orange juices stored for 10 days in contact with two different silver-based packagings [45]. Therefore, the amount provided by a glass of contaminated orange juice is comparable to the amount of silver consumed through only one confectionary silver pearl ($\pm$180 mg sugar pearl decorated with silver at 8.4 mg/kg contains 1.5 µg of silver, according to the data provided in Verleysen *et al.* [27]).

Although silver does not play any biological role in humans, it is also found in numerous dietary supplements under the label of colloidal silver (e.g., see the Danish database[2)] and the Woodrow Wilson database[7)]). One of them was characterized and the authors found that the diameter of silver nanoparticles was 10.9 ± 4.5 nm [26].

5.2.5
Miscellaneous

5.2.5.1 Other Metals (Fe, Se, Ca, etc.)
The external scientific report of EFSA [2] listed selenium, calcium, iron, and colloidal suspensions of metal particles, for example, copper, gold, platinum, silver, molybdenum, palladium, titanium, and zinc as nanoforms in fortified foods. In the Danish nanodatabase, one product was identified as a zinc source, in the form of nanoparticles of pure metallic zinc. Readers are invited to refer to the more complete review recently published on this topic [46].

5.2.5.2 Calcium Carbonate (CaCO$_3$)
Calcium carbonate (E170) belongs to the group of food additives and food colors authorized *at quantum satis*. It is used in some dairy products and analogs, confectionary (in the limit of 70 000 mg/kg), salt, in some fish and fisheries products,

7) http://www.nanotechproject.org/cpi/.

in foods intended for particular nutritional uses, and in some beverages. EFSA has recently given a scientific opinion on re-evaluation of calcium carbonate (E170) as a food additive with the conclusion that "the available data are sufficient to conclude that the current levels of adventitious nanoscale material within macroscale calcium carbonate would not be an additional toxicological concern" [28]. Calcium carbonate (E170) contains less than 1% of particles having a diameter below 100 nm.

5.2.5.3 Calcium Chloride (CaCl$_2$)

Calcium chloride (E509) is a salt used as firming agent, stabilizer, or thickener in preparations belonging to subcategories of dairy products and analogs, and fruits and vegetables (e.g., jam, marmalades). It was registered as a nanomaterial in the French register. Although the fraction of particles with one or more external dimensions in the size range 1–100 nm may be higher than 50%, this salt is highly soluble in water and will associate to any biomolecule present in the preparation, so that it is not likely to exist as a nanomaterial in processed foods.

5.2.6
Knowledge Gaps

5.2.6.1 Gold (Au)

Gold (E175) is used in confectionary for the decorations of chocolates, in spirit drinks like liqueurs, and in the external coating of others confectionaries. It was from time to time classified as nanomaterial.[8] However, in a recent re-evaluation of this food additive, the EFSA Panel noticed that the data about the mean particle size and on the particle size distribution were limited [23]. According to the manufacturer who responded to EFSAs call for data, thin sheets of gold with an approximate thickness of some tenth of a micrometer, are reduced by a milling process to get powder, crumbs, or flakes that have a size above 1 mm. Thus, if there are nanomaterials within these small parts, it is likely to be unintentional. This compound can also be found in liquid supplements under the form of "nanoparticle gold" (examples can be found in the Danish database[2]).

5.2.6.2 Aluminum (Al)

Aluminum as food additive (E173 or INS173) is produced by grinding aluminum and this process may be carried out in the presence of edible vegetable oils and/or food grade fatty acids (FAO2011). It is permitted in the external coating of sugar confectionery for decoration of cakes and pastries [47]. From the grinding process, one can reasonably expect the presence of nanosized aluminum particles in the final powder as in the case of aluminum oxides [5]. Metallic aluminum is naturally covered by a thin layer of aluminum oxide of a few millimetres that prevents the metal from reacting with water. However, aluminum is soluble in acidic media and may form a variety of complexes with ligands present in

---

[8] http://www.oekopol.de/wp-content/uploads/03_Rauscher_EFSA-Bericht-Nanoinventory.pdf.

foods (lactate, citrate) that have different physicochemical properties, such as solubility in aqueous medium, stability toward hydrolysis at different pH, electric charge, and so on [47]. Thus, the nanosized fraction of the raw powder may be rapidly converted into a variety of forms in the food products or after consumption. However, no such study has been carried out until now.

5.2.6.3 Zinc Oxide (ZnO)

Zinc oxide is not officially listed as a food additive in Europe (but is permitted in food supplements). However, it was identified as a nanomaterial in food ingredients and food additive in the EFSA Nanoinventory [2], and is listed as "generally recognized as safe" (GRAS) by the USFDA (21CFR182.8991). When authorized, it is used as zinc source and is added to cereal-based foods for example.

5.3
Characterization of Engineered Inorganic Nanomaterials

A few years ago, it was established that the lack of precise characterization of materials greatly influences the outcome of risk assessment [48–50]. For this purpose, EFSA recommended authors to characterize nanomaterials with a great care [51]. In particular, this information is needed to assess whether the material tested is representative and relevant for the exposure from the intended use. The specifications of an ENM are important, since different sizes and shapes of ENMs of the same chemical composition may have different toxicities [52]. As a consequence, an ENM should be tested with properties and characteristics falling within the specifications provided for that ENM, since subsequent approval will be based on the outcome of the risk assessment. For this purpose, EFSA established a list of 15 essential parameters (for solid or liquid samples) characterizing the physical form, morphology, and the chemical reactivity of particles. They are indicated below with some methods dedicated to get these parameters:

- Chemical composition/identity by analytical methods including UV-vis spectroscopy, high performance liquid chromatography (HPLC), gas chromatography/liquid chromatography coupled with mass spectrometry (GC/LC–MS), atomic absorption spectrometry (AAS), graphite furnace atomic absorption spectrometry (GFAAS), inductively coupled plasma mass spectrometry (ICP-MS), inductively coupled plasma - atomic (or optical) emission spectroscopy (ICP-AES or ICP-OES), Fourier transform infrared spectroscopy (FT-IR), nuclear magnetic resonance (NMR), X-ray diffraction (XRD), electron microscopy coupled to energy dispersive X-ray spectroscopy (EDX/S) or electron energy loss spectroscopy (EELS), and so on.
- Particle size (primary/secondary) accessible by field flow fractionation (FFF) such as, asymmetric flow field flow fractionation (A4F), sedimentation field flow fractionation (Sed-FFF), hydrodynamic chromatography (HDC), HPLC, analytical ultra-centrifugation (AUC), centrifugal particle sedimentation (CPS), coupled to transmission electron microscopy (TEM), scanning electron

microscopy (SEM), high angle annular dark field scanning transmission electron microscopy (HAADF-STEM), atomic force microscopy (AFM), dynamic light scattering (DLS) also named photon correlation spectrosocpy (PCS), differential mobility analyzer (DMA), single particle inductively coupled plasma mass spectrometry (sp-ICP-MS), nanoparticle tracking analysis (NTA), multi-angle light scattering (MALS), laser granulometry and so on.

- Physical form and morphology given most accurately by TEM, SEM, or STEM.
- Particle and mass concentration for dry powders.
- Specific surface area for dry powders given by gas volumetry adsorption isotherms with BET mathematical treatment.
- Surface chemistry and surface charge (by scanning probe microscopy, X-ray photoelectron spectroscopy (XPS), mass spectrometry, Raman spectroscopy, FTIR, NMR, AUC (for surface composition), capillary electrophoresis (CE), scanning probe microscopy, laser doppler electrophoresis (LDE) also named electrophoresis light scattering (ELS) with possible use of phase analysis light scattering (PALS), surface-enhanced Raman spectroscopy (SERS), and so on.
- Redox potential for inorganic nanomaterials by potentiometric methods or X-ray absorption spectroscopy (XAS).
- Solubility and partition properties.
- pH for liquid dispersions.
- Viscosity for liquid dispersions.
- Density and pour density for solids.
- Dustiness for dry powders.
- Chemical reactivity/catalytic activity.
- Photocatalytic activity for photocatalytic materials.

Since the first reviews of Tiede *et al.* [50] and Hasselov *et al.* [53] on characterization of nanomaterials, substantial progress has been made to improve them (uncertainties, limits of detection, reliability, etc.) and adapt them in complex matrices like food products. Several reviews reported at each step these progresses [54–67]. In the same time, several attempts have emerged to harmonize the procedures and improve the reliability of these methods [68–71].

5.3.1
Characterization of Engineered Inorganic Nanomaterials as Manufactured

Here are summarized the articles characterizing the raw powders of silica and titanium dioxide as well as the different methods used (Table 5.2). Silver, gold, and aluminum have not yet been characterized as raw materials but only in finished products for silver [27]. Silver and gold food additives are commercially available to consumers in contrast to aluminum.

Electron microscopy and ICP-MS are commonly used to get the particle size distribution and the elementary composition of inorganic materials. In the most recent works, advances techniques such as A4F-ICP-MS and sp-ICP-MS were used to better characterize the inhomogeneities in samples. In most cases, the

Table 5.2 Methods used to characterize food grade titanium dioxide and silica powders.

Nanoparticles	Methods	References
TiO$_2$ E171	ICP-OES, EDX fluorescence spectroscopy	[17]
	TEM, XRD, Raman spectroscopy, XPS, ICP-MS, PALS, UV-visible spectroscopy	[20]
	SEM-EDX, spICP-MS, A4F-ICP-MS, ICP-MS	[18]
SiO$_2$ E551	SdFFF, GFAAS, SEM, PCS	[8]
	PALS, STEM, TEM, SEM-EDX, XPS, XRD, ICP-MS	[14]
	TEM, DLS, ICP-MS, A4F-MALS-ICP-MS	[6]

names of suppliers are not given, so that it is difficult to carry out interlaboratory comparisons.

According to these studies, the raw powders of titanium dioxide are very pure (~100%) but may often exhibit traces of phosphorus at the surface [16,20], and sometimes alumina and silica [18,20] that are authorized in 2% weight basis. These samples have acidic isoelectric point (ranging from <2.5 to 4 [16,20]). The specific surface area of one of them was determined to be 8.8 m^2/g [37]. A study is currently under investigation in our lab to confirm and complete data about titanium dioxide.

Amorphous silica has a very high surface specific area [10] that explains its use as anticaking agent and the diameters of particles are below 100 nm (between 20.6 and 39.8 nm as determined by A4F and between 10.3 and 20.3 nm for equivalent circle diameter determined by TEM [6]). When suspended in water, silica spontaneously forms agglomerates whose sizes are several hundred nanometres [14]. It is negatively charged on a large pH range, due to a low isoelectric point (<3.2).

5.3.2
Characterization of Engineered Inorganic Nanomaterials as Present in the Food Matrices

In Table 5.3, we summarized the studies that have analyzed the content of food additives in real foods (nonspiked).

The characterization of engineered inorganic materials in food products has been performed in the most common products where they are employed (soups, confectionary, chewing gums) but also in food supplements. Here again, electron microscopy and ICP-MS-based methods were the most common techniques to characterize the engineered particles. One limitation is that engineered inorganic materials were extracted and we cannot exclude that particles may undergo alterations. This difficulty is particularly encountered for silver and iron oxides that are the most prone to chemical alterations when the physicochemical conditions are changed [42,75]. However, this is also related to the difficulty to probe particles *in situ* within food samples.

Table 5.3 Techniques used to characterize engineered inorganic particles in food products.

Nanoparticles	Food products	Techniques	References
Ag, Au, Cu, Ir, Pd, Pt, Zn, Si	Dietary supplement drinks	spICP-MS, ICP-MS, TEM-EDX, XRD, DLS, PALS	[26]
Ag	Decoration of pastry	TEM-EDX, spICP-MS, HAADF-STEM	[27]
Ag	Nutraceutical and beverage samples	A4F-ICP-MS, TEM	[25]
Al	Confectionary and all food products permitting aluminum-based silicates	ICP-MS	[72]
Au	Dietary supplements	CE-ICP-MS, TEM, UV-vis	[73]
Ca	Calcium-enriched milk	XRD	[29]
SiO_2	Coffee creamer	A4F-ICP-MS, A4F-MALS, TEM-EDX	[11]
SiO_2	Coffee creamer, instant soup, sauce, pancakes	HDC-ICP-MS	[9]
SiO_2	Not given	XRD, FTIR, TEM, EDX and DLS	[34]
SiO_2 TiO_2	Dietary supplements	SEM, TEM-EDS, DLS, XRD	[12]
SiO_2	Chocolate powder, seasoning, coffee creamer, food supplements, confectionary, pastry	High-resolution TEM, TEM-EDX, SEM-EDX, ICP-MS	[14]
TiO_2	Chewing-gum	TEM, SEM, high-resolution TEM, NTA	[15]
TiO_2	Confectionary, coffee creamer dressing, sauces, chewing-gums	ICP-OES	[21]
TiO_2	Salad dressing	n.d.	[74]
TiO_2	Chewing-gums	ICP-OES, XRD, TEM-EDX, DLS	[22]
TiO_2	Confectionary, pastry, chocolate, mayonnaise	TEM, DLS, ICP-MS SEM-EDX, A4F-ICP-MS, Sp-ICP-MS	[18]
TiO_2	Candies, sweets, and chewing gums	SEM-EDX, ICP-MS, DLS	[19]
TiO_2	Candies	ICP-MS, TEM, XRD, XPS, DLS, LDE	[16]

5.4
Conclusion and Perspectives

Food grade engineered inorganic materials containing nanoparticles are now quite well identified. They mainly belong to the food categories of coloring agents and anticaking agents. They may be daily consumed due to their widespread use as food additives in many food products. The most known engineered inorganic materials containing nanoparticles are silica and titanium dioxide, which have received more attention for the last 5 years. From their characterization, silica can be really considered as a nanomaterial whereas it is still under debate for titanium dioxide. However, one must keep in mind that the consumption of engineered inorganic nanoparticles from food represents a minor fraction of the total part of ingested inorganic particles: only 14% of silicates and 47% of titanium dioxide come from food [76].

The re-evaluation of these additives by EFSA has just been performed for iron oxides, gold, silver titanium dioxide, and is expected soon for silica. However, there are still gaps of knowledge regarding silver (needs of confirmation), gold, iron oxides, and silicates, in particular their primary particle size. For all of them, the characterization is as important as their reactivity is dependent on their surface properties [20,77]. The characterization of the nanoparticles *in situ* in food matrices and their biotransformation, decisive for risk assessment is still lacking, but several developments are under progress. Due to the applications of food-grade engineered inorganic materials containing nanoparticles in products consumed at vulnerable life stages, such as infancy (confectionary) and early development (milk powder), the fate of these nanoparticles and their evolution in organisms is also under investigation. Chapter 17 is dedicated to this topic.

References

1 Lomer, M.C.E., Thompson, R.P.H., and Powell, J.J. (2002) Fine and ultrafine particles of the diet: influence on the mucosal immune response and association with Crohn's disease. *Proc. Nutr. Soc.*, **61** (01), 123–130.

2 RIKILT and JRC (2014) Inventory of Nanotechnology applications in the agricultural, feed and food sector. pp. 1–125. Available at http://www.efsa.europa.eu/fr/supporting/pub/621e.

3 European Parliament and Council (2011) Commission recommendation of 18 October 2011 on the definition of nanomaterial. pp. 38–40. Available at http://eur-lex.europa.eu/legal-content/EN/TXT/PDF/?uri=CELEX:32011H30696&from=EN.

4 SCENIHR (Scientific Committee on Emerging and Newly Identified Health Risks) (2010) Scientific basis for the definition of the term "nanomaterial" - Pre-consultation opinion. pp. 1–43. Available at http://ec.europa.eu/health/scientific_committees/emerging/docs/scenihr_o_030.pdf.

5 Willhite, C.C., Karyakina, N.A., Yokel, R.A., Yenugadhati, N. *et al.* (2014) Systematic review of potential health risks posed by pharmaceutical, occupational and consumer exposures to metallic and nanoscale aluminum, aluminum oxides, aluminum hydroxide and its soluble salts. *Crit. Rev. Toxicol.*, **44** (4), 1–80.

6 Barahona, F., Ojea-Jimenez, I., Geiss, O., Gilliland, D., and Barrero-Moreno, J. (2016) Multimethod approach for the detection and characterisation of food-grade synthetic amorphous silica nanoparticles. *J. Chromatogr. A*, **1432**, 92–100.

7 Contado, C., Mejia, J., Lozano García, O., Piret, J.-P. *et al.* (2016) Physicochemical and toxicological evaluation of silica nanoparticles suitable for food and consumer products collected by following

the EC recommendation. *Anal. Bioanal. Chem.*, **408** (1), 271–286.

8 Contado, C., Ravani, L., and Passarella, M. (2013) Size characterization by sedimentation field flow fractionation of silica particles used as food additives. *Anal. Chim. Acta*, **788**, 183–192.

9 Dekkers, S., Krystek, P., Peters, R.J.B., Lankveld, D.P.K. et al. (2011) Presence and risks of nanosilica in food products. *Nanotoxicology*, **5** (3), 393–405.

10 Fruijtier-Pölloth, C. (2012) The toxicological mode of action and the safety of synthetic amorphous silica – a nanostructured material. *Toxicology*, **294** (2–3), 61–79.

11 Heroult, J., Nischwitz, V., Bartczak, D., and Goenaga-Infante, H. (2014) The potential of asymmetric flow field-flow fractionation hyphenated to multiple detectors for the quantification and size estimation of silica nanoparticles in a food matrix. *Anal. Bioanal. Chem.*, **406** (16), 3919–3927.

12 Lim, J.-H., Sisco, P., Mudalige, T.K., Sánchez-Pomales, G. et al. (2015) Detection and characterization of SiO_2 and TiO_2 nanostructures in dietary supplements. *J. Agric. Food Chem.*, **63** (12), 3144–3152.

13 Peters, R., Kramer, E., Oomen, A.G., Herrera Rivera, Z.E. et al. (2012) Presence of nano-sized silica during *in vitro* digestion of foods containing silica as a food additive. *ACS Nano*, **6** (3), 2441–2451.

14 Yang, Y., Faust, J.J., Schoepf, J., Hristovski, K. et al. (2016) Survey of food-grade silica dioxide nanomaterial occurrence, characterization, human gut impacts and fate across its lifecycle. *Sci. Total Environ.*, **565**, 902 912.

15 Chen, X.-X., Cheng, B., Yang, Y.-X., Cao, A. et al. (2013) Characterization and preliminary toxicity assay of nano-titanium dioxide additive in sugar-coated chewing gum. *Small*, **9** (9–10), 1765–1774.

16 Faust, J.J., Doudrick, K., Yang, Y., Westerhoff, P., and Capco, D.G. (2014) Food grade titanium dioxide disrupts intestinal brush border microvilli *in vitro* independent of sedimentation. *Cell Biol. Toxicol.*, **30** (3), 169–188.

17 Mutsuga, M., Sato, K., Hirahara, Y., and Kawamura, Y. (2011) Analytical methods for SiO_2 and other inorganic oxides in titanium dioxide or certain silicates for food additive specifications. *Food Addit. Contam. Part A*, **28** (4), 423–427.

18 Peters, R.J.B., van Bemmel, G., Herrera-Rivera, Z., Helsper, H.P.F.G. et al. (2014) Characterization of titanium dioxide nanoparticles in food products: analytical methods to define nanoparticles. *J. Agric. Food Chem.*, **62** (27), 6285–6293.

19 Weir, A., Westerhoff, P., Fabricius, L., Hristovski, K., and von Goetz, N. (2012) Titanium dioxide nanoparticles in food and personal care products. *Environ. Sci. Technol.*, **46** (4), 2242–2250.

20 Yang, Y., Doudrick, K., Bi, X., Hristovski, K. et al. (2014) Characterization of food-grade titanium dioxide: the presence of nanosized particles. *Environ. Sci. Technol.*, **48** (11), 6391–6400.

21 Lomer, M.C.E., Thompson, R.P.H., Commisso, J., Keen, C.L., and Powell, J.J. (2000) Determination of titanium dioxide in foods using inductively coupled plasma optical emission spectrometry. *Analyst*, **125** (12), 2339–2343.

22 Periasamy, V.S., Athinarayanan, J., Al-Hadi, A.M., Juhaimi, F.A. et al. (2015) Identification of titanium dioxide nanoparticles in food products: induce intracellular oxidative stress mediated by TNF and CYP1A genes in human lung fibroblast cells. *Environ. Toxicol. Pharmacol.*, **39** (1), 176–186.

23 EFSA ANS Panel (EFSA Panel on Food Additives and Nutrient Sources added to Food) (2016) Scientific opinion on the re-evaluation of gold (E 175) as a food additive. *EFSA J.*, **14** (1), 4362. Available at http://www.efsa.europa.eu/sites/default/files/scientific_output/files/main_documents/4362.pdf.

24 EFSA ANS Panel (EFSA Panel on Food Additives and Nutrient Sources Added to Food) (2016) Scientific opinion on the re-evaluation of silver (E 174) as food additive. *EFSA J.*, **14** (1), 4364. Available at http://www.efsa.europa.eu/sites/default/files/scientific_output/files/main_documents/4364.pdf.

25 Ramos, K., Ramos, L., Cámara, C., and Gómez-Gómez, M.M. (2014) Characterization and quantification of silver nanoparticles in nutraceuticals and beverages by asymmetric flow field flow fractionation coupled with inductively coupled plasma mass spectrometry. *J. Chromatogr. A*, **1371**, 227–236.

26 Reed, R.B., Faust, J.J., Yang, Y., Doudrick, K. *et al.* (2014) Characterization of nanomaterials in metal colloid-containing dietary supplement drinks and assessment of their potential interactions after ingestion. *ACS Sustain. Chem. Eng.*, **2** (7), 1616–1624.

27 Verleysen, E., Van Doren, E., Waegeneers, N., De Temmerman, P.J. *et al.* (2015) TEM and SP-ICP-MS analysis of the release of silver nanoparticles from decoration of pastry. *J. Agric. Food Chem.*, **63** (13), 3570–3578.

28 EFSA Panel on Food Additives and Nutrient Sources added to Food (ANS) (2011) Scientific opinion on reevaluation of calcium carbonate (E 170) as a food additive. *EFSA J.*, **9** (7), 2318. Available at http://www.efsa.europa.eu/sites/default/files/scientific_output/files/main_documents/2318.pdf.

29 Tsai, C.-F., Huang, L.-Y., Tseng, S.-H., and Chen, C.-H. (2012) Structural identification of nano-calcium compounds in milk powder by X-ray powder diffraction. *J. Food Drug Anal.*, **20** (2), 510–515.

30 Joint FAO/WHO Expert Committee on Food Additives (2009) Compendium of Food Additives Specifications. 71th meeting. pp. 1–120. Available at http://www.fao.org/123/a-i0971e.pdf.

31 Dekkers, S., Bouwmeester, H., Bos, P.M.J., Peters, R.J.B. *et al.* (2013) Knowledge gaps in risk assessment of nanosilica in food: evaluation of the dissolution and toxicity of different forms of silica. *Nanotoxicology*, **7** (4), 367–377.

32 European Parliament and Council (2011) Commission regulation (EU) No 1129/2011 of 11 November 2011 amending Annex II to regulation (EC) No 1333/2008 of the European Parliament and of the Council by establishing a Union list of food additives. pp. 1–177. Available at http://eur-lex.europa.eu/LexUriServ/LexUriServ.do?uri=OJ:L:2011:2295:0001:0177:en:PDF.

33 Athinarayanan, J., Alshatwi, A.A., Periasamy, V.S., and Al-Warthan, A.A. (2015) Identification of nanoscale ingredients in commercial food products and their induction of mitochondrially mediated cytotoxic effects on human mesenchymal stem cells. *J. Food. Sci.*, **80** (2), N459–N464.

34 Athinarayanan, J., Periasamy, V.S., Alsaif, M.A., Al-Warthan, A.A., and Alshatwi, A.A. (2014) Presence of nanosilica (E551) in commercial food products: TNF-mediated oxidative stress and altered cell cycle progression in human lung fibroblast cells. *Cell Biol. Toxicol.*, **30** (2), 89–100.

35 Ovissipour, M., Sablani, S.S., and Rasco, B. (2013) Engineered nanoparticle adhesion and removal from tomato surfaces. *J. Agric. Food Chem.*, **61** (42), 10183–10190.

36 Sprong, C., Bakker, M., Niekerk, M., and Vennemann, M. (2015) Exposure assessment of the food additive titanium dioxide (E 171) based on use levels provided by the industry. (RIVM Report 2015-0195), pp. 1–54. Available at www.rivm.nl/dsresource?objectid=rivmp:306146&type=org&disposition=inline&ns_nc=306141.

37 Rezwan, K., Studart, A.R., Vörös, J., and Gauckler, L.J. (2005) Change of ζ potential of biocompatible colloidal oxide particles upon adsorption of bovine serum albumin and lysozyme. *J. Phys. Chem. B*, **109** (30), 14469–14474.

38 World Health Organization - International Agency for Research on Cancer (2010) Carbon black, titanium dioxide, and talc, pp. 1–466. Available at http://monographs.iarc.fr/ENG/Monographs/vol493/mono493.pdf.

39 Jovanović, B. (2015) Critical review of public health regulations of titanium dioxide, a human food additive. *Integr. Environ. Assess. Manage.*, **11** (1), 10–20.

40 EFSA ANS Panel (EFSA Panel on Food Additives and Nutrient Sources Added to Food) (2015) Scientific opinion on the re-evaluation of iron oxides and hydroxides (E 172) as food additives. *EFSA J.*, **13**, 4317.

41 Emerton, V. (2008) *Iron Oxides and Iron Hydroxides*, Blackwell Publishing, Oxford, UK.
42 Cornell, R.M. and Schwertmann, U. (2003) *The iron oxides: Structure, Properties, Reactions, Occurrences and Uses*, Wiley-VCH Verlag GmbH, Weinheim, Germany.
43 Hadrup, N. and Lam, H.R. (2014) Oral toxicity of silver ions, silver nanoparticles and colloidal silver – a review. *Regul. Toxicol. Pharmacol.*, **68** (1), 1–7.
44 McShan, D., Ray, P.C., and Yu, H. (2014) Molecular toxicity mechanism of nanosilver. *J. Food Drug Anal.*, **22** (1), 116–127.
45 Metak, A.M., Nabhani, F., and Connolly, S.N. (2015) Migration of engineered nanoparticles from packaging into food products. *LWT – Food Sci. Technol.*, **64** (2), 781–787.
46 Hilty, F.M. and Zimmermann, M.B. (2014) Nano-structured minerals and trace elements for food and nutrition applications, in *Nano- and Microencapsulation for Foods*, John Wiley & Sons, Ltd, pp. 199–222.
47 EFSA AFC Panel (2008) Scientific opinion of the panel on food additives, flavourings, processing aids and food contact materials (AFC) on safety of aluminium from dietary intake. *EFSA J.*, **754**, 1–34.
48 Boverhof, D.R. and David, R.M. (2010) Nanomaterial characterization: considerations and needs for hazard assessment and safety evaluation. *Anal. Bioanal. Chem.*, **396** (3), 953–961.
49 Sayes, C.M. and Warheit, D.B. (2009) Characterization of nanomaterials for toxicity assessment. *Wiley Interdiscip. Rev. Nanomed. Nanobiotechnol.*, **1** (6), 660–670.
50 Tiede, K., Boxall, A.B.A., Tear, S.P., Lewis, J. *et al.* (2008) Detection and characterization of engineered nanoparticles in food and the environment. *Food Addit. Contam. Part A*, **25** (7), 795–821.
51 EFSA Scientific Committee (2011) Scientific opinion on guidance on the risk assessment of the application of nanoscience and nanotechnologies in the food and feed chain. *EFSA J.*, **9** (5), 2140.
52 George, S., Lin, S., Ji, Z., Thomas, C.R. *et al.* (2012) Surface defects on plate-shaped silver nanoparticles contribute to its hazard potential in a fish gill cell line and zebrafish embryos. *ACS Nano*, **6** (5), 3745–3759.
53 Hassellöv, M., Readman, J.W., Ranville, J.F., and Tiede, K. (2008) Nanoparticle analysis and characterization methodologies in environmental risk assessment of engineered nanoparticles. *Ecotoxicology*, **17**, (5), 344–361.
54 Adam, V., Loyaux-Lawniczak, S., and Quaranta, G. (2015) Characterization of engineered TiO_2 nanomaterials in a life cycle and risk assessments perspective. *Environ. Sci. Pollut. Res.*, **22** (15), 11175–11192.
55 Bandyopadhyay, S., Peralta-Videa, J.R., and Gardea-Torresdey, J.L. (2013) Advanced analytical techniques for the measurement of nanomaterials in food and agricultural samples: a review. *Environ. Eng. Sci.*, **30** (3), 118–125.
56 Blasco, C. and Pico, Y. (2011) Determining nanomaterials in food. *Trac-Trend Anal. Chem.*, **30** (1), 84–99.
57 Blasco, C. and Picó, Y. (2006) Nanoparticles in foods, determination of, in *Encyclopedia of Analytical Chemistry*, John Wiley & Sons, Ltd.
58 Calzolai, L., Gilliland, D., and Rossi, F. (2012) Measuring nanoparticles size distribution in food and consumer products: a review. *Food Addit. Contam. Part A*, **29** (8), 1183–1193.
59 Contado, C. (2015) Nanomaterials in consumer products: a challenging analytical problem. *Front. Chem.*, **3** (3), 48.
60 Grieger, K.D., Harrington, J., and Mortensen, N. (2016) Prioritizing research needs for analytical techniques suited for engineered nanomaterials in food. *Trends Food Sci. Technol.*, **50**, 219–229.
61 Laborda, F., Bolea, E., Cepria, G., Gomez, M.T. *et al.* (2016) Detection, characterization and quantification of inorganic engineered nanomaterials: a review of techniques and methodological approaches for the analysis of complex samples. *Anal. Chim. Acta*, **904**, 10–32.
62 Lapresta-Fernández, A., Salinas-Castillo, A., Anderson de la Llana, S.,

Costa-Fernández, J.M. et al. (2014) A general perspective of the characterization and quantification of nanoparticles: imaging, spectroscopic, and separation techniques. *Crit. Rev. Solid State*, **39** (6), 423–458.

63 Lin, P.-C., Lin, S., Wang, P.C., and Sridhar, R. (2014) Techniques for physicochemical characterization of nanomaterials. *Biotechnol. Adv.*, **32** (4), 711–726.

64 Linsinger, T.P.J., Chaudhry, Q., Dehalu, V., Delahaut, P. *et al.* (2013) Validation of methods for the detection and quantification of engineered nanoparticles in food. *Food Chem.*, **138** (2–3), 1959–1966.

65 Lopez-Serrano, A., Olivas, R.M., Landaluze, J.S., and Camara, C. (2014) Nanoparticles: a global vision. Characterization, separation, and quantification methods. Potential environmental and health impact. *Anal. Met.*, **6** (1), 38–56.

66 Singh, G., Stephan, C., Westerhoff, P., Carlander, D., and Duncan, T.V. (2014) Measurement methods to detect, characterize, and quantify engineered nanomaterials in foods. *Compr. Rev. Food Sci. Food Saf.*, **13** (4), 693–704.

67 López-Lorente, Á.I. and Valcárcel, M. (2014) Chapter 10 – determination of gold nanoparticles in biological, environmental, and agrifood samples, in *Comprehensive Analytical Chemistry* (eds V. Miguel and I. L.-L. Ángela), Elsevier, pp. 395–426.

68 OECD (2016) Report of the OECD Expert Meeting on the Physical–Chemical Properties of Nanomaterials: Evaluation of Methods Applied in the OECD - WPMN testing Programme ENV/JM/MONO, 7. pp. 1–43. Available at http://www.oecd.org/officialdocuments/publicdisplaydocumentpdf/ ?cote=env/jm/mono%282016%292297&doclanguage=en.

69 Linsinger, T.P.J., Roebben, G., Gililand, D., Calzolai, L. *et al.* (2012) Requirements on measurements for the implementation of the European Commision definition of the term "Nanomaterial," pp. 1–74. Available at http://ec.europa.eu/jrc/sites/default/files/irmm_nanomaterials_%28online%29.pdf.

70 Linsinger, T.P.J., Peters, R., and Weigel, S. (2014) International interlaboratory study for sizing and quantification of Ag nanoparticles in food simulants by single-particle ICPMS. *Anal. Bioanal. Chem.*, **406** (16), 3835–3843.

71 Peters, R., Herrera-Rivera, Z., Undas, A., van der Lee, M. *et al.* (2015) Single particle ICP-MS combined with a data evaluation tool as a routine technique for the analysis of nanoparticles in complex matrices. *J. Anal. Atom. Spectrom.*, **30** (6), 1274–1285.

72 Stahl, T., Taschan, H., and Brunn, H. (2011) Aluminium content of selected foods and food products. *Environ. Sci. Eur.*, **23** (1), 1–11.

73 Qu, H.O., Mudalige, T.K., and Linder, S.W. (2016) Capillary electrophoresis coupled with inductively coupled mass spectrometry as an alternative to cloud point extraction based methods for rapid quantification of silver ions and surface coated silver nanoparticles. *J. Chromatogr. A*, **1429**, 348–353.

74 Ma, Z. and Boye, J.I. (2013) Advances in the design and production of reduced-fat and reduced-cholesterol salad dressing and mayonnaise: a review. *Food Bioprocess Technol.*, **6** (3), 648–670.

75 Reidy, B., Haase, A., Luch, A., Dawson, K., and Lynch, I. (2013) Mechanisms of silver nanoparticle release, transformation and toxicity: a critical review of current knowledge and recommendations for future studies and applications. *Materials*, **6** (6), 2295.

76 Powell, J.J., Thoree, V., and Pele, L.C. (2007) Dietary microparticles and their impact on tolerance and immune responsiveness of the gastrointestinal tract. *Br. J. Nutr.*, **98** (Suppl 1), S59–S63.

77 Burcza, A., Gräf, V., Walz, E., and Greiner, R. (2015) Impact of surface coating and food-mimicking media on nanosilver-protein interaction. *J. Nanopart. Res.*, **17** (11), 1–15.

6
Nanostructure Characterization Using Synchrotron Radiation and Neutrons

Francois Boué[1,2]

[1] INRA-AgroParisTech-UPSay, Génie Microbiologique et Procédés Alimentaires UMR 782, 1 rue L. Brétignières, 78850 Thiverval – Grignon, France
[2] UMR 12 CNRS-CEA-IRAMIS, Laboratoire Léon Brillouin, 91191 Gif-sur-Yvette Cedex, France

6.1 Introduction

6.1.1 Observing at Nanosizes *In Situ*

When passing from micrometer sizes to nanometer sizes, a first important difference arises: using imaging methods in the most spread case, that is, involving light, or in other words, usual optical microscopy, becomes useless: The resolution is on the order of the light wavelength, that is, $\lambda = 0.5\,\mu m$, below which diffraction effects render geometrical optics impossible. We have then three possibilities:

- The first is to use smaller wavelength. In most of the case this is achieved with electronic microscopy. The electron beam travels in vacuum, which requires in practice placing the sample under vacuum – in case of food, for example, this will dry the water out, which most of the time will alter – if not totally destroy – the whole structure. For example, one will have to start from a rather dilute system (already modifying the structure), and during drying small droplets will form and reaggregate the nano-objects. An alternative is to use an environmental chamber where the sample is localized, but other complications arise, including a low thickness of the sample and also its low concentration, in addition to the fact that this equipment is costly.
- The second is to use "nanoprobes," such as AFM tips, and optical tweezers, which can be displaced at nanometer distances. Again, strong requirements are linked to the state of the sample.
- The third is taking advantage of diffraction; in other words, using scattering. Light scattering reaches tenth of micrometers (while staying sensitive to

Nanotechnology in Agriculture and Food Science, First Edition. Edited by Monique A.V. Axelos and Marcel Van de Voorde.
© 2017 Wiley-VCH Verlag GmbH & Co. KGaA. Published 2017 by Wiley-VCH Verlag GmbH & Co. KGaA.

turbidity). In order to see "truly nanoparticles," that is, object below the 100 nm scale, we can take advantage of two other types of radiation: X rays (the typical value is $\lambda = 0.1$ nm, as produced by a copper anode in a laboratory machine, although a wide range is now available owing to synchrotrons) and neutrons (0.1 nm<λ<3 nm, in nuclear reactors or spallation sources).

In short, scattering is due to the interference between the waves arising from two different scatterers, but distances much larger than λ can be reached if looking at very small scattering angles; thus, "small-angle neutron and X-ray scattering can give information up to 100–1000 times the wavelength used, that is, up to tenth of micrometers. How it works will be illustrated briefly in Section 6.2.

The huge advantage of these techniques is that they do not require such sophisticated preparation of the samples as short-scale imaging. Measurements can be made on volume lower than 1 ml (see just below) and on a wide range of concentrations, for example, à1-10% in volume). Finally, they are made *in situ*, allowing following time evolutions under various external constraints such as temperature, pH, addition of other components, all things very often met when studying food, evolution of food, and food processing.

Another particularity is that the information is gathered over all the sample volume (typically 0.1 ml for X-rays to a few milliliters for neutrons). We thus get information on an average ensemble contrary to imaging, which can induce wrong conclusions if one focuses on a nonrepresentative situation and can also require numerous and tedious recordings for complex samples. This averaging feature is so interesting that scattering is used even for sizes larger that the micrometer.

Finally, let us note that due to both ensemble averaging and the speed of the radiation, the object can move: The measurement is not sensitive to this, except if it triggers some deformation, such as under shearing or stirring all constraints to which food is often subjected to. Hence, different rheological studies can be made, looking at the effect of the flow on the structure (Figure 6.1).

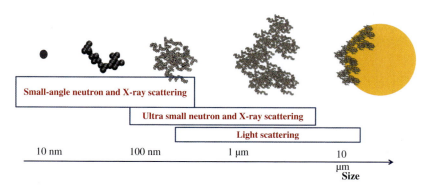

Figure 6.1 Sizes, structures, and corresponding scattering techniques.

6.1.2
Nanoparticles in Food and Agricultural Products: What is Here, What Can Be Seen

As described in other contributions of this book, many different nanosized objects or structures can be present in food, as occurring in different areas of matter, in particular soft matter, of which food can be considered a small field.

They can be present in nature, and are composed of either inorganic species, such as calcium phosphate nanoclusters in milk or other possible metallic salt aggregation or precipitation, or organic species, such as small emulsion droplets, fatty acids micelles, liposomes, periodic layers and crystals, carbohydrates macromolecules or small-scale gel structures, and proteins and protein aggregates.

They can be incorporated, or triggered, by man, due to physicochemical processes, linked with new contacts generated between different ingredients, due to heating, during the many food processes, and also due to the addition of nanoparticles such as silica, titanium oxide, and so on.

On one hand, the chemical composition – organic or not – is important because it may be linked with the possibility of detecting these objects or structures within the hosting system, among the rest of the biologic material. On the other hand, this is not relevant when considering spatial geometry: They may be considered as similar when they have the same shape, and indeed similar organizations may arise from different species. For example, SiO_2 nanoparticles and beta-lactoglobulin may form fractal aggregates owing to local interactions very similar in their physics or in their interaction potential profile $U(r)$.

The information on the following is provided by scattering:

- Size of the elementary object, symmetries, anisotropy, size distribution, internal structure often complex when made from the interaction of several components.
- At larger scale, external arrangements. Indeed, many small structures or objects tend to organize at larger scales. As already said, nano-objects are often unstable: Nanoparticles or proteins tend to aggregate over a large range of scales. We end with a multiscale system. In such a case, the basic properties are modified – for example, aggregated nanoparticles may be less dangerous since they are not as invasive, or globular proteins lose their specific role.

Another important issue is sometimes to know where nanoparticles are: Do they penetrate or not in the body cells, or interact with specific food component.

6.2
Principles

6.2.1
Scattering Process

The elementary process of scattering can be described in analogy with the well-known Young slits experience (Figure 6.2). Two narrow slits separated by a

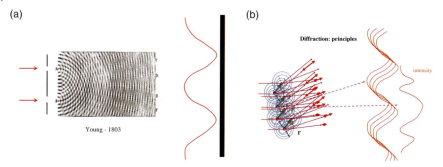

Figure 6.2 Principles of scattering: Young slits (a) and scattering from many scatterers (b).

vector **r** are illuminated by the same planar wave (the plane is parallel to the plane containing the slits). They receive the wave with the same phase, and re-emit in all direction as two spherical waves. On the detector plane located at a distance D, along a direction defined by an angle theta (θ), they interfere, producing an intensity the amplitude of which depends on their phase difference. The latter is a difference between the delay of "arrival" of the incoming beam on the two slits (if there is an angle θ_1 between the beam direction and the vector **r**) and the delay of "departure" for the outgoing beam (in a given direction $\theta_1 + \theta$). Finally, after vectorial subtraction [1,2] the phase difference is

$$4\pi \cdot r \sin(\theta/2)/\lambda = \mathbf{q} \cdot r, \tag{6.1}$$

which defines the scattering vector **q**.

If we now consider the intensity variation with respect to the angle, or along a line on the detector, we observe an oscillatory variation; in other words, some fringes. This explanation was given by Young (London, 1803) for his famous double-slit experiment.

If we now consider wider slits, or finite objects made of many scatterers (Figure 6.2b), we gather on a detector spot (a cell) an ensemble of waves, the combined oscillations of which display, along a given line, a smooth bump. The mean width or the curvature of the decreasing curve varies inversely with the size of the object.

It is also clear from this scheme that the larger the angle θ, the larger the phase differences $\mathbf{q} \cdot r_{ij}$ of two waves coming from the two points of the slits separated by a given r_{ij}. When $\mathbf{q} \cdot r_{ij}$ is large, the amplitude presents many oscillations with q, which cancel each other when averaging on many r_{ij} values. So the larger the r_{ij}, the lower the **q** to get some scattering. If the object is large, the scattering profile decreases sharply and vice versa. So, **q** is the conjugate of r_{ij}.

The analogy drawn in Figure 6.2 is all we need to understand the scattering from one isolated 3D object, made of N_{scatt} scatterers (average <Nscatt>), and corresponding to the interferences due to the N_{scatt}^2 pairs of scatterers.

To give an example at this stage, the simplest one is a collection of N_{obj} objects all of the same type (A), in vacuum and without interference between two scatterers of two distinct objects. The scattered intensity per incident beam intensity is by dimensional analysis a surface (cross section) and is proportional to

$$N_{obj} < N_{scatt} > b_A^2 \langle P(q) \rangle, \tag{6.2}$$

where b_A is a length and represents the strength of wave scattering of elementary scatterer A. We describe further below $\langle P(q) \rangle$, called the (average) form factor.

6.2.2
q and r: Orders of Magnitude

Let us give some orders of magnitude. For neutrons, λ varies between 0.2 and 2.5 nm; for hard X-rays – the most used for SAXS, we are often around 0.1 nm, but the synchrotron range is very wide. For classical SAS, the sample–detector distances are between 0.5 and 40 m, and the cell size is 5–10 mm for neutrons and 20–500 μm for X-rays. Equation (6.1) gives $q = 4\pi/\lambda \sin(\theta/2)$; after calculation of θ, this gives a range of 0.01–0.6 nm^{-1} (0.001–0.6 Å^{-1}) for SAXS and SANS. For light scattering, θ is large, but also the wavelength (500 nm) so that q covers one decade 0.0003–0.003 Å^{-1}. For ultrasmall angle, θ and therefore q values are 10 times lower. Together, this gives the values in r ($\sim 1/q$) shown in Figure 6.1.

6.2.3
Binary System: Contrast

The above simplest example was in vacuum. Usually, as in food, the objects made of molecules dispersed in a medium with other molecules. Let us consider the example of objects A dispersed in a solvent S, like water. If the scattering strength of a unit volume is the same whether it contains A or S, all volumes scatter similarly, and are equivalent to an infinite continuous medium: the wave proceeds as a planar wave, like light through glass, and there is *no macroscopic scattering*. Scattering will appear if one draws with a marker some black regions on the glass. Alternatively, if the same drawings are made by cutting away some parts of a black screen, like a photographic "negative," the scattering will be identical (Babinet theorem). This is similar for the binary system A/S, due to an important property of our soft systems, the "incompressibility": Take a volume V that could contain $n_A = V/v_A$ and have a scattering strength $V \cdot b_A/v_A$. If no scatterers A are present, $n_S = V/v_S$ scatterers S of volume v_S must be present and the scattering strength is $V \cdot b_S/v_S$. Therefore, the S spatial distribution of the scattering strength density b_S/v_S is the "negative" of the one of the scattering strengths b_A/v_A. The scattering density from pairs A and A, $(b_A/v_A)^2 \cdot S_{AA}(q)$, is identical to the squared negative $(b_S/v_S)^2 \cdot S_{SS}(q)$, and similarly the crossed terms A–S and S–A have a negative front factor $-(b_A/v_A) \cdot (b_S/v_S) \cdot S_{AA}(q)$. Finally, the sum is a squared difference

$$I(q) \sim \Phi \cdot (b_A/v_A - b_S/v_S)^2 \cdot S_{AA}(q). \tag{6.3}$$

The prefactor is called the contrast. It is the square of the difference between the "scattering length densities" $SLD_A = b_A/v_A$ and $SLD_S = b_S/v_S$ (also noted ρ_A and ρ_S, respectively). All calculations are shown in Ref. [3]. The way the SLD is determined is presented in the next section.

6.2.4
Contrast Strategies

How to Determine Scattering Length Density

In practice, since we consider the nanometer to micrometer scale, we will approximate the relevant elementary unit as a group of atoms i (lattice cell in material science; molecule or repeating unit in soft matter). We then consider the sum of all l_i by unit of volume, $\Sigma l_i / \Sigma v_i$, called the "scattering length density" (SLD in cm^{-2}, often noted ρ). The elementary scatterer depends on the radiation. For X-rays, it is the electronic cloud of one atom, so the SLD varies with the number of electrons per volume. For neutrons, it is the nucleus for which values of l_i vary in a much less systematic way as summarized in Figure 6.3. For visible light, it is a unit volume, larger than the wavelength and homogeneous so that a concentration can be defined.

How to Deal with Contrast? Contrast, Contrast Matching, and Contrast Variation

Let us first return to Eq. (6.3): The contrast, $SLD_A - SLD_B$, must be estimated. The higher, the better. It is important to check whether the cross section is large

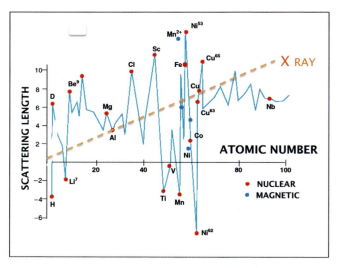

Figure 6.3 Variation of scattering length with atomic number Z. For X-rays (constant proportionality to Z) or neutrons.

enough (for the spectrometer that is used). And, if possible, this should be optimized.

For X-rays, strong differences in electron density are preferred (see Mendeleïev Table!). For example, it is interesting to replace sodium counterions, Na^+, by cesium, Cs^+, in biopolymers solutions because monovalent ions often behave very closely. This is more difficult for divalent ions: for example, complexation with carboxylic acids is different for Ca^{++} ions and for Sr^{++}, strontium. We must at least take advantage of opportunities: If a material naturally contains barium (because of its complexation potential), accurate studies are possible.

Another fascinating trick, called *anomalous scattering*, is also possible with synchrotron since the wavelength can be tuned within a large range: approaching the absorption threshold, around which the scattering length varies rapidly for a chosen atom, allowing thus a natural SLD magnification (note that the wavelength selection must be extremely reliable).

For neutrons, a very convenient option is deuteriation. The strong difference between H and D nuclei, shown in (Figure 6.3), leads to large SLD differences. This is the case, for example, for species containing a lot of protons (H) and solvent containing deuterium, often "heavy water" (D_2O). In some cases, some species may be synthesized in their deuterated version (polymers, lipids), as illustrated below.

In addition to increasing the contrast in a binary system, a wide range of availability of SLD will enable the labeling of one species when the system contains three or more species. A famous technique is "contrast matching" in a mixture of nondeuterated/deuterated (H/D) solvent: The scattering of species A in several mixtures of different deuterated fractions (contrast variation) is measured first. If only A is present, all signals are proportional to each other and to the squared difference of SLDs $(b_A/v_A - b_S/v_S)^2$ (see Eq. 6.3). At a given value of q, b_A/v_A is determined via the *H/D* percentage at which the square root of the $I(q)$ meets the zero axis (matching). A ternary mixture can thus be studied as shown in Figure 6.4: When matching A (poly(styrene sulfonate) in the example, a water-soluble polyelectrolyte – PEL, noted PSSNa), only B (lysozyme) is visible and vice versa.

6.3
The Basic Information from a SAS Profile

6.3.1
Form Factors

If the system is not dilute, the distances between objects must be accounted and influence the scattering. For undeformable and centrosymmetric objects,

$$I(q) \sim \Phi \cdot (b_A/v_A - b_S/v_S)^2 \langle V_{obj\ A} \rangle \cdot P_A(q) \cdot S_A(q), \tag{6.4}$$

where $S_A(q)$ is related to the distribution of the distances between objects.

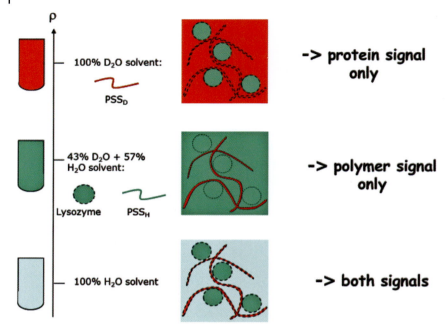

Figure 6.4 Contrast matching strategy. On the axis labeled ρ are scattering length densities for mixtures of H_2O and D_2O. On the right side of the axis, schematic representations of a mixture of d-PSSNa/PSSNa/lysozyme are given in three mixtures. For 100% D_2O (top), only the protein is visible; for 43% D_2O (middle), only the d-PSSNa is visible.

Often, $P_A(q)$ is determined directly using dilute solutions (unless using specific contrast strategies in SANS – zero average contrast [4]). The description is more complex if $P_A(q)$ varies with concentration, and even more when objects are not centrosymmetric, or deformable, such as polysaccharides, unfolded proteins, lipid tails. The right combination with structure factors has thus to be "guessed."

In any case, it is useful to have in mind a few typical form factors.

The Guinier Regime Gives the Size
The first thing is that any form factor starts by a fall from a limit 1 at $q \rightarrow 0$:

$$S_1(q) \sim M_w \cdot P(q) \sim M_w \cdot (1 - q^2 R_g^2/3), \quad q \cdot R_g \ll 1 \quad \text{(Guinier law)} \quad (6.5)$$

in the so-called Guinier regime (Figure 6.5). This defines a global average size R_g, the radius of gyration (in the mechanical meaning). Since the condition $q \cdot R_g \ll 1$ (Eq. (6.5)) is very exacting and requires a very high accuracy, measurements are done for q as high as $1/R_g$, and various strategies for estimating the R_g are possible, corresponding to known approximations that depend on the expected shape of the object. For example, a plot of $1/I(q)$ versus q^2 (*Zimm plot*) is convenient for a rod shape, and is extended to a polymer chain. Reversely, a plot of $\log(I(q))$ versus q^2 (*Guinier plot*) is suitable for polydisperse spheres.

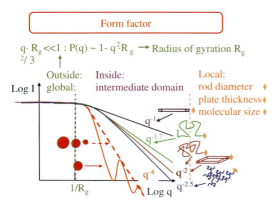

Figure 6.5 Log–log plot of examples of form factors from objects of different geometries.

The Intermediate Regime Gives the Shape

Information on the actual shape of the objects is gathered at $q \cdot R_g > 1$, the so-called intermediate regime. In this region, the inside of the object is visible.

The behavior strongly depends on the shape (Figure 6.5). A rather crude reasoning is that, for a rod, the number of scattering points of the rod inside a sphere of radius r is $n(r) \sim r$. This can be written in terms of q as $s(q) \sim 1/q$. For a platelet, it will be r^2; hence, the scattering is proportional to the number $n(q) \sim 1/q^2$.

More generally, if the object is fractal, $s(q) \sim 1/q^{D_f}$, where D_f is the assumed fractal dimension of the object: 2 for a Gaussian polymer chain (random walk), $1.8 < D_f < 2.5$ for a fractal aggregate (RCLA to DCLA [5]). Increasing D_f, we reach the limit value 3, which is equal to the one of real spaces; in other words, we have a dense object. The former law then loses significance, since the inside of a compact object is homogeneous, there is no density fluctuations, and hence no scattering "from inside."

The scattering only comes from the shape, like from a "3D" slit, and generates oscillations with maxima separated by an interval $\Delta q \sim 2\pi/R$ (Figure 6.5 (red curves) and 6.6).

However, another important point is that $P(q)$ must be calculated over an ensemble of objects. This is already the case for the fractal concept, but for an ensemble of sphere this may induce a strong change when the value of the radius R is not unique. In such case, the oscillations are blurred and the decay at large q varies as $1/q^4$ (red curves in Figure 6.5). This is the so-called Porod law, which just represents the scattering from a sharp interface between two media. For example, it will be found for a porous material, whatever the spatial size and distribution of the pores may be.

Advanced Form Factors

Many other cases can then be considered, where the form factors given above are more complex or "composite": for example, for spherical or rod-like nanoparticles coated with a shell of another material (core–shell) (Figure 6.14) or

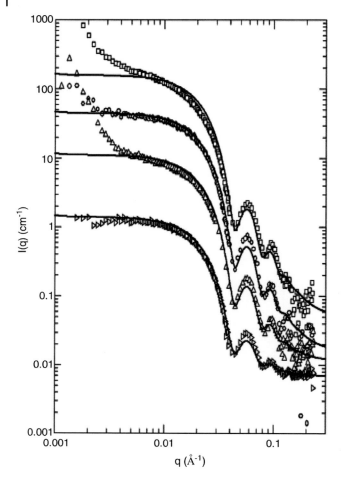

Figure 6.6 Monodisperse spheres form factor: scattering from gold particles (0.2% by weight) and for lipase-modified particles (0.02% by weight) coated step-by-step. The solid lines are fits to a smeared, polydisperse core–shell form factor (gold core radius 7.5 nm and shell thickness of the layer about 1.5 nm). The same fit is used for all cases (low density of the biotin, low effect in binding a few enzyme molecules). (Reproduced by courtesy from Ref. [6].)

with polymers (hairy spheres), spheres linked by polymer chains ("pearl necklaces" [4]).

Simple Examples of Form Factors

Before considering more complex scattered signals, in order to stay close to real world, we wish here to present two usual, simple, and very different form factors:

- The *sphere form factor of Au* nanoparticles, in dilute regime, at different steps during coating of a lipase enzyme (the aim is to monitor digestion when lipase is grafted on the NPs) (Figure 6.6). It is characterized by sharp oscillations, due

to the Fourier transform of a well-defined size, the radius R, with sharp distribution, all through the grafting steps. However, it can be noticed on the log–log plot of $I(q)$ at small q, when $q>0$, some upturns: Often, aggregates coexist with the free form, a very general problem for nano-objects.
- The form factors of a completely different shape, a polymer chain. In (Figure 6.7a while the scattering also has a Guinier range with the same q Guinier law, the shape at large q shows a succession of different power laws of q, since the conformation of the chain crosses over from rigid at short scale to self-avoiding (Figure 6.7a). The low q range on the left-hand side insert of Figure 6.7a as well as the low q range in Figure 6.7b show the Guinier regime at low q.

6.3.2
Structure Factors: Interactions between Objects

Hard Spheres
One of the most popular structure factors often applies to nanospheres and corresponds to the simplest interaction, that is to say repulsion at contact. "Hard" spheres (not penetrable) of radius R cannot have their centers at distances lower than $2R$. This repulsion kills the scattering at large q (larger than $2\pi/R$), increases the thermodynamic cost of large aggregates, while the occurrence of contact is increased at large concentration; it results in a scattering maximum at $q = 2\pi/R$. The most used is the Percus–Yevick function. There are many avatars of this S

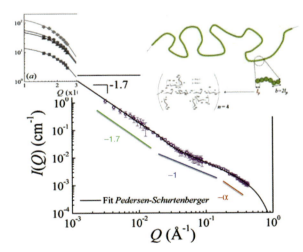

Figure 6.7 (a) Form factors of polymer chains. Form factor of a xyloglucan chain (neutral hydrophilic, low concentration) showing the different q regimes (slopes indicated on the plot) and a fit by the worm-like chain Pedersen–Schurtenberger model. *Top left:* Complementary light scattering (in semi-log plot) showing reversible aggregation. (From Ref. [7].) (b) Form factors of polymer chains. SAXS conformational tracking of amylose synthesized by amylosucrases. (From Ref. [8].)

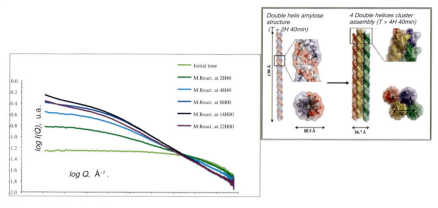

Figure 6.7 (Continued)

(q); in particular, in the case of electrically charged spheres, which repel each other at a common distance D linked to the volume fraction $\Phi = D^3/R^3$. The shape of the corresponding maximum of this function and its volume fraction dependence are also linked to the interaction energy potential $U(r)$, which represents all interactions (repulsion at contact, hard spheres or soft ones (e.g., grafted with polymer), electrostatic repulsion, van der Waals attraction, etc.), and also enables modeling of the solution rheology, important for applications. $U(r)$ also governs some other interactions, as we will see in the following section.

Aggregation
When repulsion is overbalanced by attraction, nano-objects aggregate. This is very important for applications. It modifies the genuine properties of the components: for example, nanoparticles, which may be dangerous for health, but become harmless when aggregated. It also creates new functionalities, like for food gels, turbidity, linear and nonlinear viscoelasticity, dependence over frequency, and sensorial properties. Well-known ones are fractal-like, whether aggregation is limited by diffusion (DLCA, immediate adhesion at first contact, $D_f = 1.8$) or by chemical reaction (RLCA, not fully immediate adhesion at first contact, hence more compact $D_f = 2.2$).

In such a case, the main features of the structure factor for NPs in a cluster are one of the form factors of fractal object, $S(q) \sim q^{-Df}$, down to $q \sim 1/R_{agg}$ (Figure 6.8). Exponents 4 (for compact aggregates) or ~3 (due to polydispersity or fractal surfaces) are also encountered (Figure 6.5).

Polymer Interpenetration
Until now interactions between rather compact and "hard" nano-objects have been considered. A completely orthogonal world is the one of polymers, where chains are Gaussian (random walk), or excluded volume (see above). When the available volume is restricted, these chains "interpenetrate" each other. Their form factor changes, while the structure factor does not yield the relevant

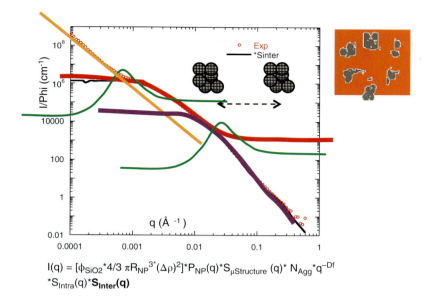

$$I(q) = [\phi_{SiO_2} * 4/3 \, \pi R_{NP}^3 * (\Delta\rho)^2] * P_{NP}(q) * S_{\mu Structure}(q) * N_{Agg} * q^{-Df} * S_{Intra}(q) * S_{Inter}(q)$$

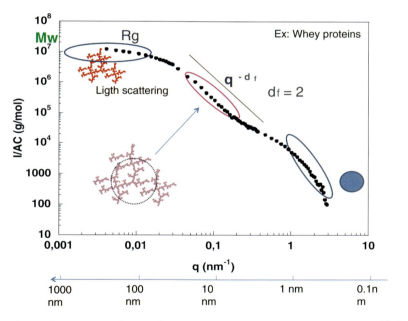

Figure 6.8 (a) Structure factors of NPs inside aggregates for tire reinforcement. The various contributions are in different colors: NP form factor, Percus–Yevick structure factor between NPs, fractal NP aggregates, structure factor between aggregates, and low q $q^{-Df=3}$ scattering from crazes. (See Ref. [9].) (b) Whey protein aggregates (WPI): at low q connection of SANS with light scattering. (See Ref. [20].)

description anymore. Rather, the new scattering function often implies a correlation size ξ (a kind of average "pore size" above which the solution is homogeneous) and recovers at large $q>1/\xi$ a fractal law akin to the one for the inside of the chain, with the corresponding D_f. Note that using mixtures of H and D chains, the form factor, in other words the polymer conformation can be directly studied in a bulk of dry amorphous polymer.

Phase Separation
Let us end by the result of a phase separation. When two species A and B (or a solvent S) in a solution are strongly immiscible, they will separate into two phases, one rich in A and the other poor in A. The resulting signal will feature a maximum size, Ξ, and a typical interfacial scattering, like a Porod one (q^{-4} for sharp interface, or softer exponent $q^{<4}$ if the separation is less complete).

6.4
A Few Examples: From Soft Matter to Agrofood

SAS has mostly developed in the agricultural domain for food (see Neutron and Food Meetings 2010–2012–2014, Delivery of Functionality, 2015), while its use is rising for biotech.

6.4.1
Proteins/Polymer: Opposite Architectures of Complexes in Mixed Systems

Using a Synthetic Polymer: An Exemplary System for Contrast Matching
This example of mixed systems [10,11] complexes made by lysozyme, a positively charged protein at pH 4.7, and sodium poly(styrene sulfonate) (PSSNa), a negatively charged polyelectrolyte, has already been presented to illustrate contrast matching (Figure 6.4): It implies a fully D_2O buffer (scattering from lysozyme only) and a 57%/43% H_2O/D_2O mixture (scattering from d-PSSNa only).

Figure 6.9 displays the effect of an excess of PSSNa: On the signal of lysozyme only, the kinetics of evolution show a vanishing of aggregation at low q, associated with the transformation at large q, of the globular protein (q^{-4}) into an unfolded one ($q^{-Df=1.7}$ like for excluded volume of polymer chains). Strikingly, when lysozyme is matched, the scattering becomes very different: (i) at low q, the "polyelectrolyte peak" characteristic of electrostatic repulsion between charged chains; (ii) at large q, PSS chains scatter as q^{-1}, due to small-scale structure rigidified by electrostatic repulsions. So, the proteins are simply unfolded and coexist with the polyelectrolyte chains.

The second example for the same phase diagram is the complexation at a lower polymer fraction (Figure 6.10). At *large* q, the two scatterings from the two species are different from each other, like above. The PSS signal (red) is akin to one of the semiflexible individual chains, and the protein signal (blue) is

In situ kinetics of athermal protein denaturation

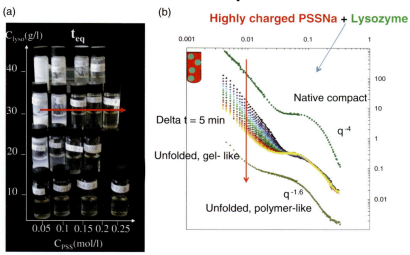

Figure 6.9 Monitoring the evolution of a form factor in a mixed system. Lysozyme is unfolded by adding PSSNa. (From Ref. [10].) (a) Photos show progressive clearing of the solutions due to protein denaturation. (b) Scattering from lysozyme only, as PSSNa is "matched."

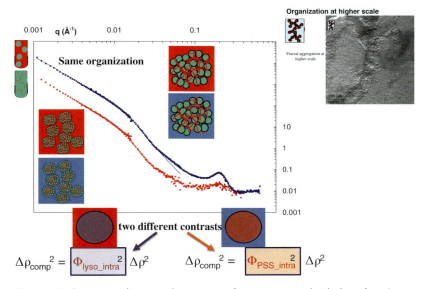

Figure 6.10 Comparison between the scattering from proteins and polyelectrolytes in coacervate complexes. *Insert:* TEM after cryofracture. (From Ref. [11].)

also identical to the individual one (q^{-4} law). At *intermediate q*, a pronounced maximum, at $q \sim 2\pi/R_{lys}$, is characteristic of close contact between two proteins inside the complexes, while the PSS scattering still shows a polyelectrolyte peak. We have thus again a quite accurate and independent description of the two species using these two different contrasts. But at *low q* a new situation arises: The signals of lysozyme (when PSS is matched) and of PSS (when lysozyme is matched) are identical! When going from large to low q, a decay in q^{-4} is observed, characteristic this time of compact objects, much larger than the lysozyme protein. At lower q, the q^{-4} signal bends down in the Guinier regime (radius R_{comp} 10 nm,), and finally at even lower q, a $q^{-2.1}$ variation. A fractal dimension $D_f = 2.1$ corresponds to a reaction-limited aggregation often seen between electrostatics species. This can be *connected with real space*, using quite sophisticated (and time-consuming) "electronic microscopy after cryo-fracture" [11], which shows grape-like aggregates (*insert* of Figure 6.10), in agreement with the precise SANS determinations of R_{comp} and its distribution variance.

The fact that both species produce the same signal in this q range means that both species are located in the spherical globules. This prompts to look at the ratio between the two signals: If the variation is identical, it is exactly equal to $\Phi^2_{lyso\_inner}/\Phi^2_{PSS\_inner}$, enabling determination of the inner charge ratio of the electrostatic complexes, $[+]/[-]_{inner}$, a very important parameter related to electrostatic neutrality inside the coacervates.

Toward Natural Systems: Striking Similarities and Quantitative Information
The above example with synthetic "deuteriable" polymer has been chosen to illustrate contrast matching but also because it finds a remarkable pendent in a system formed by natural components issued from agriculture. Solutions in D_2O with accurately tuned pH of pectin of different methylation degrees and of small globular protein (lysozyme, as well as of napin, extracted from colza with shape and mass quite close to lysozyme) were mixed together. They form coacervates complexes, the scattering of which offers a wide range of behaviors [12]. Quite interestingly, owing to the accurate work on PSSNa and lysozyme, we can recognize very resembling different trends, as seen in Figure 6.11. In this case, the polymer could not be deuterated, and we use pure D_2O (pectin matching was achieved showing the same behavior [12]), the dominant signal here being the one of lysozyme (see insert for the scattering from the free components). The experiment is not as sophisticated, but the knowledge acquired on the previous system is very useful here. Moreover, if we replace lysozyme by napin (extracted in minute quantities) we get a similar behavior.

6.4.2
Lipids: Micelles, Bilayers, Crystalline Phases

Lipids, or more generally fat, are widely spread in agronomy, agro-issued products (biodiesel and vegetal and animal fats). Metabolism of fat is also very important. Strikingly, the structures formed by fat are so strongly determined by their

Pectin–protein complex coacervates:

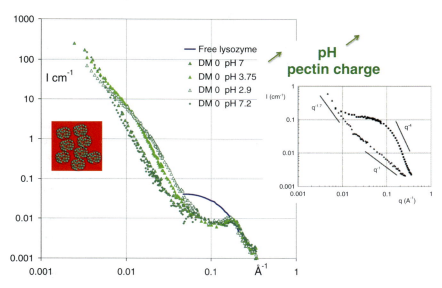

Figure 6.11 Scattering from pectin–lysozyme coacervates. *Insert*: SANS spectra from individual lysozyme and pectin DM74 10 g/l (form factors of compact protein and worm-like chain).

architecture (grossly speaking, hydrophilic head and hydrophobic tails) that the many structures they form are quite general in all types of materials, and their scattering is characteristic. The simplest one is the elementary micelles easily formed by surfactants with one lipid tail, and which often bear electrical charges: They thus display the combination of sphere form factor and structure factor with long-range repulsion. Double-tail lipids often form bilayers, hence lamellar vesicles, generally large enough to make appear a 2D behavior of the membrane over a wide enough range of size, and therefore of q. A $q^{-Df=2}$ behavior at low q is observed. Bilayers can also organize in multilamellar vesicles ("onions," or cylinders "leeks"), where the repeating distance leads to a maximum. Finally, they can also organize in specific liquid crystalline structures, such as cubosomes, and yield Bragg scattering for which the different maxima can be identified right as in a crystallographic study, with much larger lattice parameter. This occurs not at the lowest q values, but in the intermediate or large q regime of a SAS spectrometer. It is then interesting to follow the kinetics of evolution of these Bragg peaks as a function of temperature in dairy systems [13], or of time during digestion (Figure 6.12) [14]. The list of possible structures from modified fatty acids is endless! [15].

Emulsions

Emulsions are quite interesting systems, found not only in food but also in other biotechnologies, or chemical engineering using bioproducts; oil droplets can be

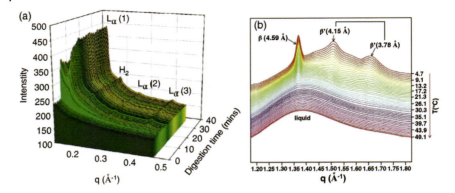

Figure 6.12 Lipids long-range structures during digestion or as function of temperature. (a) Synchrotron SAXS intensity versus scattering vector profiles during digestion for the SNEDDS formulation, indicating the three reflections for lamellar phase and single reflection for H2 phase, $q = 0.192\,\text{Å}^{-1}$. (From Ref. [14].) (b) Time-resolved synchrotron X-ray diffraction recorded on heating at 2 °C/min of the UFA-enriched TAG dispersed in 0.18 μm diameter emulsion droplets. Long-range lamellar structures and short-range hydrocarbon chain arrangements are indicated by characteristic distances. (From Ref. [13].)

stabilized using nanocellulose (pickering emulsions) and Figure 6.13 shows that the shell can be matched to the core (oil), or to external water, for different degrees of deuteration of the cellulose [16]. One could also match H/D oil mixture and H_2O/D_2O so that only the shell is visible [17].

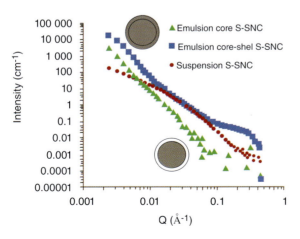

Figure 6.13 SANS results of emulsions. Emulsions are stabilized by sulfated nanocellulose contrast-matched to oil (blue) or to water (green), compared to the scattering of CNCs in aqueous suspension (red). (See Ref. [16].)

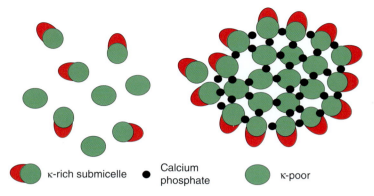

Figure 6.14 Schematic structure of casein micelle: a laboratory for scattering. (From Ref. [18].)

6.4.3
A Complex but Model Structure: Casein Micelle in Cow Milk

Cow milk, beyond molecular components such as lactose and mineral salts, is the subject of active scattering work [18,19]: It is a beautiful coexistence of fat globules of various sizes (see above) [13] – micrometer size or less (in this respect it can be considered as an emulsion), nanometer-sized globular proteins (which aggregate under heating [20]), natural mineral nanoclusters (obvious in SAXS!), and casein aggregated in giant micelles, whose size can be more than micrometer (Figure 6.14). The (many) treatments, traditional or neoindustrial, modify this structure toward systems that tend to both limit the number of components and build new complex arrangements, which resembling very much to the one studied by SAS for soft matter. This involves, for instance, aggregation/gelification of whey globular proteins or precipitation of casein large micelles.

6.4.4
Foams

Akin to emulsions are foams, involving a liquid phase and a gas phase. When the foam drains, the walls of the bubbles (Plateau borders) can have thicknesses of a few nanometers, in the presence of not only synthetic surfactants but also proteins, or even complexes of proteins and polymer such as pectin. Such coacervates studied in bulk (Figure 6.11) were also used for foaming, and the Plateau border thickness could be measured: Here SANS or SAXS results actually from the reflection of the beam on the flat walls of very drained foam cells (Figure 6.15a) [21]. A different case is where the foam is very stable, hence containing still a lot of aqueous phase, the structure of which can be characterized *in situ*. Reversible foams could be made out of a fatty acid system made of the 12 hydroxystearic acid (12-HSA) successfully dispersed in water using an organic counterion (ethanolamine). Under such conditions, one obtains self-assembled

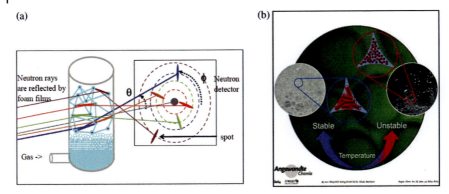

Figure 6.15 (a) SANS (reflectivity) by drained foams. (From Ref. [21].) (b) Sketch of a reversible foam containing microtubes. (From Ref. [22].)

multilayer tubes of >10 m length and 600 nm diameter (Figure 6.15b). At room temperature, the microtubes stabilize the foam for 6 months. Above 60 °C, SANS shows that it rearranges into micelles simultaneously with foam draining [22].

6.5
Other Scattering Techniques

We will quote here only some of the other radiation techniques. One has been implicitly involved: *static light scattering* (SLS), which is most of the time equivalent in its principle to SAXS and SANS except that it involves actually large angles (3–135°) but – due to the large wavelength 400–800 nm – gives access to lower q range: 3×10^{-3} to 3×10^{-2} nm^{-1}. When particles are larger than 1–10 μm, the scattering must be described by the Mie theory, not treated here. Really small-angle light scattering actually exists and gives access to even larger sizes. Another interesting potential is brought by the *dynamic light scattering* (DLS), which correlates in time (10^{-8}–10 s) the intensity variation of a speckle (spots produced by coherent light) and gives access to the dynamics of motion of nanoparticles, natural osmotic breathing, or small heterogeneities in a gel. The severe limitation of light scattering is the turbidity of the samples. Although cross-correlation gives nice improvement for samples as turbid as milk diluted 100 times, higher turbidity is difficult to address, unless samples show much higher ones, for which it is easy to use *diffusive wave scattering* (DWS) milk, or more, which gives again information on dynamics. The time range (up to 10 s) can be extended to DWS multispeckle image time correlation using a camera, which is useful to study jammed systems, very common in food and agro domains. At much shorter times, inelastic neutron scattering is available. Time of flight tackles very short times (<10^{-10} s), but the nice *neutron spin echo* technique can be pushed as far as 10^{-6} s, hence suitable for soft matter chemical physics.

Finally, let us quote *reflectivity*, which is slightly different from scattering and gives information on nanometer-sized layers, usually at interfaces, to study coatings, grafting, and thin adsorption at a surface (e.g., proteins at air–water interface, NPs deposited on a silicon surface). The variation with reflection angle (equal to the incident angle) shows decay with sometimes oscillations; this gives the variation of the SLD as a function of the altitude on the surface studied.

6.6 Recommendation and Practical: A Checklist for Scattering

This section is focused on the practice of SAXS and SANS that may require a little care in their implementations.

6.6.1 Requirements for Sample Composition and Preparation

A first requisite is the set of reasonable assumptions about a spatial structure at scale from nanometer to micrometer (reflectivity effects may also exist in SAS geometry, if flat surfaces are present).

A second requisite is that the structure creates some spatial fluctuations of the scattering length density. If we first consider a sample made of one species (e.g., polymer glass), there should be no small angle scattering except if there are large scale density fluctuations: fluid close to critical point, or some magnetic fluctuations and more prosaically crazes or pores medium (i.e., a binary system species/vacuum).

The binary system is then the basic model system built by mixing two species of different scattering length density (defined and detailed in Section 6.2.3). Porous systems are binary; however, even more frequent is the case of a solute in a solvent in the dilute regime, giving a form factor, or in semidilute regime, which implies either separation of form and structure factor, or more complex treatments and modelization. Other binary systems are objects (e.g., solid nanoparticles) dispersed in a matrix (e.g., an amorphous polymer). The SLD difference must be evaluated to aim at a sufficient contrast, ideally a maximum one. This SLD difference is natural, or artificially created:

- *For X-rays*, strong electronic densities quite different from water can be exploited when looking at silica or gold NPs solutions. This can even be built in by nature, such as $CaPO_4$ nanoclusters in milk for X-rays. The latter case corresponds to $CaPO_4$ SLD large enough to dominate the signal in the same q range as the other species, small whey globulins and caseins. They play the role of a kind of host matrix at large angle. In the same q range, the signal from casein micelles decreases strongly, while dominating all scattering at low q, due to its large size. This is true for no fat milk, but false when considering fat milk, as the scattering from fat globules interfere with the one from casein micelles.

- *For neutrons*, the simplest case is the use of deuterated solvent (D_2O), which creates contrast in most binary systems and enables contrast matching by use of mixtures of D/non-D solvent (H_2O/D_2O) to study ternary systems. Back to milk, the signal for fat can be matched with a different % of D_2O than proteins. Using D_2O is often simple, although it sometimes induces lower solubility, shifts in temperature dependences, and so on. Another way, which is more sophisticated, is to use deuterated species if they are available, which is seldom, and more costly than deuterated solvents. Soft matter studies sometimes use deuterated polymers, usually synthesized specially for this purpose from deuterated monomer (Figure 6.10).

It is also possible to attach some labeled species to object, including for X-rays, such as heavier counterions (Cs instead of Na), or introducing inorganic nanoparticles inside bilayers or cores of core–shell architecture. Finally, anomalous scattering (close to absorption threshold, see Section 3.4) is a powerful tool for SAXS, although requiring excellent stability of the beam, monochromators, and so on.

Advanced labeling can therefore become complex and require intrinsic studies during the preparation step.

6.6.2
Sample Sizes, Volumes, and Quantities

Beam sizes defined by collimation (Figure 6.16) are rather small for X-rays (below millimeters, down to less than 50 microns), hence requiring small quantities of material (mm^3). For SANS, all sizes are usually 5–10 times larger, implying volumes of order 100 mm^3 and/or larger concentrations. The thicknesses are usually 0.1–2 mm for X-rays and 1–10 mm for neutrons. Special cells are available: quartz cells for SANS, 1–2 mm capillaries for X-rays, or flat cells that can be built with mica or Capton windows. Keep in mind that the empty cell scattering is difficult to remove in case of bad signal to low q container background ratio.

Concentrations can be as low as 10 mg/l for a small protein ($R = 2$ nm), and 1 mg/l or less for a large one (or a NP with large SLD such as Au) on third- or

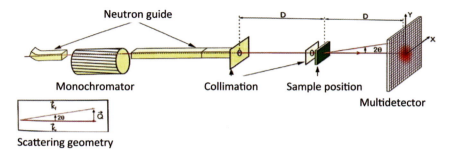

Figure 6.16 Sketch of a small-angle scattering spectrometer (here for neutron).

fourth-generation synchrotrons. This depends also on the q range: larger size requires lower q ranges, which imply much lower flux, but this is balanced in part by larger molecular scattering volumes.

Measuring durations for satisfying counting statistics depends a lot on the intensity. This is seldom a problem for synchrotron X-rays (50 ms can be enough!), while it can reach hours for SANS. Signal-to-noise ratio has to be considered carefully at large q, where the scattering become low compared to incoherent scattering (in particular, for SANS where proton incoherent scattering can be important). Due to such background effects, very low concentrations may become unreachable: Poor accuracy in transmission (error often >1%) cannot be counterbalanced by longer counting time.

6.6.3
Sample Damage

While limited to activation for neutrons, sample damage can be important for X-rays, reversibly (for surfactants) or not (proteins). A good solution is to slowly move the sample in the beam during measurement.

6.6.4
Spectrometer Setups: Sample Environment

The incident wavelength λ and the θ range (defined by detector geometry and sample–detector distance) must be chosen to define the range of $q = 4\pi/\lambda \sin(\theta/2)$, which is itself chosen such that it fits the nanostructures size to be explored. For X-rays, absorption ranges for λ is avoided, but proximity to threshold enables *anomalous scattering*.

Many sample environments are possible, starting from available sample changers, temperature-controlled cells, and pressure cells (especially with neutrons that have high penetration power) to complex injection devices sometimes built on purpose: a time-consuming task, but very rewarding for the quality of the experiment.

6.6.5
Before and After: Proposal, Data Treatment, and Fitting

In powerful radiation facilities, beam time access is very competitive. Taking care of the proposal writing is therefore important, and all technical details must be clear.

After measurements, the first step is a careful – maybe tedious – data treatment, for getting data as close to absolute values as possible. A now very widely used step is fitting: Various convenient softwares exist that enable to calculate form and structure factors of a protein, scattering from semidilute systems, and are equipped with more and efficient optimization algorithms to fit models to data.

In summary, experiments may not be so difficult if carefully prepared, at all levels.

6.7
Summary and Conclusion

Nano-world is a rich exploration field for radiation scattering, SAXS, SANS, and SLS, either via simple procedures or by advanced sample preparation. Radiation scattering can provide accurate statistical information (since it is averaged over huge numbers at variance with electronic microscopy), in particular with the help of modeling. Some examples from food and biotech fields have been developed in this chapter. Many other applications currently exist or will develop (nanoparticles in animal and human food and drugs – see Chapter 5): plant treatment and pollution of water and soils (interaction between humic acids and minerals). These few examples illustrate the generality of the approach, as well as the versatility of the models, since they are very geometric. The main burdens are the chase for contrast – and for beam time ! – meanwhile, experiments remain relatively simple, fast, and powerful.

References

1 Cotton, J.-P. (1999) Ecole SFN, Albé. *J. Phys. IV France*, **9**, 21–49. http://www-llb.cea.fr/.
2 Cotton, J.-P. (1991) Introduction to scattering experiment, in *Neutron, X-Ray and Light Scattering: Introduction to an Investigate Tool for Colloidal and Polymeric Systems* (eds P. Lindner and T. Zemb), Delta Series, North-Holland Publishing Company, Amsterdam, pp. 19–31.
3 Boué, F., Cousin, F., Gummel, J., Oberdisse, J., Carrot, G., and El Harrak, A. (2007) Small angle scattering in soft matter: application to complex mixed systems. *C.R. Phys.*, **8** (7–8), 821–844.
4 Spiteri, M.N., Williams, C.E., and Boué, F. (2007) Pearlnecklace-like chain conformation of hydrophobic polyelectrolyte. *Macromolecules*, **40** (18), 6679–6691.
5 Lin, M.Y., Lindsay, H.M., Weitz, D.A., Ball, R.C., Klein, R., and Meakin, P. (1989) Universality in colloid aggregation. *Nature*, **339**, 360–362.
6 Brennan, J., Kanaras, A., Nativo, P., Tshikhudo, P., Rees, C., Cabo-Fernandez, L., Dirvianskyte, N., Razumas, V., Skjøt, M., Svendsen, A., Jørgensen, C., Schweins, R., Zackrisson, M., Nylander, T., Brust, M., and Barauskas, J. (2010) Enzymatic activity of lipase–nanoparticle conjugates and digestion of lipid liquid crystalline assemblies. *Langmuir*, **26**, 13590–13599.
7 Muller, F., Jean, B., Perrin, P., Heux, L., Boué, F., and Cousin, F. (2013) Inherent reversible and living associations of xyloglucan. *Macromol. Chem. Phys.*, **214** (20), 2312–2323.
8 Roblin, P., Potocki-Véronèse, G., Guieysse, D., Guerin, F., Axelos, M., Perez, J., and Buléon, A. (2013) SAXS conformational tracking of amylose synthesized by amylosucrases. *Biomacromolecules*, **14** (1), 232–239.
9 Bouty, A., Petitjean, L., Degrandcourt, C., Gummel, J., Kwasniewski, P., Meneau, F., Boué, F., Couty, M., and Jestin, J. (2014) Nanofiller structure and reinforcement in model silica/rubber composites: a quantitative correlation driven by interfacial agents. *Macromolecules*, **47**, 5365–5378.
10 Cousin, F., Gummel, J., Ung, D., and Boué, F. (2005) Polyelectrolyte–protein complexes: structure and conformation of each species revealed by SANS. *Langmuir*, **21** (21), 9675–9688.
11 Gummel, J., Cousin, F., Verbavatz, J.-M., and Boué, F. (2007) Wide scale range structure in polyelectrolyte-protein dense complexes: where SANS meets freeze-

fracture microscopy. *J. Phys. Chem. B*, **111**, 8540–8546.

12 Schmidt, I., Cousin, F., Huchon, C., Boué, F., and Axelos, M.A.V. (2009) Spatial structure and composition of polysaccharide–protein complexes from small angle neutron scattering. *Biomacromolecules*, **10** (6), 1346–1357.

13 Bugeat, S., Briard-Bion, V., Pérez, J., Pradel, P., Martin, B., Lesieur, S., Bourgaux, C., Ollivon, M., and Lopez, C. (2011) Enrichment in unsaturated fatty acids and emulsion droplet size affects the crystallization behaviour of milk triacylglycerols. *Food Res. Int.*, **44**, 1314–1330.

14 Warren, D., Anby, M., Hawley, A., and Boyd, B. (2011) Real time evolution of liquid crystalline nanostructure during the digestion of formulation lipids using synchrotron small-angle X-ray scattering. *Langmuir*, **27**, 9528–9534.

15 Fay, H., Meeker, S., Cayer-Barrioz, J., Mazuyer, D., Ly, I., Nallet, F., Desbat, B., Douliez, J.-P., Ponsinet, V., and Mondain-Monval, O. (2011) Polymorphism of natural fatty acid liquid crystalline phases. *Langmuir*, **28** (1), 272–282.

16 Cherhal, F., Cousin, F., and Capron, I. (2016) Structural description of the interface of pickering emulsions stabilized by cellulose nanocrystals. *Biomacromolecules*, **17**, 496–502.

17 Jestin, J., Simon, S., Zupancic, L., and Barré, L. (2007) A small angle neutron scattering study of the adsorbed asphaltene layer in water-in-hydrocarbon emulsions: structural description related to stability. *Langmuir*, **23** (21), 10471–10478.

18 de Kruif, C., Huppertz, T., Urban, V., and Petukhov, A. (2012) Casein micelles and their internal structure. *Adv. Colloid Interface Sci.*, **171–172**, 36–52.

19 Bouchoux, A., Gesan-Guiziou, G., Perez, J., and Cabane, B. (2010) How to squeeze a sponge: casein micelles under osmotic stress, a SAXS study. *Biophys. J.*, **99**, 3754–3762.

20 Mahmoudi, N., Gaillard, C., Riaublanc, A., Axelos, M.A.V., and Boué, F. (2014) Transition from fractal to spherical aggregates of globular proteins: Brownian-like activation and/or hydrodynamic stress? *Curr. Top. Med. Chem.*, **14**, 630–639.

21 Schmidt, I., Novales, B., Boué, F., and Axelos, M.A.V. (2010) Foaming properties of protein/pectin electrostatic complexes and foam structure at the nanoscale. *J. Colloid Interface Sci.*, **345** (2), 316–324.

22 Fameau, A.L., Saint-Jalmes, A., Cousin, F., Houinsou-Houssou, B., Navailles, L., Nallet, F., Gaillard, C., Novales, B., Boué, F., and Douliez, J.P. (2011) Smart foams: switching reversibly between ultrastable and unstable foams. *Angew. Chem., Int. Ed.*, **50**, 8264–8269.

**Part Two
Opportunities, Innovations, and New Applications
in Agriculture and Food Systems**

7
Nanomaterials in Plant Protection

Angelo Mazzaglia,[1,3] Elena Fortunati,[2] Josè Maria Kenny,[2] Luigi Torre,[2] and Giorgio Mariano Balestra[1,3]

[1]*University of Tuscia, Department of Agricultural and Forestry Sciences (DAFNE), Via S. Camillo De Lellis snc, 01100 Viterbo, Italy*
[2]*University of Perugia, Civil and Environmental Engineering Department, Materials Engineering Center, UdR INSTM, Strada di Pentima, 4-05100 Terni, Italy*
[3]*Phytoparasites Diagnosticssrl. Via S. Camillo De Lellis snc, 01100 Viterbo, Italy*

7.1
Introduction

Nanotechnology can change the entire scenario of current agricultural and food industry by the development of new tools able to minimize production inputs and maximize agricultural production outputs for meeting the increasing needs of the world sustainability. Moreover, nanotechnology has the potential to conceive products based on environment-friendly natural polymers, which in addition of being biodegradable can also be obtained from natural biowaste. Nanotechnology offers, in fact, substantial prospects for the development of innovative systems and applications and it is expected that the number of products and production volumes involving nanotechnology will increase in the near future. Specifically, the application of nanoscale-based systems usually ranging from 1 to 100 nm is an emerging area of nanoscience and nanotechnology [1,2]. Nanomaterials (NMs) often show unique and considerably changed physical, chemical, and biological properties compared to their macroscale counterparts. These unique functionalities are being used by many industries, and they provide novel solutions to technological and environmental challenges in the areas of solar energy conversion, catalysis, medicine, and water treatment [3]. Moreover, multifunctional nanocomposites with specific properties can be obtained by adding specific nanomaterials and/or nanoparticles into the selected thermoplastic and/or thermosetting polymer matrix [4].

However, due to the dimension characteristics of nanomaterials, there are concerns about their safety and potential toxicity, especially because of our limited knowledge of the human health. In response to this uncertainty, more

Nanotechnology in Agriculture and Food Science, First Edition. Edited by Monique A.V. Axelos and Marcel Van de Voorde.
© 2017 Wiley-VCH Verlag GmbH & Co. KGaA. Published 2017 by Wiley-VCH Verlag GmbH & Co. KGaA.

up-to-date information is required on the state-of-the-art applications of nanotechnology as pesticides. Nanotechnology applications are currently being studied, developed, and in some cases already used in the different parts of the agri/feed/food sector. If these issues are taken care of, nanotechnological intervention in farming has bright prospects to improve the efficiency of surveillance and control of pests and plant diseases and developing of new-generation pesticides: the nanopesticides (NPs).

Plant-based agricultural production is the basis of the broad agriculture systems providing food, feed, fiber, and fuels. While the demand for crop yield will rapidly increase in the future, the agriculture and natural resources such as land, water, and soil fertility are limited. Modern agricultural practices, associated with the Green Revolution, have greatly increased the global food supply. However, they have had a detrimental impact on the environment, highlighting the need for more sustainable agricultural methods. It is well documented that excessive and inappropriate use of pesticides has increased toxins in groundwater and surface water. Moreover, production inputs, including synthetic pesticides, are predicted to be much more expensive due to the constraints of known petroleum reserve.

Nanotechnology in agriculture has played and will play a key role in food production, food security, and food safety throughout the world. The broad range of applications in agriculture also includes NPs to control crop pests and plant pathogens. Over the past decade, patents and products incorporating nanomaterials for agricultural practices (e.g., nanopesticides) have been developed. It has been previously reported that in 2011 over 3000 patent applications dealing with nanopesticides were submitted [5]. The collective goal of all of these approaches is to enhance the efficiency and sustainability of agricultural practices by requiring less input and generating less waste than conventional products and approaches [6]. Plant diseases are caused by bacteria, fungi, insects, nematodes, phytoplasmas, and viruses, and they are responsible for billions of dollars in agricultural crop loss each year in the United States alone (USDA); in addition, over $600 million is spent annually on fungicides in an attempt to control pathogens [7].

It is clear that new strategies for plant disease management are greatly needed and will be a critical component to any long-term plan for sustaining or increasing agricultural productivity. The potential of nanotechnology, and specifically the use of NPs, in this effort has been a topic of discussion for several years. Current conventional pesticides often have active ingredients (AIs) with poor water solubility, and availability to crops can be quite limited. Consequently, larger volumes/quantities of these formulations must be used to control pathogenic diseases effectively. Moreover, metal ion-based pesticide formulations are subject to processes such as leaching, volatilization, and precipitation by soil minerals. The result has been an exceptionally inefficient and costly approach to pathogen/pest control. However, the solubility and effectiveness of these agrichemical formulations could be greatly increased through the use of nanoenabled additives or carriers as well as by the incorporation of NPs themselves as the active ingredient [8].

The perspective applicability of different and novel nanotechnological systems as pest control materials, the optimization of their processing procedures and strategies (i.e., powders, emulsions or suspensions, etc.), and their final market response are deeply discussed in this chapter. Finally, the most relevant results in the literature and some recent advances developed in our laboratories on multifunction-based formulations, are here reported.

7.2
Nanotechnology and Agricultural Sector

Nanotechnology is defined by the US Environmental Protection Agency [9] as the science of understanding and control of matter at dimensions of 1–100 nm. Other challenges to define nanoparticles from the point of view of agriculture include "particulate based formulations between 10 and 1000 nm in size that are simultaneously colloidal particulate." The burgeoning applications of nanotechnology in agriculture will continue to rely on the problem-solving ability of the material and are unlikely to adhere very rigidly to the upper limit of 100 nm. This is because nanotechnology for agricultural sector should address the large-scale inherent imperfections and complexities of farm productions (e.g., extremely low-input use efficiency) that might require nanomaterials with flexible dimensions, characteristics, and quantity. However, this is in contrast with nanomaterial concept that might be working well in well-knit factory-based production systems.

Nanotechnology design and development is, furthermore, usually represented by two different approaches: *top-down* and *bottom-up*. Top-down refers to making nanoscale structures from smallest structures by machining, templating, and lithographic techniques, for example, photonics applications in nanoelectronics and nanoengineering, whereas bottom-up approach refers to self-assembly or self-organization at molecular level, which is applicable in several biological processes [10]. Biologists and chemists are actively engaged in the synthesis of inorganic, organic, hybrid, and metal nanomaterials, including different kinds of nanoparticles that, due to relevant optical, physical, biological properties, have enormous applications in many fields such as electronic, medicine, pharmaceuticals, engineering, and agriculture [11].

Nanotechnology has the potential to change the entire scenario of the current agricultural and food industry with the help of new tools developed for the treatment of plant diseases, rapid detection of pathogens using nano-based kits, improvement of the ability of plants to absorb nutrients, and so on. Biosensors, at the nanoscale, and other smart delivery systems will also help the agricultural industry to fight against different crop pathogens. In the near future, in fact, nanostructure-based catalysts will be available in order to increase the efficacy of commercial pesticides and insecticides actually used and to reduce the dose levels required for crop plant protection [12]. The current global population is about 7 billion with 50% living in Asia. A large proportion of those living in developing countries face daily food shortages as a result of environmental impacts on agriculture, including storms, droughts, and flood [12]. Similarly, agricultural production continues to be constrained by a number of biotic and abiotic factors. For instance, pests, diseases, and weeds cause considerable damage to potential agricultural production. Evidences indicate that pests and plant pathogens cause 25% loss in rice, 5–10% in wheat, 30% in pulses, 35% in oilseeds, 20% in sugarcane, and 50% in cotton [13]. An example of remarkable economic losses caused worldwide by a plant pathogen is the bacterium *Pseudomonas syringae* pv. *actinidiae* (Psa) causal agent of kiwifruit bacterial canker [14–16]. There have been a few studies on the economic cost of Psa. The Agribusiness and Economics Research Unit of Lincoln University estimated a cost, only for New Zealand, of between $310 and $410 million in net present value terms between its discovery (2010) and 2016. Over a longer time period (15 years), they estimated this could increase toward $740 and $855 million [17].

Nanotechnology applications are currently being studied, developed, and in some cases already used in different parts of the agri/feed/food sector:

Agriculture

- Nanocapsules for more efficient delivery of pesticides, fertilizers, and other agrochemicals.
- Nanomaterials for detection of animal and plant pathogens.
- Nanomaterials for identity preservation and tracking and tracing.

Food and feed

- Nanocapsules to improve dispersion, bioavailability, and absorption of nutrients.
- Nanomaterials as color enhancers.
- Nanoencapsulated flavor enhancers.
- Nanotubes and nanoparticles as gelation and viscosifying agents.
- Nanoparticles for selective binding and removal of chemicals and pathogens from food.

Food packaging

- Nanoparticles to detect chemicals of foodborne pathogens.
- Biodegradable nanosensors for temperature and moisture monitoring.
- Nanoclays and nanofilms as barrier materials to prevent spoilage and oxygen absorption.
- Nanoparticles for antimicrobial and antifungal surface coatings.

Food supplements

- Nanoparticle suspensions as antimicrobials.
- Nanoencapsulation for targeted delivery of nutraceuticals.

Finally, in Figure 7.1 are summarized the most relevant applications of nanotechnology in different fields of agriculture.

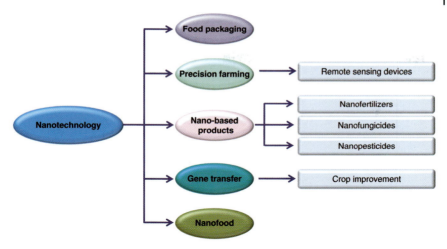

Figure 7.1 Applications of nanotechnology in different industrial and agricultural fields. (With permission from Ref. [12]. © 2012, Springer.)

7.2.1
Nanomaterials

Nanomaterials can be divided into three main categories: organic nanomaterials, inorganic nanomaterials, and combined organic/inorganic (surface-modified) nanomaterials, as shown in Figure 7.2.

In recent years, all these materials have attracted great interest. Organic nanomaterials are developed as nanoencapsulates or surface-modified systems for vitamins, antioxidants, colors, flavors, preservatives, and active ingredients in

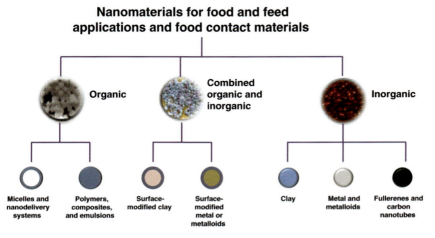

Figure 7.2 Types of nanomaterials for (future) food and feed applications and food contact materials. (With permission from Ref. [18].).

general. The production of organoclay is of interest for making polymer–clay nanocomposites with improved physical and mechanical properties. Inorganic nanomaterials are used in food and food supplements, while embedded in polymer matrices they are used in food packaging applications [4]. The use of these different nanomaterials in the agricultural and food sectors is described briefly in the following sections.

7.2.1.1 Organic Nanomaterials

Organic nanomaterials are normally used as potential carriers of additives, pesticides, and veterinary drugs. These so-called nanoencapsulates generally consist of micelles, liposomes, or nanospheres, and are composed of materials that are generally recognized as safe (GRAS). In general, three types of organic nanomaterials can be distinguished: lipid-, protein-, and polysaccharide-based nanomaterials. Lipid-based nanomaterials are among the most applied organic nanomaterials since they can be produced by using natural ingredients at an industrial scale and have the ability to encapsulate compounds with different solubilities. Protein-based nanomaterials are often prepared by using a bottom-up approach, where structures are built from molecules capable of self-assembly. Polysaccharides are naturally occurring compounds in plants (e.g., pectin, guar gum), animals (e.g., chitosan, chondroitin sulfate), algae (e.g., alginates), and microorganisms (e.g., dextran). Until now, chitosan, a naturally biodegradable and biocompatible polysaccharide, cellulose, the most abundant natural polymer, and starch, appear to be the most used polysaccharides for nanomaterials in biomedicine and packaging, but have also found application in the agricultural field.

Among natural polymers, native starch is considered one of the most promising materials because of its attractive combination of availability, price, performance, and good film-forming ability. At the same time, the film-forming capability of chitosan allows the realization of films, coating, and capsules, having excellent properties as biodegradability, biocompatibility, and nontoxicity [19]. Furthermore, it is inexpensive and commercially available, whereas the antimicrobial activity of chitosan has been demonstrated against many bacteria, yeasts, and a wide range of fungi guaranteeing defensive mechanisms in plants and fruit against infections caused by several pathogens.

Furthermore, lignocellulosic materials, and in particular cellulose and lignin, are being considered as potential nanocarriers for active ingredients or in combination in agri/food sector applications [20,21].

7.2.1.2 Inorganic Nanomaterials

A limited number of inorganic nanomaterials are known to be used in agro-food applications as nanopesticides (NPs), although some of them might be used in high volumes and several different applications. These include nanomaterials of noble metals (silver (Ag), iron (Fe), copper (Cu), titanium dioxide (TiO_2), zinc oxide (ZnO), etc.) and calcium (Ca), magnesium (Mg), selenium (Se), and silicates (silicon (Si)-based materials).

Nanoparticles of noble metals have been studied with growing interest, due to their great physical, chemical, and biological properties if compared with the bulk counterparts. Discoveries in the past decade have demonstrated that the electromagnetic, optical, and catalytic properties of noble metal nanoparticles such as gold, silver, and platinum are strongly influenced by shape and size. The size-dependent properties of small metal particles are known to yield particular optical, electrochemical, and electronic properties. This has motivated an upsurge in research on the synthesis routes that allow better control of shape and size [4]. Metal nanoparticles, for example, gold nanoparticles, are largely involved in the agricultural sector as biosensor for diagnosis of pests and pesticide residues [22]. Moreover, silver (Ag) nanoparticles display a strong inhibitory activity to microorganisms, and are currently used in many industrial sterilization processes. Even though the bactericidal mechanisms of Ag nanoparticles are still not fully understood, researchers have hypothesized that bacteria are killed through a combination of elemental Ag nanoparticles and released Ag^+ ions. Silver ions interact with cysteine-containing proteins and enzymes on bacterial membranes, causing structural deformation and biochemical imbalance of the cell membrane. Subsequently, Ag^+ ions penetrate the damaged cell membrane to inactivate cytoplasmic enzymes leading to inhibition of cell replication and finally cell death [23].

Among metals, copper deserves a specific mention due to its fundamental role in the control of plant diseases. Cupric salts are, in fact, still largely used, being allowed even in organic agriculture. Unfortunately, copper derivatives cannot be degraded or destroyed in the environment, and thus treatments with copper derivatives used as fungicides and bactericides contribute to its accumulation in soils more than any other agricultural activity. The input of copper into European agricultural soils is estimated about 504 tons per year (included that coming from manure and sludge), with a net input of about 400 tons per year, when considering output through crops [24]. This tendency to accumulate in soil and water poses a serious threat to a wide range of organisms and microorganisms and to their ecosystems. Moreover, copper-contaminated soils were proved containing higher percentages of antibiotic-resistant bacteria than noncontaminated soils, making them a sort of dangerous reservoir of genes for antibiotic resistance, possibly transmittable to animals and human pathogenic bacteria. Accordingly, severe restrictions have been established within the EU countries concerning the use of copper compounds in plant protection and their maximum residue levels in food and feed (Directive 2009/37/EC; Regulation 396/2005/EC; Council Directive 91/414/EEC; Commission Regulation 149/2008/EC).

Again, the chance to obtain nanoparticles will be of great help. In fact, stated that the biological activity, as well as the ability to penetrate into the organism, of Cu^{++} depends on its physical and chemical characteristics [25], it is mandatory to have in-depth knowledge of these parameters to standardize copper nanomaterials for their subsequent use as antibacterial agents. In such sense, the use of cupric nanoparticles to control dangerous plant pathogens to replace and reduce current amount of copper compounds, is wholly in accordance with EU Regulation.

Considering silicates (also called clays), they are abundant and environmentally benign, providing a greener alternative to the conventional crop protection strategies. Kaolin, a white nonswelling plate-shaped aluminosilicate mineral, is widely used in agriculture as a biopesticide and a repellent. These clay particles protect plants against various pests and insects [22] by masking the visual cues. Another clay, at the nanoscale, that has shown potential in agriculture is a three-layered clay called montmorillonite (MMT). While the naturally available MMT clays are hydrophilic, their high cation exchange capacity makes it possible for modification with a cationic surfactant to render them hydrophobic (see Section 7.2.1.3 about organic/inorganic nanomaterials). Encapsulation of various pesticides has been accomplished in both hydrophilic and hydrophobic MMT clays and due to this modification strategy, the efficiency of, for example, the insecticides diazinon and chlorpyrifos has been shown to improve from 4 weeks in the commercial formulation to as high as 20 weeks by using the MMT clay [26].

7.2.1.3 Combined Organic/Inorganic Nanomaterials

Combined organic and inorganic nanomaterials, also called surface-modified nanomaterials, are nanomaterials that add certain types of functionality to the matrix, such as antimicrobial activity or a preservative action through absorption of oxygen. Surface-modified clays, metal, or metalloids represent an example of these specific nanomaterials. The most known and used organic/inorganic nanomaterials in agriculture are the modified nanoclays. It was proved, for example, that surface-functionalized hydrophobic silica can act as an insecticide by getting adsorbed into the cuticular lipids of insects and causing their death by desiccation [27]. A similar effect was shown recently by Debnath *et al.* against rice weevil *Sitophilus oryzae* [28]. Water-soluble pesticide validamycin encapsulated in porous hollow silica nanoparticles is shown to last for 800 min instead of the instantaneous release of free validamycin [29]. Encapsulation of the pesticides inside the hollow silica spheres has also been shown to protect them from environmental factors such as ultraviolet degradation [30]. But these organic/inorganic nanomaterials normally find practical applications also in the polymer science as reinforcement phase of thermoplastic matrices. The surface of clays is usually organically modified in order to enhance their adhesion to and interaction with the polymer matrices in which they are embedded and, in such way, to increase the final properties of the obtained polymer/nanoclay-based formulations. In food packaging, for example, functionalized nanomaterials are incorporated in the polymer matrix to guarantee increased mechanical strength or barrier properties against gases, volatile components (e.g., flavors), or moisture. One example is the use of functionalized nanoclays in food packaging sector to develop materials with enhanced gas barrier properties [31].

7.2.2
Functionalization of Nanomaterials (NMs): Development of Novel Nanoformulations for Pests and Plant Pathogens Control

In the last few years, a number of strategies have been developed to functionalize nanostructured materials for drugs delivery in pharmaceutical and biomedical fields promoting their controlled releases. This mainly involved organic materials, for example, polymer-based and in particular biopolymer-based systems, used in the nanostructured form (nanoparticles, nanomicelles, film base systems), as possible carriers for active organic ingredients (AIs). In this context, the functionalization of their surfaces and/or the encapsulation procedures of specific AI represent important challenges in order to modulate and control the release while different strategies are usefully applied and studied.

Covalent strategies, for example, contemplate stable bonds between nanomaterials and AIs and are achieved linking native or modified hydroxyl groups present on the nanocarriers to the reactive functional groups of the selected additives. The esterification reaction is an important tool in order to introduce a covalent bond between the alcoholic groups of nanomaterials and the carboxylic groups of AIs. In addition, hydroxyl groups will be converted into carboxylic and amino groups. In this context, (2,2,6,6-tetramethylpiperidine-1-oxyl, TEMPO)-mediated oxidation of nanomaterials will be useful to convert the hydroxymethyl groups to their carboxylic form. This oxidation reaction, highly selective for primary hydroxyl groups, is also "'green'" and simple to implement. At the same time, (3-aminopropyl)triethoxysilane (APTES) modification is a useful way to convert hydroxyl groups into amino groups. Carboxylic and amino groups represent reactive functionalities useful to link AIs with the appropriate functionalities onto modified surface of nanomaterials. Moreover, amidation is also an efficient reaction to covalently bond selected compounds to the nanocarriers. Finally, click chemistry represents a powerful and versatile tool for nanocarriers functionalization owing to its simplicity, selectivity, efficiency, and tolerance of various functional groups.

Noncovalent surface modifications of nanomaterials are achieved via adsorption of surfactants, oppositely charged entities, or polyelectrolites. Electrostatic interactions and van der Waals forces are responsible for the interactions with the nanomaterials. Alternatively, multivalent metal salts will be adsorbed on the surface of the nanomaterials by complexation with hydroxyl groups of nanobiopolymers that represent the coordination sites [32]. Another strategy to functionalize nanocarriers is the polymer grafting based on the "grafting onto" strategy that involves attachment of presynthesized polymer chains, carrying reactive end groups onto native or modified hydroxyl groups of the nanomaterial surface. Moreover, hydroxyl groups are strategic in order to modify the chemical properties of nanomaterials. For example, the esterification of nanomaterials with organic fatty acid chlorides having aliphatic chains of different lengths (C2 up to C18) confer to the corresponding nanomaterials a hydrophobic character, when necessary.

Chemicophysical interactions between the AI and the nanomaterials used as carriers are responsible for the stability of the resulting system. The effect a nanoformulation has on the fate of an AI may be multiple and depends on the AI–nanocarrier combination under consideration and on the functionalization strategy adopted to associate the AI with the polymeric nanostructures. However, the common objective of the nanoformulations is to achieve the slow release of the AI, to protect it from premature degradation or to enhance a poor solubility. Each functionalized nanocarrier is therefore expected to have a different impact in terms of kinetic, persistence of the AIs, and ultimately of efficacy against pathogens. Determining the extent to which a nanoformulation will affect processes such as AI release and bioavailability is critical to the development of a robust technology. Published literature on pharmaceutical delivery systems based on biodegradable polymers indicates that the release mechanisms can be controlled by diffusion and/or dissolution (erosion), depending on the polymer properties, the way the AI is distributed, and the AI loading and solubility. The processes leading to polymer degradation are thus expected to greatly influence the release profiles. However, it cannot be excluded that, if the complete release of the AI from the nanocarrier occurs much more rapidly than the degradation processes, the impact of the nanoformulation on degradation is likely to be negligible. In contrast, if release and degradation proceed over similar timescales, the overall persistence of the AI will depend on the release rate, the persistence of the released AI, and the persistence of the AI in the AI–nanocarrier form. Bioavailability of the AI–nanocarrier certainly depends on the carrier properties and target cells/organisms considered. Bioavailability may likewise depend on the location of the AIs within the polymer matrix. A number of polymer characteristics have been shown to influence release profiles, including the length of the polymer chains, nanoparticle size, and environmental physicochemical and biological factors. Finally, the possibility that a proportion of the AI may remain associated with the nanocarrier for extended periods should also be considered, in fact this might be the case when covalent conjugations of the AI to the polymer matrix are contemplated, and as a matter of fact it has been observed for some pharmaceutical formulations from which the release of a proportion of the AI was triggered by severe degradation of the polymer matrix.

A number of analytical methods have been developed over the last decade (e.g., liquid chromatography with mass spectrometry and matrix-assisted laser desorption/ionization time-of-flight mass spectrometry) and data are now available for a wide variety of polymers. It remains, however, very difficult to predict how fast and by which mechanisms polymer-based nanoformulations, consisting of complex mixtures, will be eliminated following their release into the environment. Understanding how the characteristics of a nanocarrier evolve with time is, however, necessary in order to assess the fate of the AI. It is also important for real applications in order to ensure, for instance, a practical shelf-life or to support the selection of the most suitable nanocarrier for field applications.

7.3
Applications of Nanomaterials against Plant Pathogens and Pests

The main scientific studies related to NMs utilized in plant protection strategies to reduce damages caused by plant pathogens and pests (mainly bacteria, fungi, insects and virus) are deeply discussed in the following sections.

7.3.1
Bacteria

Bacteria are particularly problematic to control in plant pathology and this difficulty is even increased by restriction in those countries where the use of antibiotics is banned, such as in the entire European community. This makes the search for alternative control methods particularly demanding, and nano biotechnology especially promising. Results obtained with metal nanoparticles are very encouraging. Jayaseelan et al. [33] demonstrated that biosynthesized ZnO nanoparticles (25 lg/ml) yielded high suppression of pathogenic bacteria (*Pseudomonas aeruginosa*). In a field study, Cui et al. used TiO_2 sol, which is a neutral, viscous aqueous colloid of TiO_2 with the size of 10–50 nm, to create an adhesive layer on the plant surface that prevents pathogens infection. They demonstrated that TiO_2 nanoparticles treatment determines a 69% reduction of the infection of cucumber by *P. syringae* pv. *lachrymans* (and 91% against the oomycetes *Pseudoperonospora cubensis*), and also increases photosynthetic activity (30%) [34].

Fortunati et al. [35] proved that novel poly(DL-lactide-*co*-glycolide) (PLGA) copolymer-based biopolymeric nanoparticles and cellulose nanocrystals (CNCs) are useful as nanocarriers in developing innovative plant protection formulations. CNC were evaluated as possible nanostructured formulation to be directly applied in plant protection treatment. The effect of both, PLGA nanoparticles and CNC, was investigated with respect to their influence on the survival of the causal agent of bacterial speck disease (*P. syringae* pv. *tomato*, Pst), on plant development and damages (phytotoxicity effects), and on tomato plant. The proposed nanocarriers proved able in covering, by a uniform distribution, the tomato vegetal surfaces without any damage, thus allowing a regular development of the tomato-treated plants. In addition, after the distribution of the starch-PLGA nanoparticle-based formulation, no Pst (CFU/cm^2) were detectable on the tomato leaves.

Microfabricated xylem vessels with nanosize features have been shown very useful in gaining an appreciation of the mechanisms and kinetics of bacterial colonization in xylem vessels such that novel disease control strategies may be developed [36]. Also, Henrnádez-Montelongo et al. showed by the application of nanofilms of hyaluronan/chitosan (HA/CHI) assembled layer-by-layer a potential antimicrobial material for the phytopathogen *Xylella fastidiosa* [37]. Alternatively, metal oxides such as TiO_2 have been used as nanoscale amendments in agriculture due to their photocatalytic properties. Field studies indicated that TiO_2 nanoparticles (10–50 nm) inhibit pathogenic infection in cucumber by

P. syringae pv. *lachrymans* (68.6%) and *P. cubensis* (90.6%) [34]; importantly, the authors also reported significant increases in photosynthetic activity (30%) compared to control plants. Paret *et al.* [38,39] showed that after photoactivation, TiO_2 nanoparticles amendment reduced bacterial spots (*Xanthomonas* sp.) on both roses and tomato at levels equivalent to or better than conventional approaches. Huang *et al.* [40] used MgO nanoparticles as an effective bactericide due to its strong interactions with negatively charged bacteria and spores.

7.3.2
Fungi

Several studies have already demonstrated that nanoparticles, mostly inorganic such as Ag, ZnO, TiO_2, Mg, and Si [41], can be used to counteract important fungal diseases of plants. In one of these, Jo *et al.* [42] studied the inhibitory effect of Ag nanoparticles on two plant-pathogenic fungi that infect perennial ryegrass (*Lolium perenne*). A 50% inhibition of colony formation was observed at 200 mg Ag NPs per liter. Similarly, Lamsal *et al.* [43] reported that Ag nanoparticles could suppress anthracnose pathogen *Colletotrichum* spp. in field trials. Both studies [42,43] reported that Ag nanoparticles attached to and penetrated into the microbial membrane, thereby reducing the magnitude of infection. Additionally, the later study reported that Ag nanoparticles reduces plant diseases more effectively when applied prophylactically, suggesting that the efficiency of suppression may be greatly influenced by treatment timing and/or the induction of resistance mechanisms. Similarly, Gajbhiye *et al.* [44] assessed the antifungal activity of Ag nanoparticles in combination with the fungicide fluconazole by a disk diffusion assay against several pathogenic fungi, including *Pleospora herbarum*, *Trichoderma* sp., *Candida albicans*, *Phoma glomerata*, and *Fusarium*. Results showed that combined fluconazole–Ag nanoparticles had the greatest antifungal activity, achieving maximum zone inhibition against *C. albicans*, followed by *P. glomerata* and *Trichoderma* sp. Another approach involving the extracellular synthesis of Ag nanoparticles by culture supernatant of the bacterium *Serratia* sp. strain BHU-S4 reveals its effective use for the control of spot blotch disease in wheat, caused by the pathogenic fungus *Bipolaris sorokiniana* [45]. The authors demonstrated that the defensive mechanism induced by this compound involves lignin deposition in vascular system of the plant.

The potential of ZnO for controlling pathogen growth was also recently demonstrated. In this case, given the lower overall metal phytotoxicity and the secondary benefits on soil fertility, ZnO nanoparticles are likely a more appropriate choice for fungal pathogen control than is Ag [46]. For example, ZnO nanoparticles demonstrated higher inhibition (26%) against *Fusarium graminearum* compared to bulk ZnO and controls (~47%) for mung bean in broth agar medium [46]. Other studies with postharvest pathogenic fungi showed that at 3–12 mmol/l, ZnO nanoparticles caused significant inhibition of *Botrytis cinerea* (from 63–80%) and *Penicillium expansum* (61–91%) in microbiological plating [47]. The authors reported that ZnO nanoparticles inhibited growth by affecting broad cellular function within *B. cinerea* and *P. expansum*, thereby resulting in hyphal deformation and death, respectively. In a similar study, biosynthesized ZnO nanoparticles (25 μg/ml) displayed high inhibition rates against the fungal species *Aspergillus flavus* (inhibition halo: 9 ± 1.0 mm) [33]. Also Wani and Shah [48] investigated the potential of ZnO and MgO nanoparticles to inhibit pathogens such as *Alternaria alternata, Fusarium oxysporum, Rhizopus stolonifer,* and *Mucor plumbeus*. The authors reported a high inhibition rate in the germination of fungal spores upon exposure to the metal oxide NPs; the highest spore inhibition in germination (93.6%) was reported for *Mucor plumbeus* at 100 mg/l MgO nanoparticles.

Concerning nanoparticles of copper, Giannousi *et al.* [49] tested the antifungal activity of Cu_2O nanoparticles toward *Phytophtora infestans* on tomato and reported that foliar application resulted in significantly greater protection (73.5%) from the pathogen, compared to currently available non-nano Cu formulations (57.8%) Given the lack of nanoparticle-induced phytotoxicity, the potential dual use of nanoscale amendments for both suppressing disease and promoting nutrient status becomes a topic worthy of further investigation [49]. Servin *et al.* proved that foliar application of CuO solution onto tomato shoots results in significant inhibition of disease (*Fusarium* spp.) progression, as well as increased Cu content in the roots, compared to bulk CuO and untreated controls. This suggests that nanoscale size does indeed yield enhanced phloem-based shoot–root translocation and supports the hypothesis that nanoparticle metal oxides may not only directly inhibit pathogens but also indirectly affect disease by improving plant nutritional status. In fact, in recently completed field studies by our group with transplanted egg plant in *Verticillium dahliae*-infested soil, CuO nanoparticle treatment of the seedlings at planting resulted in significantly increased (17–31% in comparison with bulk treatment) marketable yield in 2013 and 2014 (unpublished). Importantly, corresponding bulk CuO had no such effect [50].

Published studies have also reported that the application of quantum dots (QDs) can significantly enhance plant growth and may offer selective activity against plant pathogens. Rispail *et al.* [51] discussed the use of QDs and superparamagnetic nanoparticles for the detection and labeling of the fungal pathogen *Fusarium oxysporum*. The authors demonstrated intracellular internalization of nanoparticles, suggesting that this technique could be applied for both detection and control platforms with this pathogen. Separately, the authors observed that 500 nM QDs reduced both fungal germination (20%) and hyphal growth (15.4%) [51].

7.3.3
Insects

Avermectin is a known pesticide that blocks neurotransmission in insects by inhibiting chloride channel; it is however inactivated by ultraviolet when applied in field, and its half-life is of 6 h only. Interestingly, when encapsulated by nanoparticle-based carrier, avermectin is released much slower so that it can be reported for about 30 days [52]. Similarly, a commercial product Karate® ZEON is a quick release microencapsulated formulation containing lambda-cyhalothrin that breaks open upon contact with leaves. In contrast, the gutbuster microencapsules containing pesticide breaks open to release its contents upon coming in contact with alkaline environments, including the stomach of certain insects [53]. A series of polyethylene glycol (PEG)-based insecticide formulations were found to release active compounds at slower rate compared to commercial formulations comprising imidacloprid [54], carbofuran [55], and thiram [56]. The release of insecticide was noted to be dependent on PEG molecular weight. The release of β-cyfluthrin from the nanoformulation was recorded over a period of 1–20 days, whereas release from a commercial formulation was found within 4–5 days [57].

Yang et al. demonstrated the insecticidal activity of polyethylene glycol-coated nanoparticles loaded with garlic essential oil against adult *Tribolium castaneum* insect found in stored products. It has been observed that the control efficacy against adult *T. castaneum* was about 80%, presumably due to the slow and persistent release of the active components from the nanoparticles [58].

Stadler et al. for the first time studied the insecticidal activity of nanostructured alumina against two insect pests, namely, *S. oryzae* L. and *Rhyzopertha dominica* (F.), which are major insect pests in stored food supplies across the world. They reported significant mortality after 3 days of continuous exposure to nanostructured alumina-treated wheat. Therefore, compared to commercially available insecticides, inorganic nanostructured alumina may provide a cheap and reliable alternative for control of insect pests, and such studies may expand the frontiers for nanoparticle-based technologies in pest management [59]

7.3.4
Virus

Goswami et al. studied the application of different kinds of nanoparticles as silver nanoparticles, aluminum oxide, zinc oxide, and titanium dioxide in the control of rice weevil and grasserie disease in silkworm (*Bombyx mori*) caused by *Sitophilus oryzae* and baculovirus BmNPV (*B. mori* nuclear polyhedrosis virus), respectively. In their study, they performed bioassay, in which they prepared solid and liquid formulations of the above-mentioned nanoparticles; later, they applied these formulations on rice and kept in a plastic box with 20 adults of *S. oryzae* and observed the effects for 7 days. It was reported that hydrophilic silver nanoparticles were most effective on the first day. On day 2, more than 90% mortality was obtained with silver and aluminum nanoparticles. After 7 days of exposure, 95 and 86% mortality were reported with hydrophilic and hydrophobic silver nanoparticles and nearly 70% of the insects were killed when the rice was treated with lipophilic silver nanoparticles. However, 100% mortality was observed in case of aluminum nanoparticles. Similarly, in another bioassay carried for grasserie disease in silkworm (*B. mori*), a significant decrease in viral load was reported when leaves of *B. mori* were treated with ethanolic suspension of hydrophobic aluminosilicate nanoparticles [60].

7.4 Conclusions

Conventional plant protection methods have affected both the environment and economy of farmers, as 90% of the applied pesticides are lost to the air during application and as runoff. Additionally, indiscriminate usage of pesticide increases pathogen and pest resistance, reduces soil biodiversity, diminishes nitrogen fixation, contributes to bioaccumulation of pesticides pollinator decline, and destroys habitat for birds [52]. In US agriculture, soilborne pathogens reduce average crop yield by 10–20%, resulting in billions of dollars of losses. Although a number of disease management options exist for many crop species, all strategies suffer from significant shortcomings. This fact, including the building pressure for greater food production and the potential challenges posed by a warming climate, highlights the need for new plant disease management tools.

Nanoscaled delivery system with active compound (often organic) can be applied only when necessary in the field. The overall goal of this imaging NPs is to reduce the number of unnecessary problems in agriculture in relation to pest and plant pathogens. Moreover, an economic evaluation is likely in order. The high cost of pesticide formulations and secondary costs related to environmental contamination concerns have stimulated interest and research in novel approaches for plant protection.

During primary production, nanoformulated agrochemicals are employed to increase the efficacy of the agrochemicals compared to conventional formulations. For example, nanoencapsulated and solid lipid nanoparticles have been explored for the delivery of agrochemicals [61]. The application of novel nanotechnology techniques in agriculture has recently been reviewed [62]. Nanomaterials come in many diverse forms (surprisingly often >100 nm), from solid-doped particles to (often nonpersistent) polymer and oil–water-based structures.

Pesticides are chemical substances that are meant to prevent, destroy, or mitigate pests. Nanotechnology shows great promise to improve pest management strategies through the formulations of nanomaterial-based pesticides. Nanoencapsulates are used for delivering pesticides and other agrochemicals [5,63]. A surfactant-based nanoemulsion has been described as a delivery system for the pesticide beta-cypermethrin [64] as are porous hollow silica and surface-modified nanomaterials for controlled release of the pesticide validamycin and the herbicide 2,4-dichlorophenoxyacetate [29]. The interesting results obtained in academic researches over the last few decades have been closely followed by several companies. Some companies over the last decade have already deposited several different patents comprising a wide range of protocols for production and application of encapsulated formulations, which can be used to produce nanoinsecticides [65]. Despite the hard work and heavy investment, no commercial nanoinsecticide formulation has been extensively commercialized up to 2012 [65]. Some nanoproducts are already being applied, for example, Nanocid®-based pesticides [66] and chitosan [67]; however, most of the applications are still in the developmental stage. For example, the preparation of the commonly used conventional chemical pesticide permethrin on a nanoscale by Suresh Kumar *et al.* [68] showed that the resulting nanopermethrin could be potentially used as a safe and effective alternative. Also the use of the pheromone methyl eugenol as a nanogel proved to be a low-cost approach [69], due to lower applied doses without losing efficacy [70]. Another group of nanoagents with insecticidal, antimicrobial, or antifungal properties are naturally occurring ashes and green produced inorganic metal nanoparticles [59,71,72]

Different commercial companies are currently producing nanoenabled pesticides that are more soluble, better dispersed, less persistent, and more specific with regard to target [12]. In this context, nanomaterials and nanotechnologies offer a great opportunity to develop NPs against pests and plant pathogens reducing drastically the agrochemicals use, as well as for effectiveness, organic active compounds, and preserving the environment [73–79].

References

1 Curtis, A. and Wilkinson, C. (2001) Nanotechniques and approaches in biotechnology. *Trends Biotechnol.*, **19** (3), 97–101.

2 Armentano, I., Arciola, C.R., Fortunati, E. *et al.* (2014) The interaction of bacteria with engineered nanostructured polymeric materials: a review. *Sci. World J.*, **2014**, 410423.

3 Anselme, K., Davidson, P., Popa, A.M. *et al.* (2010) The interaction of cells and bacteria with surfaces structured at the nanometre scale. *Acta Biomater.*, **6** (10), 3824–3846.

4 Armentano, I., Dottori, M., Fortunati, E. *et al.* (2010) Biodegradable polymer matrix nanocomposites for tissue engineering: a review. *Polym. Degrad. Stab.*, **95** (11), 2126–2146.

5 Kah, M., Beulke, S., Tiede, K., and Hofmann, T. (2013) Nanopesticides: state of knowledge, environmental fate, and exposure modeling. *Crit. Rev. Environ. Sci. Technol.*, **43** (16), 1823–1867.

6 Chinnamuthu, C.R. and Boopathi, P.M. (2009) Nanotechnology and agroecosystem. *Madras Agric. J.*, **96**, 17–31.

7 González-Fernández, R., Prats, E., and Jorrín-Novo, J.V. (2010) Proteomics of plant pathogenic fungi. *J. Biomed. Biotechnol*, **2010** 932527.

8 Naderi, M.R. and Danesh-Shahraki, A. (2013) Nanofertilizers and their roles in sustainable agriculture. *Int. J. Agric. Crop Sci.*, **5**, 2229–2232.

9 US EPA (2007) Nanotechnology White Paper, EPA 100/B-07/001(February), p. 136.

10 Mukhopadhyay, S.S. (2014) Nanotechnology in agriculture: prospects and constraints. *Nanotechnol. Sci. Appl.*, **7** (2), 63–71.

11 Salata, O. (2004) Applications of nanoparticles in biology and medicine. *J. Nanobiotechnol.*, **2** (1), 3.

12 Rai, M. and Ingle, A. (2012) Role of nanotechnology in agriculture with special reference to management of insect pests. *Appl. Microbiol. Biotechnol.*, **94** (2), 287–293.

13 Dhaliwal, G., Jindal, V., and Dhawan, A. (2010) Insect pest problems and crop losses: changing trends. *Indian J. Ecol.*, **37** (1), 1–7.

14 Mazzaglia, A., Studholme, D.J., Taratufolo, M.C. *et al.* (2012) *Pseudomonas syringae* pv. *actinidiae* (PSA) isolates from recent bacterial canker of kiwifruit outbreaks belong to the same genetic lineage. *PLoS One*, 7 (5), 11.

15 Renzi, M., Copini, P., Taddei, A.R. *et al.* (2012) Bacterial canker on kiwifruit in Italy: anatomical changes in the wood and

in the primary infection sites. *Phytopathology*, **102** (9), 827–840.
16 Ciarroni, S., Gallipoli, L., Taratufolo, M.C. *et al.* (2015) Development of a multiple loci variable number of tandem repeats analysis (MLVA) to unravel the intrapathovar structure of *Pseudomonas syringae* pv. *actinidiae* populations worldwide. *PLoS One*, **10** (8), 24.
17 Bagrie, C., Williams, C., and Croy, D. (2015) Kiwifruit revival. New Zealand Economics ANZ Agri Focus, June, pp. 2–13.
18 RIKILT and JRC (2014) Inventory of nanotechnology applications in the agricultural, feed and food sector. EFSA supporting publication 2014, EN-621, 125 pp.
19 Barikani, M., Oliaei, E., Seddiqi, H., and Honarkar, H. (2014) Preparation and application of chitin and its derivatives: a review. *Iran. Polym. J.*, **23** (4), 307–326.
20 Yang, W., Fortunati, E., Dominici, F. *et al.* (2016) Synergic effect of cellulose and lignin nanostructures in PLA based systems for food antibacterial packaging. *Eur. Polym. J.*, **79**, 1–12.
21 Yang, W., Fortunati, E., Dominici, F. *et al.* (2016) Effect of cellulose and lignin on disintegration, antimicrobial and antioxidant properties of PLA active films. *Int. J. Biol. Macromol.*, **89**, 360–368.
22 Narayanan, A., Sharma, P., and Moudgil, B.M. (2012) Applications of engineered particulate systems in agriculture and food industry. *KONA Powder Part. J.*, **30**, 221–235.
23 Ocsoy, I., Paret, M.L., Ocsoy, M.A. *et al.* (2013) Nanotechnology in plant disease management: DNA-directed silver nanoparticles on graphene oxide as an antibacterial against *Xanthomonas perforans*. *ACS Nano*, **7** (10), 8972–8980.
24 SCHER (2009) Voluntary Risk Assessment Report on Copper and Its Compounds: Environmental Part, Scientific Committee on Health and Environmental Risks, pp. 1–15.
25 Mamonova, I.A., Matasov, M.D., Babushkina, I.V. *et al.* (2013) Study of physical properties and biological activity of copper nanoparticles. *Nanotechnol. Russ.*, **8** (5–6), 303–308.
26 Choudary, B.M., Prasad, B.P., and Kantam, M.L. (1989) New interlamellar pesticide–metal–montmorillonite complexes: a novel technique for controlled release. *J. Agric. Food Chem.*, **37** (5), 1422–1425.
27 Mewis, I. and Ulrichs, C. (2001) Effects of diatomoceous earth on water content of *Sitophilus granarius* (L.) (Col., Curculionidae) and its possible use in stored product protection. *J. Appl. Entomol.*, **125** (6), 351–360.
28 Debnath, N., Das, S., Seth, D. *et al.* (2011) Entomotoxic effect of silica nanoparticles against *Sitophilus oryzae* (L.). *J. Pest Sci.*, **84** (1), 99–105.
29 Liu, F., Wen, L.-X., Li, Z.-Z. *et al.* (2006) Porous hollow silica nanoparticles as controlled delivery system for water-soluble pesticide. *Mater. Res. Bull.*, **41** (12), 2268–2275.
30 Li, Z.Z., Chen, J.F., Liu, F. *et al.* (2007) Study of UV-shielding properties of novel porous hollow silica nanoparticle carriers for avermectin. *Pest Manag. Sci.*, **63** (3), 241–246.
31 Armentano, I., Bitinis, N., Fortunati, E. *et al.* (2013) Multifunctional nanostructured PLA materials for packaging and tissue engineering. *Prog. Polym. Sci.*, **38** (10–11), 1720–1747.
32 Ciesielski, W. and Tomasik, P. (2004) Complexes of amylose and amylopectins with multivalent metal salts. *J. Inorg. Biochem.*, **98** (12), 2039–2051.
33 Jayaseelan, C., Rahuman, A.A., Kirthi, A.V. *et al.* (2012) Novel microbial route to synthesize ZnO nanoparticles using *Aeromonas hydrophila* and their activity against pathogenic bacteria and fungi. *Spectrochim. Acta A Mol. Biomol. Spectrosc.*, **90**, 78–84.
34 Cui, H., Zhang, P., Gu, W., and Jiang, J. (2009) Application of anatase TiO_2 sol derived from peroxotitannic acid in crop diseases control and growth regulation. *Nanotech*, **2**, 286–289.
35 Fortunati, E., Rescignano, N., Botticella, E. *et al.* (2016) Effect of poly (DL-lactide-co-glycolide) nanoparticles or cellulose nanocrystals based formulations on *Pseudomonas syringae* pv. *tomato* (Pst)

and tomato plant development. *J. Plant Dis. Prot.*. doi: 10.1007/s41348-016-0036-x.

36 Zaini, P.A., De La Fuente, L., Hoch, H.C., and Burr, T.J. (2009) Grapevine xylem sap enhances biofilm development by *Xylella fastidiosa*. *FEMS Microbiol. Lett.*, **295** (1), 129–134.

37 Hernández-Montelongo, J., Nascimento, V.F., Murillo, D. *et al.* (2016) Nanofilms of hyaluronan/chitosan assembled layer-by-layer: an antibacterial surface for *Xylella fastidiosa*. *Carbohydr. Polym.*, **136**, 1–11.

38 Paret, M.L., Palmateer, A.J., and Knox, G.W. (2013) Evaluation of a light-activated nanoparticle formulation of titanium dioxide with zinc for management of bacterial leaf spot on rosa "Noare." *Hortscience*, **48** (2), 189–192.

39 Paret, M.L., Vallad, G.E., Averett, D.R. *et al.* (2013) Photocatalysis: effect of light-activated nanoscale formulations of TiO_2 on *Xanthomonas perforans* and control of bacterial spot of tomato. *Phytopathology*, **103** (3), 228–236.

40 Huang, L., Li, D.Q., Lin, Y.J. *et al.* (2005) Controllable preparation of nano-MgO and investigation of its bactericidal properties. *J. Inorg. Biochem.*, **99** (5), 986–993.

41 Ram, P., Vivek, K., and Kumar, S.P. (2014) Nanotechnology in sustainable agriculture: present concerns and future aspects. *Afr. J. Biotechnol.*, **13** (6), 705–713.

42 Jo, Y.-K., Kim, B.H., and Jung, G. (2009) Antifungal activity of silver ions and nanoparticles on phytopathogenic fungi. *Plant Dis.*, **93** (10), 1037–1043.

43 Lamsal, K., Kim, S.W., Jung, J.H. *et al.* (2011) Application of silver nanoparticles for the control of *Colletotrichum* species *in vitro* and pepper anthracnose disease in field. *Mycobiology*, **39** (3), 194–199.

44 Gajbhiye, M., Kesharwani, J., Ingle, A. *et al.* (2009) Fungus-mediated synthesis of silver nanoparticles and their activity against pathogenic fungi in combination with fluconazole. *Nanomedicine*, **5** (4), 382–386.

45 Mishra, S., Singh, B.R., Singh, A. *et al.* (2014) Biofabricated silver nanoparticles act as a strong fungicide against *Bipolaris sorokiniana* causing spot blotch disease in wheat. *PLoS One*, **9** (5), 11.

46 Dimkpa, C.O., Hansen, T., Stewart, J. *et al.* (2015) ZnO nanoparticles and root colonization by a beneficial pseudomonad influence essential metal responses in bean (*Phaseolus vulgaris*). *Nanotoxicology*, **9** (3), 271–278.

47 He, L., Liu, Y., Mustapha, A., and Lin, M. (2011) Antifungal activity of zinc oxide nanoparticles against *Botrytis cinerea* and *Penicillium expansum*. *Microbiol. Res.*, **166** (3), 207–215.

48 Wani, A.H. and Shah, M.A. (2012) A unique and profound effect of MgO and ZnO nanoparticles on some plant pathogenic fungi. *J. Appl. Pharm. Sci.*, **2**, 40–44.

49 Giannousi, K., Avramidis, I., and Dendrinou-Samara, C. (2013) Synthesis, characterization and evaluation of copper based nanoparticles as agrochemicals against *Phytophthora infestans*. *RSC Adv.*, **3** (44), 21743–21752.

50 Servin, A., Elmer, W., Mukherjee, A. *et al.* (2015) Nanoscale micronutrients suppress disease. VFRC Report 2015/2, 33 pp.

51 Rispail, N., De Matteis, L., Santos, R. *et al.* (2014) Quantum dot and superparamagnetic nanoparticle interaction with pathogenic fungi: internalization and toxicity profile. *ACS Appl. Mater. Interfaces*, **6** (12), 9100–9110.

52 Ghormade, V., Deshpande, M.V., and Paknikar, K.M. (2011) Perspectives for nano-biotechnology enabled protection and nutrition of plants. *Biotechnol. Adv.*, **29** (6), 792–803.

53 Lyons, R.E., Wong, D.C.C., Kim, M. *et al.* (2011) Molecular and functional characterisation of resilin across three insect orders. *Insect Biochem. Mol. Biol.*, **41** (11), 881–890.

54 Adak, T., Kumar, J., Shakil, N.A., and Walia, S. (2012) Development of controlled release formulations of imidacloprid employing novel nano-ranged amphiphilic polymers. *J. Environ. Sci. Health. B*, **47** (3), 217–225.

55 Pankaj, Shakil, N.A. Kumar, J. *et al.* (2012) Bioefficacy evaluation of controlled release formulations based on amphiphilic nano-polymer of carbofuran against

Meloidogyne incognita infecting tomato. *J. Environ. Sci. Health B*, **47** (6), 520–528.
56 Kaushik, P., Shakil, N.A., Kumar, J. *et al.* (2013) Development of controlled release formulations of thiram employing amphiphilic polymers and their bioefficacy evaluation in seed quality enhancement studies. *J. Environ. Sci. Health B*, **48** (8), 677–685.
57 Loha, K.M., Shakil, N.A., Kumar, J. *et al.* (2012) Bio-efficacy evaluation of nanoformulations of β-cyfluthrin against *Callosobruchus maculatus* (*Coleoptera: Bruchidae*). *J. Environ. Sci. Health B*, **47** (7), 687–691.
58 Yang, F.L., Li, X.G., Zhu, F., and Lei, C.L. (2009) Structural characterization of nanoparticles loaded with garlic essential oil and their insecticidal activity against *Tribolium castaneum* (Herbst) (*Coleoptera: Tenebrionidae*). *J. Agric. Food Chem.*, **57** (21), 10156–10162.
59 Stadler, T., Buteler, M., and Weaver, D.K. (2010) Novel use of nanostructured alumina as an insecticide. *Pest Manag. Sci.*, **66** (6), 577–579.
60 Goswami, A., Roy, I., Sengupta, S., and Debnath, N. (2010) Novel applications of solid and liquid formulations of nanoparticles against insect pests and pathogens. *Thin Solid Films*, **519** (3), 1252–1257.
61 Frederiksen, H.K., Kristensen, H.G., and Pedersen, M. (2003) Solid lipid microparticle formulations of the pyrethroid gamma-cyhalothrin: incompatibility of the lipid and the pyrethroid and biological properties of the formulations. *J. Control. Release*, **86** (2–3), 243–252.
62 Gogos, A., Knauer, K., and Bucheli, T.D. (2012) Nanomaterials in plant protection and fertilization: current state, foreseen applications, and research priorities. *J. Agric. Food Chem.*, **60** (39), 9781–9792.
63 Pérez-de-Luque, A. and Rubiales, D. (2009) Nanotechnology for parasitic plant control. *Pest Manag. Sci.*, **65** (5), 540–545.
64 Wang, L., Li, X., Zhang, G. *et al.* (2007) Oil-in-water nanoemulsions for pesticide formulations. *J. Colloid Interface Sci.*, **314** (1), 230–235.

65 Perlatti, B., Bergo, P.D.S., Fernandes, J., and Forim, M. (2013) Polymeric nanoparticle-based insecticides: a controlled release purpose for agrochemicals, in *Insecticides: Development of Safer and More Effective Technologies* (ed. S. Trdan), InTech Publishers, pp. 523–550.
66 Alavi, S.V. and Dehpour, A.A. (2010) Evaluation of the nanosilver colloidal solution in comparison with the registered fungicide to control greenhouse cucumber downy mildew disease in the North of Iran. *Acta Hortic.*, **877**, 1643–1646.
67 Cota-Arriola, O., Onofre Cortez-Rocha, M., Burgos-Hernández, A. *et al.* (2013) Controlled release matrices and micro/nanoparticles of chitosan with antimicrobial potential: development of new strategies for microbial control in agriculture. *J. Sci. Food Agric.*, **93** (7), 1525–1536.
68 Suresh Kumar, R.S., Shiny, P.J., Anjali, C.H. *et al.* (2013) Distinctive effects of nano-sized permethrin in the environment. *Environ. Sci. Pollut. Res.*, **20** (4), 2593–2602.
69 Bhagat, D., Samanta, S.K., and Bhattacharya, S. (2013) Efficient management of fruit pests by pheromone nanogels. *Sci. Rep.*, **3**. doi: 10.1038/srep01294
70 Chin, C.P., Wu, H.S., and Wang, S.S. (2011) New approach to pesticide delivery using nanosuspensions: research and applications. *Ind. Eng. Chem. Res.*, **50** (12), 7637–7643.
71 Gunalan, S., Sivaraj, R., and Rajendran, V. (2012) Green synthesized ZnO nanoparticles against bacterial and fungal pathogens. *Prog. Nat. Sci. Mater. Int.*, **22** (6), 693–700.
72 Debnath, N., Das, S., and Goswami, A. (2011) Novel entomotoxic nanocides for agro-chemical industry. 11th IEEE International Conference Nanotechnology, pp. 53–56.
73 Caraglia, M., De Rosa, G., Abbruzzese, A., and Leonetti, C. (2011) Nanotechnologies: new opportunities for old drugs. The case of aminobisphosphonates. *J. Nanomed. Biotherapeutic Discov.*, **1** (103e), 2.

74 Begum, N., Sharma, B., and Ravi, S.P. (2010) Evaluation of insecticidal efficacy of *Calotropis Procera* and *Annona Squamosa* ethanol extracts against *Musca Domestica*. *J. Biofertil. Biopestic.*, **1** (101), 6.

75 Miralles, P., Church, T.L., and Harris, A.T. (2012) Toxicity, uptake, and translocation of engineered nanomaterials in vascular plants. *Environ. Sci. Technol.*, **46** (17), 9224–9239.

76 Mittal, A.K., Chisti, Y., and Banerjee, U.C. (2013) Synthesis of metallic nanoparticles using plant extracts. *Biotechnol. Adv.*, **31** (2), 346–356.

77 Wang, W.N., Tarafdar, J.C., and Biswas, P. (2013) Nanoparticle synthesis and delivery by an aerosol route for watermelon plant foliar uptake. *J. Nanopart. Res.*, **15**, 13.

78 Wang, Z., Xie, X., Zhao, J. *et al.* (2012) Xylem- and phloem-based transport of CuO nanoparticles in maize (*Zea mays* L.). *Environ. Sci. Technol.*, **46** (8), 4434–4441.

79 Khot, L.R., Sankaran, S., Maja, J.M. *et al.* (2012) Applications of nanomaterials in agricultural production and crop protection: a review. *Crop Prot.*, **35**, 64–70.

8
Nanoparticle-Based Delivery Systems for Nutraceuticals: Trojan Horse Hydrogel Beads

Benjamin Zeeb[1] and David Julian McClements[2]

[1]Institute of Food Science and Biotechnology, Department of Food Physics and Meat Science, Garbenstrasse 25, 70599 Stuttgart, Germany
[2]University of Massachusetts Amherst, Department of Food Science, 240 Chenoweth Laboratory, 102 Holdsworth Way, Amherst, MA 01003, USA

8.1
Introduction

Recently, the food industry has laid considerable emphasis on the development of functional food and beverage products specifically designed to improve human health and well-being. Many of these products consist of foods that are enriched with bioactive food components (nutraceuticals) that have health benefits over and above their normal nutritional role, such as carotenoids, curcumin, flavonoids, coenzyme Q10, peptides, and ω-3 fatty acids [1–3]. For example, regular consumption of these nutraceuticals may help prevent cancer, heart disease, hypertension, or diabetes, or they may help ensure good eye health, improved attention, or brain development. Consequently, regular consumption of functional foods or beverages enriched with these bioactive agents may lead to an overall improvement in human health and reduction in associated health care costs. Nutraceuticals are often isolated from a natural source and then converted into functional ingredients that can be introduced into food products. Some nutraceuticals can simply be mixed with existing food products, for example, lipophilic bioactives can be simply dissolved in the fat phase of oily products (e.g., cooking oils, margarine, and butter), whereas hydrophilic bioactives can simply be dissolved in the water phase of aqueous products (e.g., beverages, dressings, desserts, or sauces). However, there are problems associated with incorporating nutraceuticals into certain types of products, for example, the addition of lipophilic bioactives into aqueous-based products [4]. Moreover, there are also various other challenges associated with the utilization of certain types of nutraceuticals into functional foods and beverages, such as chemical instability, poor food matrix compatibility, and low oral bioavailability. Finally, there is often a need to deliver nutraceuticals to specific regions within the human gastrointestinal tract (GIT), such as the mouth, stomach, small intestine, or colon [5]. As a consequence, there has been considerable interest in the development of nanostructured colloidal delivery systems to encapsulate, protect, and deliver nutraceuticals [5–8].

8.2
Overview of Nanoparticles-Based Colloidal Delivery Systems

There has been a large increase in the amount of research carried out on the development and characterization of nanoparticle-based colloidal delivery systems for bioactive food components in the past decade. Much of this research has been covered in some detail in recent comprehensive review articles and books, and the reader is referred to these sources for more details on the formation, properties, and applications of colloidal delivery systems [5,6,9–12]. Consequently, only a brief overview of nanoparticle-based colloidal delivery systems suitable for utilization within the food industry will be given here.

8.2.1
Microemulsions

Microemulsions are thermodynamically stable colloidal dispersions [13]. Oil-in-water microemulsions typically consist of small particles ($d = 5$–50 nm) mainly composed of surfactant and oil molecules that are dispersed in water (Figure 8.1). The surfactant molecules are arranged so that the nonpolar tails are clustered together into the hydrophobic interior, whereas the polar heads are located at the hydrophilic exterior where they are exposed to water. The oil molecules may form a separate hydrophobic core or they may be located between the nonpolar tails of the surfactant molecules. The primary driving force for the formation of

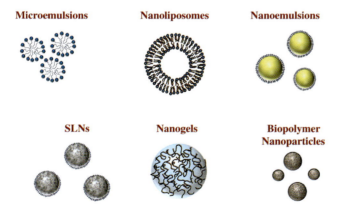

Figure 8.1 Examples of nanoscale colloidal delivery systems that can be used to encapsulate, protect, and deliver lipophilic nutraceuticals.

microemulsions is the hydrophobic effect, for example, the tendency for the system to reduce the thermodynamically unfavorable contact area between nonpolar groups and water. Lipophilic bioactive molecules can be solubilized within the hydrophobic interior of the microemulsions. Microemulsions can often be formed spontaneously by simply mixing specific surfactants, oils, and water together at an appropriate composition and environmental conditions (e.g., temperature). Typically, microemulsions lead to the formation of optically transparent colloidal dispersions that are suitable for utilization in clear foods and beverages. The major disadvantage is that a relatively high surfactant concentration is required to formulate them, and typically synthetic surfactants are required to form them [14]. Lipophilic nutraceuticals may be used as the oil phase themselves, or they may be mixed with an oil phase.

8.2.2
Nanoliposomes

Nanoliposomes are typically fabricated from phospholipid molecules that have a polar head group and two nonpolar tails [15]. These molecules have an optimum curvature close to unity, which means that they tend to self-assemble into bilayer structures with layers of phospholipid molecules aligned head-to-head and tail-to-tail (Figure 8.1). Nonpolar bioactive molecules can be trapped into the nonpolar regions between the tail groups, whereas polar bioactives can be trapped in the hydrophilic inner water region [16].

8.2.3
Nanoemulsions

Nanoemulsions are thermodynamically unstable colloidal dispersions that are typically fabricated from oil, emulsifier, and water [17]. Oil-in-water nanoemulsions consist of small lipid-rich particles dispersed in water (Figure 8.1). In this case, the nanoparticles consist of a nonpolar core of lipid molecules coated by a layer of emulsifier molecules (which may be surfactants, phospholipids, proteins, polysaccharides, or particles). Lipophilic bioactive molecules can be solubilized within the lipid interior of the nanoemulsion droplets. Nanoemulsions may be formed using low- or high-energy methods [18–21]. Low-energy methods rely on the spontaneous formation of tiny oil droplets when certain types of surfactant, oil, and water are mixed together under appropriate conditions. High-energy methods utilize mechanical devices known as homogenizers to generate intense disruptive forces that break up oil and water phases and form tiny oil droplets.

8.2.4
Solid Lipid Nanoparticles

Solid lipid nanoparticles (SLNs) consist of small crystalline lipid particles that are dispersed in water (Figure 8.1). SLNs are usually fabricated by forming an oil-in-water nanoemulsion at an elevated temperature, and then cooling it below the crystallization point of the lipid phase [22]. SLNs may have advantages over liquid lipid nanoparticles (LLNs) for certain applications, because the solid core may restrict molecular diffusion inside the particle. This may help to prevent the release of encapsulated components, or prevent undesirable chemical degradation reactions from occurring [23]. Alternatively, the solid-to-liquid phase transition may be utilized as a trigger mechanism for releasing an encapsulated lipophilic nutraceutical. It should be noted that SLNs must be carefully designed to avoid particle aggregation and nutraceutical expulsion during their preparation [24].

8.2.5 Biopolymer Nanoparticles and Nanogels

Nanoparticles can be fabricated from various types of proteins and polysaccharides using a number of different fabrication methods, including antisolvent precipitation, injection, coacervation, and templating methods [10]. In this case, the bioactive component is usually mixed within the biopolymers prior to nanoparticle fabrication (Figure 8.1). Biopolymer nanoparticles predominantly consist of tightly packed biopolymer molecules, whereas biopolymer nanogels contain relatively high amounts of water trapped in a biopolymer network.

8.3 Designing Particle Characteristics

The functional performance of nanostructured colloidal delivery systems depends on the characteristics of the particles they contain [5,6,9–12]. Consequently, food researchers can rationally design nanoparticles with different functional attributes, such as encapsulation efficiency, light scattering properties, texture-modifying properties, and physicochemical stability by carefully controlling nanoparticle properties. A detailed discussion of the relationship between nanoparticle characteristics and the functional performance of colloidal delivery systems has been given elsewhere [5]. In this section, only a brief overview of important nanoparticle properties is therefore given (Figure 8.2).

8.3.1 Composition

Nanoparticles can be fabricated from various types of food constituents, with the most commonly used being lipids, surfactants, proteins, polysaccharides, minerals, and water. The composition of a nanoparticle will influence its physicochemical properties (refractive index, density, physical state, and rheology), which will determine its functional attributes (optical properties, stability, and release characteristics). For example, the tendency of a nanoparticle to dissolve, dissociate, or disintegrate under different environmental conditions depends on

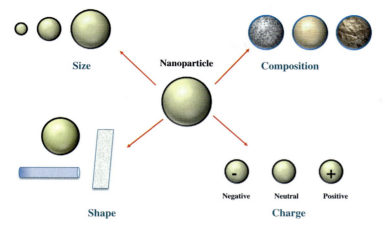

Figure 8.2 Nanoparticles can be designed to have different particle properties, such as composition, size, shape, and charge, which will alter their functional attributes.

its composition. This can be used to create delivery systems that will release encapsulated components in response to different environmental triggers, such as pH, ionic strength, temperature, or enzyme activity.

8.3.2
Particle Size

Nanoparticles typically have mean diameters within the range of about 5–200 nm depending upon the components and fabrication method used to assemble them [19]. The size of the nanoparticles affects their optical properties, physicochemical stability, and release characteristics. For example, nanoparticles must be below a certain critical diameter (around 70 nm) before the overall colloidal dispersions appear transparent [22]. In addition, the stability of nanoparticles to aggregation and gravitational separation usually improves as their size decreases [17]. On the other hand, encapsulated substances tend to be more susceptible to chemical degradation and rapid release for smaller particles, which may limit their use for certain applications.

8.3.3
Particle Charge

The electrical charge on nanoparticles may range from highly positive to neutral to highly negative depending on the nature of the charged groups on their surfaces. The electrical properties of nanoparticles can be altered by fabricating them from building blocks with different charge characteristics, such as emulsifiers, proteins, polysaccharides, or minerals. The charge on nanoparticles is important because it influences their chemical and physical stability, as well as their interactions with charged surfaces inside and outside the human body.

8.3.4
Particle Structure

Nanoparticles may have a variety of different structures, depending on the components and preparation methods used to fabricate them: spherical or nonspherical, homogeneous, dispersed, or core–shell, and individual or clustered [5]. The structure of the nanoparticles will impact their functional performance and can often be designed to obtain particular functional attributes.

Typically, a food formulator needs to define the properties of the nutraceutical ingredient for the purpose of encapsulation (e.g., polarity, charge, chemical stability, and physical state) and then define the characteristics of the food that they would like to incorporate the nutraceutical into (e.g., optical properties, rheology, and stability). The formulator must then select particle characteristics that will enable the nutraceutical to be effectively encapsulated, protected and delivered, and that is compatible with the food matrix.

8.4
Trojan Horse Nanoparticle Delivery Systems

As mentioned earlier, there have already been a large number of review articles, book chapters, and books on various kinds of nanostructured colloidal delivery systems suitable for nutraceutical encapsulation. For this reason, we focus on a relatively new approach that has been applied in the food industry recently: Trojan horse nanoparticle delivery systems. These systems consist of lipophilic nutraceuticals dissolved within nanoparticles that are themselves trapped inside hydrogel beads. The hydrogel beads are specifically designed to retain the nanoparticles under one set of conditions, but then release them under another set of conditions. A number of physiochemical mechanisms can be utilized to trigger the release of the encapsulated nutraceutical-loaded nanoparticles, including weakening of attractive interactions between the nanoparticles and hydrogel beads; swelling of the beads to increase the hydrogel pore size; or disintegration of the beads (Figure 8.3). These trigger mechanisms can be designed to occur in response to a particular change in environmental conditions (e.g., pH, ionic strength, temperature, or enzyme activity). In this section, we begin by discussing the main building blocks that can be used to assemble hydrogel beads and then highlight the potential application of Trojan horse nanoparticle delivery systems using curcumin as an example.

8.4.1
Biopolymers as Building Blocks to Form Hydrogel Beads

The hydrogel beads act as the "Trojan horse" that contains the nanoparticles inside, hides them from their immediate environment, but releases them when they experience another environment. Food-grade hydrogel beads are typically

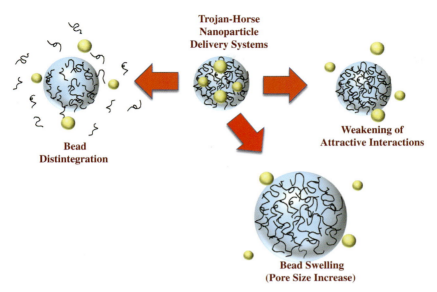

Figure 8.3 Nanoparticles can be trapped inside hydrogel beads to form Trojan horse nanoparticle delivery systems that release the nanoparticles in response to an external trigger.

assembled from food biopolymers, such as proteins and polysaccharides [10,25]. These biopolymers are naturally produced by living organisms such as plants, animals, or microorganisms, and then are isolated, purified, and sometimes modified before they are used as functional ingredients.

Animal-derived proteins from milk, egg, fish, or meat have been widely used as building blocks to fabricate various kinds of structures in foods. The functional properties of these proteins (e.g., solubility, surface activity, thickening, and gelling) are determined by their molecular characteristics (e.g., molecular weight, conformation, flexibility, charge, and polarity). Plant-derived proteins have gained great interest among food scientists and manufacturers recently as structural elements due to ethical and ecological reasons. Plant proteins are suitable for vegetarian or vegan diets, and can usually be produced in a more sustainable fashion than animal proteins. Proteins from potato, rice, pea, sunflower, or canola are potential candidates to fabricate hydrogel beads with tailor-made performances. However, their functional properties vary considerably depending on their origin as well as on the isolation, fractionation, and purification methods used to convert them into functional ingredients. For example, differences in the pH dependence of the water solubility of different plant and animal proteins are shown in Figure 8.4. These results clearly show that there are major differences between their solubility characteristics, which will impact their ability to be used as building blocks for hydrogel beads and other structures. For example, protein ingredients isolated from peas show vastly different solubility in water, even though they are nominally from the same origin (Figure 8.4). One major reason

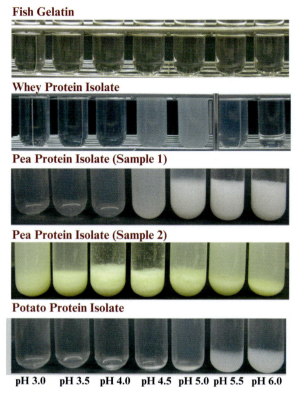

Figure 8.4 Solubility of animal and plant proteins derived from various origins as a function of pH (3.0–6.0).

for these differences is that many vegetable proteins mainly function as storage units in plant tissues, whereas many animal proteins function as structural units in the skin, hair, or bones (e.g., collagen) or as nutrient carriers (e.g., caseins or whey proteins).

Various types of polysaccharides can also be used as building blocks to construct hydrogel beads, such as starch, cellulose, carrageenan, pectin, alginate, locust bean gum, and gum arabic. These polysaccharides also vary in their molecular weight, branching, polarity, and charge characteristics, which alter their ability to form hydrogel beads [25]. The fabrication of hydrogel beads typically involves the use of either single or mixed biopolymer systems. Knowledge of the nature of the forces acting between biopolymers can be used to facilitate the design of Trojan horse nanoparticle delivery systems. In general, biopolymers may interact with each other through a variety of interaction forces, which may be physical (electrostatic, hydrophobic, hydrogen bonding, excluded volume effects) or chemical (covalent) in origin [10].

8.4.2
Fabrication Methods for Hydrogel Beads

Numerous fabrication methods have been developed to assemble hydrogel beads suitable for use in Trojan horse nanoparticle delivery systems. Typically, a biopolymer particle is formed first using a suitable approach (e.g., phase separation, electrostatic complexation, extrusion, or templating), and then the biopolymer particle is gelled using a suitable cross-linking approach (e.g., temperature change, ion addition, or enzymes) [10,26,27]. Hydrogel beads tailored for specific applications can be created by selecting the most appropriate building blocks and fabrication method. In particular, the size, structure, composition, interfacial properties, loading, and release behavior of the hydrogel template structure can be specifically tuned so that the hydrogel beads swell or disintegrate under a specific set of different environmental conditions, or change their molecular interactions with the nanoparticles. The following section gives a brief overview of some of the most commonly used fabrication methods for assembling biopolymers into hydrogel beads. All of these structures can be potentially used as Trojan horse nanoparticle delivery systems by the food industry. However, one should mention that many biopolymer-based systems are held together by relatively weak attractive forces, and may therefore have a tendency to dissociate or degrade during storage. This problem can be overcome by covalently cross-linking the biopolymers within the hydrogel beads using physical, chemical, or enzymatic methods [4,28].

8.4.3
Thermodynamic Incompatibility

Hydrogel beads can often be formed based on the phase separation that occurs in aqueous solutions containing mixtures of different biopolymers that are not strongly attracted to each other. In general, when two biopolymer solutions are mixed together, a variety of different structures may be produced depending on the nature of the biopolymers, the solution composition, and the environmental conditions [29]. The Flory Huggins theory has been used to explain the phase separation phenomena that may be encountered when combining similarly charged or uncharged polymers. Restrictions in the freedom of orientation of the biopolymer molecules reduce the entropy contribution to the overall free energy of the system, thereby favoring phase separation rather than mixing – a fact that is known as *thermodynamic incompatibility* [30]. This process leads to the formation of two separate phases that have different compositions: one phase that is rich in biopolymer 1 and depleted in biopolymer 2, and another phase that is rich in biopolymer 2 and depleted in biopolymer 1. These two phases may be treated like oil and water, and when subjected to mild shear stresses, they form dispersions of one in the other, for example, water-in-water emulsions. As such, whey protein isolate (WPI) and pectin solutions – if mixed at a neutral pH – tend to phase separate over time. A mild homogenization step can then be applied to form water-in-water dispersions consisting of WPI-rich dispersed phase surrounded by a pectin-rich continuous phase [31,32]. If the pH is then reduced, core–shell-like structures composed of an inner WPI-rich core stabilized by WPI–pectin aggregates may be formed. Nanoparticles may be trapped within the hydrogel beads formed using this method by mixing the biopolymers and nanoparticles together and then adjusting solution conditions to promote thermodynamic incompatibility.

8.4.4
Complex Coacervation

The phase separation of mixed biopolymer solutions may also occur when the biopolymer molecules are attracted to each other [30]. In this case, the system separates into one phase that is rich in both biopolymers, and into another phase that is depleted in both biopolymers. The most common driving force for the formation of the biopolymer-rich phase is electrostatic attraction between oppositely charged biopolymers. The electrostatic complexes formed may be soluble or insoluble depending on their size, density, and charge characteristics, which depend on the nature of the biopolymers, solution properties, and environmental conditions [29,33]. Electrostatic complexes have been used to stabilize individual lipid droplets from aggregation [34], but they may also be used to trap a number of nanoparticles inside [10]. In this case, the nanoparticles to be encapsulated are mixed with the biopolymer solutions and then the conditions are changed to favor the formation of the electrostatic complexes.

8.4.5
Antisolvent Precipitation

Antisolvent precipitation methods rely on changes in the solvent quality of biopolymer solutions to promote nanoparticle or microparticle formation. These particles can be used as protein nanoparticles to encapsulate nutraceuticals, or they can be used to form hydrogel beads to encapsulate nanoparticles. When the solvent quality surrounding the biopolymer molecules is reduced, they spontaneously form particles through a nucleation and growth mechanism [26,35,36]. Manipulating system composition and production parameters can control the size of the biopolymer particles formed. The key to producing ultrafine particles by antisolvent precipitation is to create conditions that favor rapid particle nucleation and slow particle growth. Particle growth can be inhibited using polymers and/or surfactants that are dissolved in the antisolvent phase [37,38]. Both hydrophilic [39] and lipophilic [40–42] molecules have been encapsulated in particles produced by antisolvent precipitation. The entrapment of molecules is usually improved if the polarity of the matrix material (the biopolymer) and the molecule of interest (the micronutrient) closely match [41]. Thus, hydrophobic proteins (e.g., gliadin and zein) are suitable for encapsulation of hydrophobic bioactive molecules [41,42], while hydrophilic proteins (e.g., gelatin or whey protein isolate) are more suitable for entrapment of hydrophilic compounds [39,43]. Nanoparticles may be trapped inside particles formed using this method by mixing the nanoparticles and biopolymers together in a good solvent, and then adding this mixture into a poor solvent for the biopolymers, which leads to particle formation.

8.4.6
Electrospinning

Electrospinning is a manufacturing technology capable of producing solid biopolymer fibers from a biopolymer solution by applying a strong electric field to a spinneret equipped with a small capillary orifice [44,45]. In general, electrospun biopolymer fibers can range in diameter from around 10 to 1000 nm. Biopolymer fibers may exhibit unusual functionalities in terms of release and retention of nanoparticles incorporated within them due to the high surface-to-volume ratio. Typically, a small drop of a biopolymer solution is formed at the tip of a syringe that is connected to a high-voltage generator. A strong electrostatic field causes the biopolymers in the solution to be so strongly charged that it leads to distorting the droplet shape to form a so-called Taylor cone. As a consequence, a very thin charged polymers solution jet is expelled at the tip of the Taylor cone, which is then accelerated toward the grounded target [46]. Charge instabilities cause the biopolymer jet to bend and stretch during its travel to the collector, which causes the jet to become even thinner. On the way to the collector plate, the solvent evaporates and the jet solidifies to form a single-fiber strand. The fiber strand forms a film on the surface of the collector plate [46]. In this case, the nanoparticles to be encapsulated would be mixed with the biopolymer solution to be electrospun.

8.4.7
Extrusion Techniques

Injection or extrusion methods are one of the most common approaches for forming hydrogel beads. They involve the injection of single- or mixed-biopolymer solutions containing the nanoparticles to be encapsulated into gelling solutions. The cross-linking of biopolymers into hydrogel beads can be induced by ions, temperature, and/or enzymes, leading to beads with different sizes, shapes, and porosities [47]. Typically, ionic gelation occurs between charged biopolymers (e.g., alginate or pectin) and multivalent ions (e.g., calcium or barium), although monovalent ions can be used to induce cross-linking in some cases (e.g., potassium ions and carrageenan). In addition, temperature might be utilized as a cross-linking agent to induce heat-set or cold-set gelation [48]. The kinetics of bead formation depends on the biopolymers used, the gelling conditions, and the mechanical setup utilized. The resulting differences in the properties of the hydrogel beads formed will lead to differences in the retention, protection, and release behavior of Trojan horse delivery systems. In general, filled hydrogel beads produced by extrusion provide a cost-effective and simple to scale-up method that might be easily implemented in food-processing lines [49].

8.4.8
Fibril Formation

Biopolymer fibrils may also be useful constructs for forming Trojan horse nanoparticle delivery systems. A number of studies have shown that protein fibrils can be formed by controlled self-assembly of globular proteins [50,51]. Globular proteins from various sources (e.g., whey, egg, or bovine) can be subjected to an acid-induced hydrolysis under elevated temperatures (80–95 °C), which leads to the formation of peptides that assemble into fibril structures. Fibril formation depends on peptide composition, degree of hydrophobicity, ionic strength, and temperature. The protein fibrils generated through this process can be used to form hydrogels through a cold-setting mechanism by altering the electrostatic interactions between them (e.g., by adding salt or adjusting pH). Thus, Trojan horse nanoparticle delivery systems can be fabricated by mixing the nanoparticles with fibrils and then altering solution conditions to promote hydrogel formation.

8.5
Case Study: Alginate Hydrogel Beads as Trojan Horse Nanoparticle Delivery Systems for Curcumin

Curcumin is a highly nonpolar molecule that has many biological activities, including antioxidant, anti-inflammatory, anticancer, and antimicrobial properties, which make it an interesting candidate as a nutraceutical for application in functional foods [52]. However, its utilization as a bioactive agent in its pure form in food matrices is limited due to its crystalline character, low water solubility, and chemical instability [53,54]. Recent studies have demonstrated that the aqueous solubility and bioavailability of curcumin can be greatly increased when it is incorporated into nanoemulsion-based delivery systems consisting of curcumin trapped within lipid nanoparticles. Nevertheless, the curcumin in nanoemulsions may still be susceptible to chemical degradation or metabolism, because the small size of the lipid droplets means that the curcumin is still in relatively close proximity to pro-oxidants in the aqueous phase [17,55].

Recent research has therefore focused on the development of Trojan horse delivery systems to improve the stability and bioavailability of curcumin, as well as to control its release within the human gastrointestinal tract (Figure 8.5). In particular, our laboratory has carried out studies to establish the major factors influencing the retention and release of curcumin-loaded nanoemulsion droplets

8.5 Case Study: Alginate Hydrogel Beads as Trojan Horse Nanoparticle Delivery Systems for Curcumin

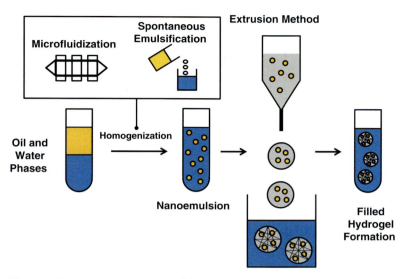

Figure 8.5 Schematic representation of the formation of Trojan horse delivery systems to deliver curcumin-loaded lipid nanoparticles [49,56].

from alginate hydrogel beads [49,56]. Initially, curcumin-loaded nanoemulsions were prepared by two emulsification techniques based on different physicochemical principles, namely, spontaneous emulsification (low-energy method) and microfluidization (high-energy method) [20]. Filled alginate beads were then generated by an extrusion technique, whereby nanoemulsions were mixed with alginate solutions and then injected into gelling solutions containing divalent calcium ions. This study showed that the curcumin was mainly located within the hydrogel beads when the nanoemulsions were formed using high-energy emulsification, but that the curcumin was distributed between the interior and exterior of the hydrogel beads when the nanoemulsions were formed using low-energy emulsification [49]. The difference was attributed to the fact that both methods produced lipid droplets that were trapped inside the hydrogels by the biopolymer network, but the low-energy method also produced an appreciable amount of curcumin-loaded micelles due to the high levels of surfactants used. These micelles are so small (<15 nm) that they rapidly diffuse through the hydrogel beads and into the surrounding liquid. The same study showed that the size, shape, and permeability of the hydrogel beads depended on the calcium and alginate concentrations, which could therefore be manipulated to control the release rate of curcumin (Figure 8.6). Greater lipid droplet release occurred at low calcium and alginate levels, which was attributed to larger pore sizes in the hydrogel matrix. Moreover, the release behavior of the lipid droplets also depended on solution pH, which was attributed to pH-dependent changes in hydrogel pore size and droplet–biopolymer interactions. For example, protein-coated lipid nanoparticles are positively charged at pH values below their

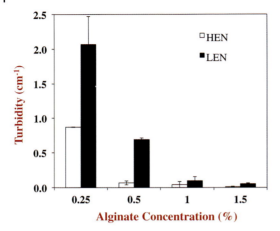

Figure 8.6 Turbidity (at 600 nm) of aqueous solution surrounding lipid nanoparticle-loaded alginate hydrogel beads after 24 h incubation (500 mM calcium chloride bath). An increase in turbidity is a measure of the release of lipid nanoparticles from the beads. Increasing the alginate concentration reduces the release of the beads because of the smaller pore size. HEN: high-energy nanoemulsions; LEN: low-energy nanoemulsions [49,56].

isoelectric point, but negatively charged at pH values above their isoelectric point. Consequently, they will tend to be electrostatically attracted to the anionic alginate molecules in the hydrogel beads at low pH (and therefore be retained), but electrostatically repelled from the beads at high pH (and therefore be released). It was therefore suggested that these Trojan horse beads might be designed to encapsulate and release nutraceutical-loaded lipid droplets in response to specific triggers in foods or the human body [56].

As mentioned, the release properties of the curcumin also depended on the nature of the colloidal particles it was dissolved within (i.e., micelles versus lipid droplets). Curcumin-loaded micelles rapidly diffused out of the hydrogel beads regardless of alginate or calcium concentrations (Figure 8.6). On the other hand, curcumin-loaded lipid droplets were retained within the hydrogel beads at sufficiently high alginate and calcium levels due to the formation of pores that were small enough to restrict droplet movement. These results indicated that the modulation of the hydrogel bead composition might be used to create controlled-release profiles. Moreover, a burst release might be obtained by encapsulating curcumin in small surfactant micelles that rapidly diffuse through the hydrogel matrix, whereas a high retention could be achieved by encapsulating curcumin within larger lipid droplets. Alginate beads are resistant to degradation within the upper gastrointestinal tract, but are degraded by microbes within the lower gastrointestinal tract, and so they may be useful as colonic delivery systems for lipophilic bioactive molecules.

8.6 Conclusions

There has been considerable progress in the design, fabrication, and characterization of food-grade nanoparticles for utilization as delivery systems to encapsulate, protect, and release nutraceuticals. These nanoparticles can be fabricated from a variety of food-grade ingredients (including lipids, surfactants, proteins, polysaccharides, and minerals) using a variety of different production methods. The functionality of these edible nanoparticles can often be improved by encapsulating them within hydrogel beads to form Trojan horse nanoparticle delivery systems. The hydrogel beads can be designed to retain and protect the nutraceutical-loaded nanoparticles under one set of conditions, but release them under another set of conditions. This kind of delivery system may therefore have a number of applications in the food industry. For example, they could be used to encapsulate and protect nanoparticles in the upper gastrointestinal tract, but release them in the colon.

References

1 Ramaa, C.S., Shirode, A.R., Mundada, A.S., and Kadam, V.J. (2006) Nutraceuticals: an emerging era in the treatment and prevention of cardiovascular diseases. *Curr. Pharm. Biotechnol.*, **7** (1), 15–23.

2 Espin, J.C., Garcia-Conesa, M.T., and Tomas-Barberan, F.A. (2007) Nutraceuticals: facts and fiction. *Phytochemistry*, **68** (22–24), 2986–3008.

3 Wildman, R.E.C. and Kelley, M. (2007) Nutraceuticals and functional foods, in *Handbook of Nutraceuticals and Functional Foods*, 2nd edn (ed. R.E.C. Wildman), CRC Press, Boca Raton, FL, pp. 1–22.

4 Zeeb, B., Fischer, L., and Weiss, J. (2014) Stabilization of food dispersions by enzymes. *Food Funct.*, **5** (2), 198–213.

5 McClements, D.J. (2014) *Nanoparticle- and Microparticle-Based Delivery Systems: Encapsulation, Protection and Release of Active Components*, CRC Press, Boca Raton, FL.

6 Velikov, K.P. and Pelan, E. (2008) Colloidal delivery systems for micronutrients and nutraceuticals. *Soft Matter*, **4** (10), 1964–1980.

7 Sagalowicz, L. and Leser, M.E. (2010) Delivery systems for liquid food products. *Curr. Opin. Colloid Interface Sci.*, **15** (1–2), 61–72.

8 McClements, D.J. (2012) Advances in fabrication of emulsions with enhanced functionality using structural design principles. *Curr. Opin. Colloid Interface Sci.*, **17** (5), 235–245.

9 Spernath, A. and Aserin, A. (2006) Microemulsions as carriers for drugs and nutraceuticals. *Adv. Colloid Interface Sci.*, **128**, 47–64.

10 Matalanis, A., Jones, O.G., and McClements, D.J. (2011) Structured biopolymer-based delivery systems for encapsulation, protection, and release of lipophilic compounds. *Food Hydrocoll.*, **25** (8), 1865–1880.

11 Singh, H., Thompson, A., Liu, W., and Corredig, M. (2012) Liposomes as food ingredients and nutraceutical delivery systems, in *Encapsulation Technologies and Delivery Systems for Food Ingredients and Nutraceuticals* (eds N. Garti and D.J. McClements), Woodhead Publishing, Oxford, UK, pp. 287–318.

12 Ezhilarasi, P.N., Karthik, P., Chhanwal, N., and Anandharamakrishnan, C. (2013)

Nanoencapsulation techniques for food bioactive components: a review. *Food Bioprocess Technol.*, **6**, 628–647.

13 Flanagan, J. and Singh, H. (2006) Microemulsions: a potential delivery system for bioactives in food. *Crit. Rev. Food Sci. Nutr.*, **46** (3), 221–237.

14 Anton, N., Benoit, J.-P., and Saulnier, P. (2008) Design and production of nanoparticles formulated from nano-emulsion templates: a review. *J. Control. Release*, **128** (3), 185–199.

15 Mozafari, M.R., Johnson, C., Hatziantoniou, S., and Demetzos, C. (2008) Nanoliposomes and their applications in food nanotechnology. *J. Liposome Res.*, **18** (4), 309–327.

16 Gibis, M., Zeeb, B., and Weiss, J. (2014) Formation, characterization, and stability of encapsulated hibiscus extract in multilayered liposomes. *Food Hydrocoll.*, **38**, 28–39.

17 McClements, D.J. and Rao, J. (2011) Food-grade nanoemulsions: formulation, fabrication, properties, performance, biological fate, and potential toxicity. *Crit. Rev. Food Sci. Nutr.*, **51** (4), 285–330.

18 Wooster, T.J., Golding, M., and Sanguansri, P. (2008) Impact of oil type on nanoemulsion formation and Ostwald ripening stability. *Langmuir*, **24** (22), 12758–12765.

19 Qian, C. and McClements, D.J. (2011) Formation of nanoemulsions stabilized by model food-grade emulsifiers using high-pressure homogenization: factors affecting particle size. *Food Hydrocoll.*, **25** (5), 1000–1008.

20 Saberi, A.H., Fang, Y., and McClements, D.J. (2013) Fabrication of vitamin E-enriched nanoemulsions: factors affecting particle size using spontaneous emulsification. *J. Colloid Interface Sci.*, **391** (1), 95–102.

21 Zeeb, B., Herz, E., McClements, D.J., and Weiss, J. (2014) Impact of alcohols on the formation and stability of protein-stabilized nanoemulsions. *J. Colloid Interface Sci.*, **433**, 196–203.

22 Helgason, T., Salminen, H., Kristbergsson, K., McClements, D.J., and Weiss, J. (2015) Formation of transparent solid lipid nanoparticles by microfluidization: influence of lipid physical state on appearance. *J. Colloid Interface Sci.*, **448**, 114–122.

23 Salminen, H., Gömmel, C., Leuenberger, B.H., and Weiss, J. (2015) Influence of encapsulated functional lipids on crystal structure and chemical stability in solid lipid nanoparticles: towards bioactive-based design of delivery systems. *Food Chem.*, **190**, 928–937.

24 Weiss, J., Decker, E.A., McClements, D.J., Kristbergsson, K., Helgason, T., and Awad, T. (2008) Solid lipid nanoparticles as delivery systems for bioactive food components. *Food Biophys.*, **3** (2), 146–154.

25 Jones, O.G. and McClements, D.J. (2010) Functional biopolymer particles: design, fabrication, and applications. *Compr. Rev. Food Sci. Food Safety*, **9** (4), 374–397.

26 Joye, I.J. and McClements, D.J. (2013) Production of nanoparticles by anti-solvent precipitation for use in food systems. *Trends Food Sci. Technol.*, **34** (2), 109–123.

27 Zhang, Z.P., Zhang, R.J., Chen, L., Tong, Q.Y., and McClements, D.J. (2015) Designing hydrogel particles for controlled or targeted release of lipophilic bioactive agents in the gastrointestinal tract. *Eur. Polym. J.*, **72**, 698–716.

28 Zeeb, B., Gibis, M., Fischer, L., and Weiss, J. (2012) Crosslinking of interfacial layers in multilayered oil-in-water emulsions using laccase: characterization and pH-stability. *Food Hydrocoll.*, **27** (1), 126–136.

29 McClements, D.J., Decker, E.A., Park, Y., and Weiss, J. (2009) Structural design principles for delivery of bioactive components in nutraceuticals and functional foods. *Crit. Rev. Food Sci. Nutr.*, **49** (6), 577–606.

30 De Kruif, C.G., Weinbreck, F., and De Vries, R. (2004) Complex coacervation of proteins and anionic polysaccharides. *Curr. Opin. Colloid Interface Sci.*, **9** (5), 340–349.

31 Kim, H.J., Decker, E.A., and McClements, D.J. (2006) Preparation of multiple emulsions based on thermodynamic incompatibility of heat-denatured whey protein and pectin solutions. *Food Hydrocoll.*, **20** (5), 586–595.

32 Thongkaew, C., Gibis, M., Hinrichs, J., and Weiss, J. (2015) Shear-induced morphological changes in associative and segregative phase-separated biopolymer systems. *Food Hydrocoll.*, **51**, 414–423.

33 Davidov-Pardo, G., Joye, I.J., and McClements, D.J. (2015) Food-grade protein-based nanoparticles and microparticles for bioactive delivery: fabrication, characterization, and utilization. *Adv. Protein Chem. Struct. Biol.*, **98**, 293–325.

34 Salminen, H. and Weiss, J. (2013) Electrostatic adsorption and stability of whey protein–pectin complexes on emulsion interfaces. *Food Hydrocoll.*, **35**, 410–419.

35 Oppenheim, R.C., Marty, J.J., and Speiser, P. (1978) Injectable compositions, nanoparticles useful therein, and process of manufacturing same. U.S. Patent 4107288A.

36 Duclairoir, C., Nakache, E., Marchais, H., and Orecchioni, A.M. (1998) Formation of gliadin nanoparticles: influence of the solubility parameter of the protein solvent. *Colloid. Polym. Sci.*, **276** (4), 321–327.

37 Kim, S., Ng, W.K., Dong, Y., Das, S., and Tan, R.B.H. (2012) Preparation and physicochemical characterization of *trans*-resveratrol nanoparticles by temperature-controlled antisolvent precipitation. *J. Food Eng.*, **108** (1), 37–42.

38 Zhang, X.P., Le, Y., Wang, J.X., Zhao, H., and Chen, J.F. (2013) Resveratrol nanodispersion with high stability and dissolution rate. *LWT Food Sci. Technol.*, **50** (2), 622–628.

39 Khan, S.A. and Schneider, M. (2013) Improvement of nanoprecipitation technique for preparation of gelatin nanoparticles and potential macromolecular drug loading. *Macromol. Biosci.*, **13**, 455–463.

40 Ezpeleta, I., Irache, J.M., Stainmesse, S., Chabenat, C., Gueguen, J., Popineau, Y., and Orecchioni, A.M. (1996) Gliadin nanoparticles for the controlled release of all-*trans*-retinoic acid. *Int. J. Pharm.*, **131**, 191–200.

41 Duclairoir, C., Orecchioni, A.M., Depraetere, P., Osterstock, F., and Nakache, E. (2003) Evaluation of gliadins nanoparticles as drug delivery systems: a study of three different drugs. *Int. J. Pharm.*, **253**, 133–144.

42 Patel, A.R., Hu, Y.C., Tiwari, J.K., and Velikov, K.P. (2010) Synthesis and characterisation of zein-curcumin colloidal particles. *Soft Matter*, **6**, 6192–6199.

43 Gunasekaran, S., Ko, S., and Xiao, L. (2007) Use of whey proteins for encapsulation and controlled delivery applications. *J. Food Eng.*, **83** (1), 31–40.

44 Zhang, Y., Chwee, T.L., Ramakrishna, S., and Huang, Z.M. (2005) Recent development of polymer nanofibers for biomedical and biotechnological applications. *J. Mater. Sci.*, **16** (10), 933–946.

45 Kriegel, C., Arrechi, A., Kit, K., McClements, D.J., and Weiss, J. (2008) Fabrication, functionalization, and application of electrospun biopolymer nanofibers. *Crit. Rev. Food Sci. Nutr.*, **48** (8), 775–797.

46 Taylor, G. (1969) Electrically driven jets. *Proc. R. Soc. Lond. A Math. Phys. Sci.*, **313** (1515), 453–475.

47 Gombotz, W.R. and Wee, S.F. (1998) Protein release from alginate matrices. *Adv. Drug Deliv. Rev.*, **31** (3), 267–285.

48 Lee, B.B., Ravindra, P., and Chan, E.S. (2013) Size and shape of calcium alginate beads produced by extrusion dripping. *Chem. Eng. Technol.*, **36** (10), 1627–1642.

49 Zeeb, B., Saberi, A.H., Weiss, J., and McClements, D.J. (2015) Formation and characterization of filled hydrogel beads based on calcium alginate: factors influencing nanoemulsion retention and release. *Food Hydrocoll.*, **50**, 27–36.

50 Veerman, C., Sagis, L.M.C., and van der Linden, E. (2003) Gels at extremely low weight fractions formed by irreversible self-assembly of proteins. *Macromol. Biosci.*, **3** (5), 243–247.

51 Adamcik, J. and Mezzenga, R. (2011) Adjustable twisting periodic pitch of amyloid fibrils. *Soft Matter*, **7** (11), 5437–5443.

52 Heger, M., van Golen, R.F., Broekgaarden, M., and Michel, M.C. (2014) The molecular basis for the pharmacokinetics and pharmacodynamics of curcumin and

its metabolites in relation to cancers. *Pharmacol. Rev.*, **66** (1), 222–307.

53 Ahmed, K., Li, Y., McClements, D.J., and Xiao, H. (2012) Nanoemulsion- and emulsion-based delivery systems for curcumin: encapsulation and release properties. *Food Chem.*, **132** (2), 799–807.

54 Aggarwal, B.B., Gupta, S.C., and Sung, B. (2013) Curcumin: an orally bioavailable blocker of TNF and other pro-inflammatory biomarkers. *Br. J. Pharmacol.*, **169** (8), 1672–1692.

55 Zou, L.Q., Zheng, B.J., Liu, W., Liu, C.M., Xiao, H., and McClements, D.J. (2015) Enhancing nutraceutical bioavailability using excipient emulsions: influence of lipid droplet size on solubility and bioaccessibility of powdered curcumin. *J. Funct. Food*, **15**, 72–83.

56 Zeeb, B., Saberi, A.H., Weiss, J., and McClements, D.J. (2015) Retention and release of oil-in-water emulsions from filled hydrogel beads composed of calcium alginate: impact of emulsifier type and pH. *Soft Matter*, **11** (11), 2228–2236.

9
Bottom-Up Approaches in the Design of Soft Foods for the Elderly

José Miguel Aguilera[1] and Dong June Park[2]

[1]*Pontificia Universidad Católica de Chile, Department of Chemical and Bioprocess Engineering, Escuela de Ingeniería, Vicuña Mackenna, 4860 Macul, Santiago 7820436, Chile*
[2]*Korea Food Research Institute, Division of Strategic Food Research, Anyangpangyo-ro 1201beon-gil, Bundang-gu, Seongnam-si, Gyeonggi-do, Seoul 13539-62, Republic of Korea*

9.1 Foods and the Elderly

9.1.1 An Aging Society

By 2020, more than 700 million people will be over 65 years of age; life expectancy at birth will rise from the current 70 years to 77 years by 2045 [1]. It is expected that by 2050, around 2 billion people will be aged 60 and over and in many countries (e.g., Japan, Germany, and South Korea) about 15% of their population will be over 80 years old by that time (Table 9.1). These predictions mean that worldwide by 2050, there will be over 400 million seniors older than 80 years.

Since extreme longevity is new to humankind, food needs and nutritional requirements for elderly people are largely unknown. However, one of the most important challenges to be faced by our society is to provide them with a good quality life, as is recognized by the European Union and several countries alike.

9.1.2 The Elderly and Food-Related Issues

The aging phenomenon has important implications in the types of foods and nutritional needs of older consumers. Within the scope of this presentation, there are some recurrent conditions that make elderly people different in their needs from younger individuals [2]:

- Increasing difficulties in mastication and swallowing the oral contents (dysphagia) caused by a diminishing strength in the jaw and neck muscles, and changes

Nanotechnology in Agriculture and Food Science, First Edition. Edited by Monique A.V. Axelos and Marcel Van de Voorde.
© 2017 Wiley-VCH Verlag GmbH & Co. KGaA. Published 2017 by Wiley-VCH Verlag GmbH & Co. KGaA.

Table 9.1 Percentage of population in age groups over 60 and over 80 years old, by country.

Country	2105		2050	
	+60	+80	+60	+80
Australia	20.4	3.9	28.3	8.3
Austria	24.2	5.1	37.1	12.9
Canada	22.3	4.2	32.4	10.6
China	15.2	1.6	36.5	8.9
France	25.2	6.1	31.8	11.1
Germany	27.6	5.7	39.3	14.4
Japan	33.1	7.8	42.5	15.1
Mexico	9.6	1.5	24.7	5.4
The Netherlands	24.5	4.4	33.2	11.8
Spain	24.4	5.9	41.4	14.0
Republic of Korea	18.5	2.8	41.5	15.9
The United States of America	20.7	3.8	37.9	8.3
World	12.3	1.7	21.5	4.5

Source: Ref. [1].

in the neck anatomy and physiology [3]. This condition implies that foods consumed must yield into a bolus that is soft, cohesive, and moist [4].
- As a consequence of the above, aging people tend to eat foods having a low content of dietary fiber (i.e., fruits and vegetables), making constipation a common disorder and causing changes in the composition of the colonic microbiota [5,6].
- A progressive loss of taste and smell, which is quite unfortunate since flavor is the most important determinant for food choice in the elderly [7].
- A gradual decline in skeletal muscle mass and strength (sarcopenia), meaning that older adults require greater amounts of dietary protein than younger people [8].
- The loss of bone mass (or mineral density) and bone strength (osteoporosis) leading to an increased risk of fracture, particularly among women [8].
- An increased risk of nutritional deficiencies due to a reduced appetite, decreased food intake, and, consequently, malnutrition [9].

Although food products that take into account these requirements are an opportunity for the food industry and a wealth of scientific and technological knowledge is available today for tailor-made products, companies are moving slowly into this market [10]. Figure 9.1 shows how the drivers of food choices – convenience, health and nutrition, and pleasure – may be fulfilled by products aimed at the market of foods for the elderly with special needs. Obviously, the silver target is the black zone in the figure.

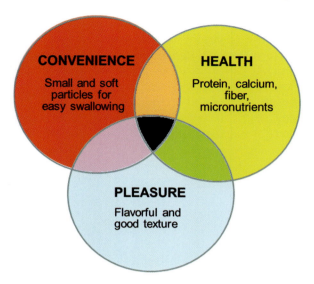

Figure 9.1 Drivers of food choices and requirements of the elderly.

9.1.3
Special Foods for the Elderly: Texture-Modified Foods

The term texture-modified (TM) foods refers to foods that are minced, pureed, or liquidized as well as liquids that are thickened to various extents intended for people with difficulties in swallowing [4,11]. Around 8% of the world's population suffer from an impairment when swallowing foods and need some kind of TM foods [12]. For example, in long-term care facilities for seniors, around 40% of the meals have TM foods and 10% are special-type diets [13]. The subject of texture modification of food for elderly people has been recently reviewed [14]. Initiatives are currently aimed at standardizing and classifying the different textures of TM foods from a sensorial as well as a rheological viewpoint [15]. Relevant to the topic of this chapter are soft, small-particle foods that can thicken liquids, be used as carriers of nutrients and flavor, or added to regular meals to improve their acceptability and specific nutrient contents.

9.2
Rational Design of Soft and Nutritious Gel Particles

9.2.1
Structure and Food Properties

The unique need for tailor-made foods for old people opens the opportunity to "design" foods for the first time in history. The culinary formats and dishes that we normally eat were never really designed, but were the result of trial-and-error attempts of those in charge of food preparation or cooking [16]. In the early 1990s, food scientists started to relate (micro)structural aspects of foods to their desirable (and undesirable) properties, a subject now called food materials science. Soon it became evident that the microstructure (i.e., the structure at scales <100 μm) played a crucial role in the quality of foods and their functionality, that is, texture and appearance, the release of flavors, and the bioavailability of nutrients, among others [17–20]. Recently, it has been demonstrated that the nanolevel (e.g., dimensions <100 nm) conditions the final structure of many food products of our daily life in the form of macromolecular arrangements, aggregates, colloidal networks, interfaces, and nanoparticles [21]. Hence, the concept of a rational design of foods has emerged to address the challenge of linking the structure of foods to their function and impact on pleasure, health, and well-being [22,23]. In particular, products developed for promoting health must ensure that the macro- and micronutrients in the food ingredients are effectively digested and absorbed [24].

9.2.2
Molecular Gastronomy: An Example of Food Design

By the end of the past century, some famous chefs were attracted by the multitude of new ingredients available on the market and the possibility of incorporating scientific knowledge into their dishes. It was the time of molecular gastronomy, a term coined in 1988 by the French chemist Hervé This, which refers to the application of scientific principles in cooking and the preparation of dishes [25]. The practical side was molecular cooking or the development of novel formats and dishes by cooks using refined ingredients (gelling agents, lecithin, etc.), new tools (e.g., syringes, temperature-controlled baths), and novel methods (e.g., freezing with liquid nitrogen, vacuum impregnation). Molecular cooking had two characteristics that set it apart from the traditional (and historical) way used by cooks to make new dishes. First, the dish was projected from a concept or a formalized discourse and it was not the result of chance or serendipity [26]. Second, the design had always in mind the structure to be attained by the molecules, so it was built "bottom-up" from the molecular scale. A good example is the desire by *el Bulli* restaurant to simulate the delicate texture of caviar or the membrane/liquid structure of egg yolk. The answer was found in the interaction of long molecules of alginate with calcium ions to form a soft gelled structure in a process called "spherification" [27]. Interestingly, in the search for foods that are easily chewable and swallowable, Korean scientists have reported that techniques used in molecular gastronomy (e.g., gelation, foam formation, and carbonation) produced novel foods with a very high preference among the elderly, particularly, through the gelation and spherification methods [28].

9.2.3
Nanotechnology and Foods for the Elderly

The subject of nanotechnology and their applications to foods is covered in other sections of this book. Suffice to say that an enormous amount of work has been carried out on food nanotechnology and the subject has been recently discussed in the United States at a prestigious Gordon Conference on Nanoscale Science and Engineering for Agriculture and Food Systems (visit www.grc.org).

Table 9.2 presents examples in which some of the nanotechnologies and food nanomaterials presently available (although at limited stages of practical applications) may be used to accomplish some of the targets for TM foods (Figure 9.1). The most recurrent food nanostructures proposed are solid nanoparticles

Table 9.2 Nanostructures that may be used in the design of TM foods.

Impact	Example	Reference
Better taste	Nanoencapsulaton of drugs and bitter substances in *micelles* and *liposomes* can mask their taste	[30]
Flavor protection	Thin, amorphous *films* (50 nm or less) were applied as a coating to confectionary products to improve shelf life and flavor	[31]
Rheology control	*Nanoemulsions* (220 nm) encapsulated in deformable alginate microgels (40–60 μm) led to a free-flowing suspension	[32]
Microstructure stabilization	*Nanogels* stabilized macroemulsions by the Pickering mechanism	[33]
Protein delivery	Protein *nano- and microparticles* may be added in high concentration (e.g., 20%) to beverages needing heat treatment	[34]
Improved digestibility	*Hydrolyzed* whey protein stimulated protein gains in elderly people due to faster digestion and absorption	[35]
Aid to constipation	Cellulosic *nanofibers* (diameter 100–800 nm) produced by electrospinning may be used as a laxative and to increase fecal mass	[36]
Improved vitamin stability	Whey protein *nanoparticles* protected vitamin D3 to be added to beverages	[37]
Increase in bioavailability	*Nanoemulsions* containing an anti-inflammatory and encapsulated in *nanotubes* (20 nm diameter, micrometer long) showed an increase in oral bioavailability	[38]
Carrier for nutraceuticals	Casein micelles were used as a natural *nanocapsular carrier* for protein, Ca, and the fat-soluble vitamin D2	[39]
Osteoporosis	*Nanoparticles* of Ca carbonate and Ca citrate enhanced the serum calcium concentration and maintained bone mineral density (in rats)	[40]
Cancer prevention	Lycopene in particle sizes of 100 nm was added for color and possible protection against prostate cancer	[31]

(e.g., from protein, polysaccharides, and lipids), nanogels and nanofibers, association colloids (micelles, vesicles, liposomes, liquid crystals, etc.), liquid nanodispersions (e.g., nanoemulsions, nanostructured emulsions, nanosuspensions), and nanocomposites and thin films. Most applications delve into effective encapsulation and delivery systems that contain, protect, and deliver nutrients, functional ingredients, and drugs in their specific sites of action. A particular concern nowadays is bioavailability or the fraction of an ingested nutrient or bioactive substance that is available for utilization in normal physiological functions. Some common foods present a surprisingly low nutrient bioavailability, often improved by processing and formulation. It has been suggested that the man-made matrices can be manipulated at the nano- and microstructural levels to effectively protect nutrients, bioactives, and beneficial microorganisms during digestion and improve their absorption in the gut [29].

9.2.4
Building-Up Healthy Gels with Soft Textures

Dairy products are an excellent example where molecules (whey proteins, enzymes), a nanosized protein aggregate (casein micelles) and a micrometer-sized structural unit (fat globules), suffice to generate a wide variety of products and textures [41]. Equally important, specific characteristics of dairy products may be linked to structures at different scales from nanometers to centimeters. Conceivably, food scientists will become inspired by how these "natural" food structures (i.e., butter, cheese, yogurt, whipped cream, etc.) have come into being when they have to develop novel and functional products "bottom-up."

Microgels are soft, stable, and small particles (e.g., less than a few hundred micrometers in equivalent diameter) that contain a significant amount of water (e.g., >80 by weight) that swells a biopolymer network and whose structure can be adjusted within a wide range of shapes and textural properties. In contrast to bulk gels, a dilute microgel suspension may behave as a free-flowing liquid or as a paste if the volume fraction is high, so it can be mixed with other foods or consumed as such, much in the way cooked starch granules do (e.g., tapioca pearls) [42]. Moreover, microgel particles can be built as carriers of protein and amino acids, lipids, dietary fiber, and other nutrients as well as delivery systems for bioactives and micronutrients (e.g., minerals and vitamins).

Bottom-up approaches for the design and manufacture of soft gel microparticles are based on the controlled assembly of molecules and/or smaller particles or aggregates, as is also the case in the design of nanoscale delivery systems [43]. The sequence and levels of a bottom-up approach for the design of soft gel microparticles is presented in Figure 9.2. All start with the identification of those consumer needs and key benefits that are drivers of product development and added value – in this case, those of an aging population [44]. These requirements condition the selection of ingredients that must contain the structure-building molecules, the nutrients, and required additives (flavorings and flavor potentiators, colorants, preservatives, etc.). Most natural biopolymeric

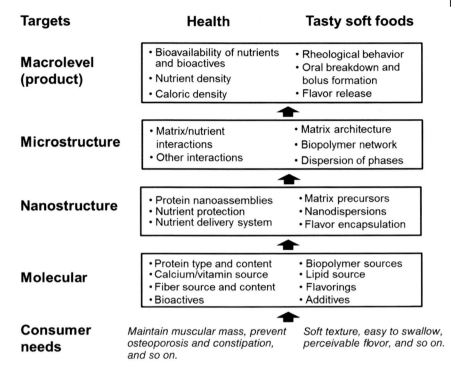

Figure 9.2 Scheme of a bottom-up planning and design of soft gel particles for texture-modified foods.

structurants of foods such as globular proteins and some polysaccharides already have nanodimensions (e.g., β-lactoglobulin in whey is about 3.6 nm in size in the native state). Proteins (e.g., casein, whey proteins, gelatin, etc.) and hydrocolloids (e.g., alginate, agar, gellan gum), alone or in combination, are the preferred food biopolymers to form the matrices of soft gels and a vast knowledge is available regarding their gelling properties and performance [42,45].

At the next level, molecules associate with a multitude of nanostructures at the colloidal scale by self-assembly mechanisms or are driven by physical interactions or chemical reactions. Some of these transient or permanent entities may include protein aggregates and nanofibrils, protein–polysaccharide complexes and coacervates, micelles, vesicles, liposomes and nanoemulsions from surfactants, and nanoplatelets and lipid nanoparticles assembled from triacylglycerol molecules, among others. Think of the passage from casein micelles of about 300 nm in size to the long strands forming the gel network of cheese curd or the association between denatured globular proteins with casein micelles to form coaggregates that are responsible for a stiff yogurt.

Arrangements at the microstructural level (in a scale of micrometers) are responsible for stability and mechanical and flow properties of the particles, and largely determine their textural and sensorial properties as well as their applications. Possible chemical or steric interactions may occur at this level between nutrients and the matrix or between bioactives and other components of the dispersed phase (e.g., fiber particles), which may hinder their bioavailability. Further processing may be necessary at the macroscale to ensure food safety (e.g., thermal treatments) or to control particle size and shape. At the top of Figure 9.2 emerges the target product with its properties: good texture and great taste as well as a beneficial nutritional value.

9.3
Technological Alternatives for the Design of TM Foods

As already mentioned, TM foods range from foods that are soft *per se* (yogurt, mousses, mashed potatoes, etc.) to liquids like thin juices. Several known technologies, beyond the scope of this paper, such as high-pressure processing, pulsed electric fields, freeze-thawing, and so on have been proposed to soften the structure of tissue foods such as meats, fruits, and vegetables, while largely preserving the color and flavor.

Figure 9.3 shows some of the technologies that have been used to make soft gel particles. A detailed description of several of these technologies is available elsewhere [46]. Enzyme hydrolysis of proteins not only changes the mechanism of gel formation and modulates the gel properties but also improves the gastrointestinal absorption of peptides compared to the intact protein. Although extensive protein breakdown normally impairs gelation, limited hydrolysis of whey protein followed by mixing with a hydrocolloid resulted in enhanced gel strength [47]. It was also proven that gelation of extensively hydrolyzed whey proteins is possible if the selected enzyme is alcalase [48]. Interesting from the

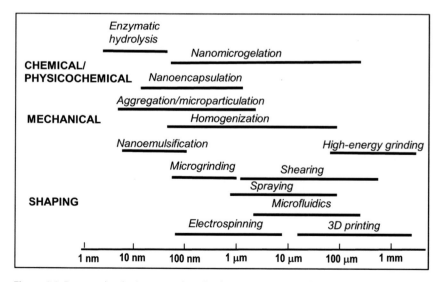

Figure 9.3 Some technologies to produce food nanostructures and microstructures.

viewpoint of increasing the fiber content in soft foods, strong gels were formed by a network of highly entangled cellulose nanofibrils produced following a mild enzymatic hydrolysis of cellulose combined with mechanical shearing and high-pressure homogenization [49].

Microparticulated protein based on heating-aggregation–high-shear procedures has been around commercially for some time and used to modulate the texture of high-protein foods, especially that of heat-processed beverages, and as fat replacers. Size (100 nm to micrometers) and morphology (spherical, fibrillar strands) of microparticles depend on protein concentration, heat treatment, pH, and the ionic environment [50]. Novel applications of protein micro- and nanoparticles are in high-protein drinks as an ingredient to modify the texture of semisolid foods and stabilize emulsions, and as a carrier and delivery medium for bioactives.

Food nanogels and microgels belong to the class of gel particulates, the frontier between them being a size of around 100 nm. The subject of formation and characterization of both types of gels has been reviewed [50,51]. Among the various mechanisms to form protein gel particles are (i) homogenization of a biopolymer solution into a water-in-oil emulsion followed by gelation of the biopolymer phase by ions or temperature [52]; (ii) exposing a spray of a biopolymer solution (e.g., alginate) into a countercurrent mist of a $CaCl_2$ solution [53]; (iii) self-aggregation into spherical units (200–400 nm) after heating a isoelectric globular protein solution beyond denaturation temperature and subsequent cooling to 4 °C [33], and (iv) size reduction of bulk gels by high-shear mechanical means [54]. Hydrocolloid gel particles of micrometer and submicrometer size are particularly attractive due to their soft texture and are finding uses as structuring agents to strengthen dispersed phases and as thickeners in soups and sauces. These microparticles are usually formed while gelling under shear or by gelation of preformed droplets [55].

Among the top-down methods to produce soft particles in the nano- and microrange is size reduction by mechanical means, commonly referred to as homogenization. There are several high-shear devices that accomplish this task such as rotor–stator mixers (e.g., the Ultra-turrax), high and ultrahigh pressure valve homogenizers, ultrasonicators, and microfluidizers. These types of equipment are routinely used to prepare macroemulsions and those that deliver a high-energy input may be used to form nanoemulsions. There are also low-energy approaches to form nanoemulsions that rely on the spontaneous formation of tiny oil droplets by phase inversion or spontaneous emulsification [56].

Mention should be made of emerging technologies that may find innovative applications in the design and fabrication of TM foods. Microfluidics utilize devices where small amounts of fluids flow in channels with at least one dimension on the order of a few hundred micrometers or less. Systems have been developed to generate emulsions, foams, and even produce monodisperse alginate gel droplets of about 100 μm [57,58]. Three-dimensional printing refers to rapid prototyping techniques based on digitally controlled depositing of materials layer-by-layer. Both liquid deposition and powder binding are capable of building geometrically complex food structures from "printable" mixtures of carbohydrates, proteins, and lipids [59]. Electrospinning uses a high-voltage electric field to produce electrically charged jets of a biopolymer solution that becomes nanofibers upon evaporation of the solvent [60]. Thin electrospun protein fibrils (e.g., <1 μm in diameter) are finding applications in the encapsulation of probiotics and bioactives, as dietary supplements, and to impart texture and mouthfeel to foods [61].

9.4 Conclusions

The elderly population and the life expectancy will continue to increase worldwide, as will our knowledge of food-related problems derived from aging. By 2050, around 400 million people will be over 80 years old. At present, our understanding of the aging process is limited, but links between foods and the quality of late life start to appear. Seniors needing special foods due to physiological dysfunctions (e.g., dysphagia, constipation) and/or requirements for special nutrition should be offered soft foods that are palatable and tasty as well as nutritious and healthy. This circumstance gives food technologists the unique opportunity to design texture-modified foods bottom-up, as molecular gastronomy chefs did. Abundant science knowledge and technology expertise is already available and accumulating at a fast pace in food nanoscience and food materials science. Soft gel microparticles with sizes <500 μm are an attractive option as TM foods since they could harbor in their interior key nutrients and bioactives as well as adequate amounts of odorants and tastants (Figure 9.4). Small soft and tasty particles may be used to thicken liquids, modify the texture of sauces and

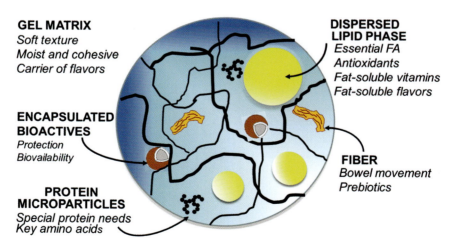

Figure 9.4 Idealized scheme of a gel microparticle designed to satisfy nutritional and sensory needs of elderly people.

purees, or simply be added to season regular dishes. An interface between food and pharma is conceivable when functional ingredients, nutritional supplements, and perhaps some drugs become added to these microgel carriers. In a futuristic perspective, we may imagine elder people formulating and making these soft particles at home in a 3D printer!

Acknowledgments

Financial support from the Korea Food Research Institute (KFRI) through a contract with the Pontificia Universidad Católica de Chile is appreciated.

References

1. United Nations (2015) World Population Prospects: The 2015 Revision, Key Findings and Advance Tables. Department of Economic and Social Affairs, Population Division. Working Paper No. ESA/P/WP.241.
2. Raats, M.M., de Groot, C.P.G.M., and van Staveren, W.A. (2009) *Food for the Ageing Population*, Woodhead Publishers, Cambridge.
3. Humbert, I.A. and Robbins, J. (2008) Dysphagia in the elderly. *Phys. Med. Rehabil. Clin. N. Am.*, **19**, 853–866.
4. Cichero, J.A.Y. (2015) Texture-modified meals for hospital patients, in *Modifying Food Texture, Volume 2: Sensory Analysis, Consumer Requirements and Preferences* (eds J. Chen and A. Rosenthal), Woodhead Publishing, Cambridge, pp. 135–162.
5. Gallegos-Orozco, J.F., Foxx-Orenstein, A.E., Sterler, S.M., and Stoa, J.M. (2012) Chronic constipation in the elderly. *Am. J. Gastroenterol.*, **107**, 18–25.
6. Donini, L.M., Savina, C., and Cannella, C. (2009) Nutrition in the elderly: role of fiber. *Arch. Gerontol. Geriatr.*, **49** (Suppl. 1), 61–69.
7. Nordin, S. (2009) Sensory perception of food and ageing, in *Food for the Ageing Population* (eds M.M. Raats, C.P.G.M. de Groot, and W.A. van Staveren), Woodhead Publishing, Cambridge, pp. 73–94.
8. Bauer, J., Biolo, G., Cederholm, T., Cesari, M., and Cruz-Jentoft, A.J. *et al.* (2013) Evidence-based recommendations for optimal dietary protein intake in older people: a position paper from the PROT-AGE Study Group. *J. Am. Med. Dir. Assoc.*, **14**, 542–559.
9. Cederholm, T. (2009) Nutrition and bone health in the elderly, in *Food for the Ageing Population* (eds M.M. Raats, C.P.G.M. de Groot, and W.A. van Staveren), Woodhead Publishing, Cambridge, pp. 252–270.
10. Scott-Thomas, C. (2012) http://www.foodnavigator.com/Science/R-D-challenge-Developing-texture-modified-foods-for-the-elderly (accessed September 21, 2015).
11. International Dysphagia Diet Standardisation Initiative (2016) www.iddsi.org (accessed February 16, 2016).
12. Cichero, J.A.Y., Steele, C., Duivestein, J., Clavé, P., Chen, J., Kayashita, J., Dantas, R., Lecko, C., Speyer, R., Lam, P., and Murray, J. (2013) The need for international terminology and definitions for texture-modified foods and thickened liquids used in dysphagia management: foundations of a global initiative. *Curr. Phys. Med. Rehabil. Rep.*, **1**, 280–291.
13. The International Union of Food Science and Technology (2014) http://www.iufost.org/iufostftp/IUF.SIB.Meeting%20the%20Food%20Needs%20of%20the%20Ageing%20Population.pdf (accessed October 14, 2014).

14 Rothenberg, E. and Wendin, K. (2015) Texture modification of food for elderly people, in *Modifying Food Texture, Volume 2: Sensory Analysis, Consumer Requirements and Preferences* (eds J. Chen and A. Rosenthal), Woodhead Publishing, Cambridge, pp. 164–185.

15 Wendin, K., Ekman, S., Bülow, M., Ekberg, O., Johansson, D., Rothenberg, E., and Stading, M. (2010) Objective and quantitative definitions of modified food textures based on sensory and rheological methodology. *Food Nutr. Res.*, **54**, 5134–5145.

16 Aguilera, J.M. (2013) *Edible Structures: The Basic Science of What We Eat*, CRC Press, Boca Raton, FL.

17 Aguilera, J.M. (2005) Why food microstructure? *J. Food Eng.*, **67**, 3–11.

18 Aguilera, J.M. and Stanley, D.W. (1999) *Microstructural Principles of Food Processing and Engineering*, 2nd edn, Aspen Publishers Inc., Gaithersburg, MD.

19 Axelos, M. (2013) Du nano au macro quelles structures pour quelles fonctionnalites? in *Conception Raisonnée des Aliments* (eds C. Michon and J.-P. Canselier), Société Chimique de France, Paris, pp. 102–111.

20 Heertje, I. (2014) Structure and function of food products: a review. *Food Structure*, **1**, 3–23.

21 Aguilera, J.M. (2014) Where is the nano in our foods? *J. Agric. Food Chem.*, **62**, 9953–9956.

22 Lesmes, U. and McClements, D.J. (2009) Structure–function relationships to guide rational design and fabrication of particulate food delivery systems. *Trends Food Sci. Technol.*, **20**, 448–457.

23 Palzer, S. (2009) Food structures for nutrition, health and wellness. *Trends Food Sci. Technol.*, **20**, 194–200.

24 Lundin, L. and Golding, M. (2009) Structure design for healthy food. *Aust. J. Dairy Technol.*, **64** (1), 68–74.

25 This, H. (2008) *Molecular Gastronomy: Exploring the Science of Flavor*, Columbia University Press, New York.

26 Opazo, M.P. (2012) Discourse as driver of innovation in contemporary haute cuisine: the case of elBulli restaurant. *Int. J. Gastron. Food Sci.*, **1** (2), 82–89.

27 Vega, C. and Castells, P. (2012) Spherification, in *The Kitchen as the Laboratory* (eds C. Vega, J. Ubbink, and E. van der Linden), Columbia University Press, New York, pp. 25–32.

28 Kim, S. and Joo, N. (2015) The study on development of easily chewable and swallowable foods for elderly. *Nutr. Res. Pract.*, **9** (4), 420–424.

29 Parada, J. and Aguilera, J.M. (2007) Food microstructure affects the bioavailability of several nutrients. *J. Food Sci.*, **72**, R21–32.

30 Neethirajan, S. and Jayas, D.S. (2011) Nanotechnology for the food and bioprocessing industries. *Food Bioprocess Technol.*, **4**, 39–47.

31 Chaudhry, Q., Scotter, M., Blackburn, J., Ross, B., Boxall, A., Castle, L., Aitken, R., and Watkins, R. (2008) Applications and implications of nanotechnologies for the food sector. *Food Addit. Contam.*, **25** (3), 241–258.

32 Ching, S.H., Bansal, N., and Bhandari, B. (2016) Rheology of emulsion filled alginate microgel suspension. *Food Res. Intern.*, **80**, 50–60.

33 Schmitt, C., Bovay, C., and Rouvet, M. (2014) Bulk self-aggregation drives foam stabilization properties of whey protein microgels. *Food Hydrocolloid*, **42**, 139–148.

34 Saglam, D., Venema, P., van der Linden, E., and de Vries, R. (2014) Design, properties and applications of protein micro- and nanoparticles. *Curr. Opin. Colloid Interface*, **19**, 428–437.

35 Pennings, B., Boirie, Y., Senden, J.M.G., Gijsen, A.P., Kuipers, H., and van Loon, L.J.C. (2011) Whey protein hydrolysate stimulates postprandial muscle protein accretion more effectively than do casein and casein hydrolysate in older men. *Am. J. Clin. Nutr.*, **93**, 997–1005.

36 Rezaei, A., Nasirpour, A., and Fathi, M. (2015) Application of cellulosic nanofibers in food science using electrospinning and its potential risk. *Compr. Rev. Food Sci. Food Saf.*, **15**, 269–284.

37 Abassi, A., Emam-Djomeh, Z., Mousavi, M.A.E., and Davoodi, D. (2014) Stability of vitamin D_3 encapsulated in nanoparticles of whey protein isolate. *Food Chem.*, **143**, 379–383.

38 Graveland-Bikker, J.F. and de Kruif, C.G. (2006) Unique milk protein based nanotubes: food and nanotechnology meet. *Trends Food Sci. Technol.*, **175**, 196–203.

39 Semo, E., Kesselman, E., Danino, D., and Livney, Y.D. (2007) Casein micelle as a natural nanocapsular vehicle for nutraceuticals. *Food Hydrocolloid*, **21**, 936–942.

40 Huang, S., Chen, J.C., Hsu, C.W., and Chang, W.C. (2009) Effects of nano calcium carbonate and nano calcium citrate on toxicity in ICR mice and on bone mineral density in an ovariectomized mice model. *Nanotechnology*, **20** (37), 375102.

41 Aguilera, J.M. (2006) Food product engineering: building the right structures. *J. Sci. Food Agric.*, **86**, 1147–1155.

42 Dickinson, E. (2015) Microgels: an alternative colloidal ingredient for stabilization of food emulsions. *Trends Food Sci. Technol.*, **43**, 178–188.

43 McClements, D.J. (2015) Nanoscale nutrient delivery systems for food applications: improving bioactive dispersibility, stability, and bioavailability. *J. Food Sci.*, **80**, N1602–1611.

44 Costa, A.I.A. and Jongen, W.M.F. (2010) Designing new meals for an ageing population. *Crit. Rev. Food Sci.*, **50**, 489–502.

45 Funami, T., Ishihara, S., Nakauma, M., Kohyama, K., and Nishinari, K. (2012) Texture design for products using food hydrocolloids. *Food Hydrocolloid.*, **26**, 412–420.

46 Re, M.I., Andrade Santana, M.H., and d'Avila, M.A. (2010) Encapsulation technologies for modifying food performance, in *Innovations in Food Engineering* (eds M.L. Passos and C.P. Ribeiro), CRC Press, Boca Raton, FL, pp. 223–270.

47 Rocha, C., Teixeira, J.A., Hilliou, L., Sampaio, P., and Gonçalves, M.P. (2009) Rheological and structural characterization of gels from whey protein hydrolysates/locust bean gum mixed systems. *Food Hydrocolloid.*, **23**, 1734–1745.

48 Doucet, D., Gauthier, S.F., and Foegeding, E.A. (2001) Rheological characterization of a gel formed during extensive enzymatic hydrolysis. *J. Food Sci.*, **66** (5), 711–715.

49 Pääkkö, M., Ankerfors, M., Kosonen, H., Nykänen, A., Ahola, S., Österberg, M., Ruokolainen, J., Laine, J., Larsson, P.T., Ikkala, O., and Lindström, T. (2007) Enzymatic hydrolysis combined with mechanical shearing and high-pressure homogenization for nanoscale cellulose fibrils and strong gels. *Biomacromolecules*, **8** (6), 1934–1941.

50 Saglam, D., Venema, P., van der Linden, E., and de Vries, R. (2014) Design, properties and applications of protein micro- and nanoparticles. *Curr. Opin. Colloid Interface Sci.*, **19**, 428–437.

51 Stokes, J.R. (2012) Food biopolymer gels, microgel and nanogel structures, formation and rheology, in *Food Materials Science and Engineering* (eds B. Bhandari and Y. Roos), Wiley-Blackwell, Chichester, pp. 151–176.

52 Egan, T., Jacquier, J.-C., Rosenberg, Y., and Rosenberg, M. (2013) Cold-set whey protein microgels for the stable immobilization of lipids. *Food Hydrocolloid.*, **31**, 317–324.

53 Sohail, A., Bhandari, B., Turner, M.S., and Coombes, A. (2012) Direct encapsulation of small molecule hydrophilic and hydrophobic actives in micron size alginate microspheres using a novel impinging aerosol method. *J. Drug Deliv. Sci. Technol.*, **22**, 139–143.

54 Leon, A.M., Medina, W.T., Park, D.J., and Aguilera, J.M. (2016) Mechanical properties of whey protein/Na alginate gel microparticles. *J. Food Eng.*, **188**, 1–7.

55 Burey, P., Bhandari, B.R., Howes, T., and Gidley, M.J. (2008) Hydrocolloid gel particles: formation, characterization and application. *Crit. Rev. Food Sci. Nutr.*, **48**, 361–377.

56 McClements, D.J. and Rao, J. (2011) Food-grade nanoemulsions: formulation, fabrication, properties, performance, biological fate, and potential toxicity. *Crit. Rev. Food Sci. Nutr.*, **51**, 285–330.

57 Amici, E., Tetradis-Meris, G., Pulido de Torres, C., and Jousse, F. (2008) Alginate

gelation in microfluidic channels. *Food Hydrocolloid.*, **22**, 97–104.

58 Skurtys, O. and Aguilera, J.M. (2008) Applications of microfluidic devices in food engineering. *Food Biophys.*, **3**, 1–15.

59 Godoi, F.C., Prakash, S., and Bhandari, B.S. (2016) 3D printing technologies applied for food design: status and prospects. *J. Food Eng.*, **179**, 44–54.

60 Nieuwland, M., Geerdink, P., Brier, P., van den Eijnden, P., Henket, J.T.M.M., Marloes, M.L.P., Stroeks, N., van Deventer, H.C., and Martin, A.H. (2013) Food-grade electrospinning of proteins. *Innov. Food Sci. Emerg. Technol.*, **20**, 269–275.

61 Ghorani, B. and Tucker, N. (2015) Fundamentals of electrospinning as a novel delivery vehicle for bioactive compounds in food nanotechnology. *Food Hydrocolloid.*, **51**, 227–240.

10
Barrier Nanomaterials and Nanocomposites for Food Packaging

Jose M. Lagaron,[1] Luis Cabedo,[2] and Maria J. Fabra[1]

[1]IATA-CSIC, Novel Materials and Nanotechnology Group, Av. Agustin Escardino 7, 46980 Paterna, Valencia, Spain
[2]Polymers and Advanced Materials Group (PIMA), Universidad Jaume I, 12071, Castellon, Spain

10.1
Introduction

The principal function of packaging is protection and preservation of the content from external spoiling factors. This function involves retardation of deterioration, extension of shelf life, and maintenance of quality and safety of packaged foods. Packaging protects foods from environmental influences such as heat, light, the presence or absence of moisture, oxygen, pressure, enzymes, spurious odors, microorganisms, insects, dirt and dust particles, gaseous emissions, and so on.

Using polymers to produce food packages presents many advantages. They are light and cheap materials and can even be microwavable. The optical properties of the package can also be adapted to the specific requirements of each product. Transparent packages allow the consumer to see the product. In addition, many polymers are printable. Furthermore, polymeric packages can be produced as part of integrated processes where they are formed, filled, and sealed in the same production line, making the process quick and cheap and can also be formed into a myriad of sizes and shapes. Polymers have however some drawbacks such as waste management issues and transport of low-molecular-weight components. These environmental drawbacks are mainly related to the extremely slow degradation rate of the most commonly used polymers and to the fact that most of them are petroleum-derived products. Although package stability during the shelf life of a product is an advantage, it turns into a disadvantage when the packages are rarely reused or recycled and the used containers generate huge volumes of residues. This problem has been attenuated by the creation and improvement of recycling systems and by the implementation of biodegradable/compostable materials. The other main drawback of polymers for food packaging applications is that they are all permeable to the transport of low-molecular-weight compounds. Multilayer structures and blending with other polymers and fillers have long been used as solutions to this problem. Until recently, the most interesting plastic packaging technology based on blending to generate barrier properties was the so-called active oxygen scavengers. This technology is known to lead to relatively low levels of oxygen in contact with the food because it traps permeated oxygen from both the headspace and the outside. However, in carbonated beverages, for instance, a barrier to carbon dioxide is also a requirement. As most commodity plastic packaging materials, for example, PET and its main sustainable counterpart the PLA, are not sufficient barrier to these gases, multilayer structures had to be devised in which one layer (made of EVOH, MXD6, PEN, nanocomposites of PA6, etc.) needs to be high barrier to carbon dioxide and to oxygen while the scavenger reduces oxygen levels at the packaging headspace. In this respect, the latest more interesting developments are in our view coming from the recent scaling up of nanocomposites making use of, for instance, clay-based nanofillers and electrospun nanostructured layers and biolayers.

10.2
Nanocomposites

Combining two different materials in order to obtain a composite with enhanced behavior for a particular application is a very convenient technique that can be easily found in nature, and has been used by our ancestors almost since the beginning of humanity. However, the use and development of composite materials did not progress until last century, when polymers and new processing methods made possible their blossom. The composite technology, during the last 40 years, has enabled the development of materials with very interesting combinations of properties that cannot be achieved by any material on its own. This fact has supposed a revolution in materials science, pushing the boundaries of what could be expected in materials performance until then.

A traditional definition of a composite material is a material artificially obtained and formed by at least two chemically different materials separated by an interface. In a traditional composite, there is a continuous phase known as matrix and a discontinuous or disperse phase, and the final properties of the composite are generally, in the best case, a trade-off between the properties of the components. However, it has been widely reported that the interaction at the interface between the different phases present in a composite is a key factor in the performance of the resulting composite. Moreover, it is also known that the size of the disperse phase can affect the final properties of the composite. In this sense, a reduction in size of the disperse phase can lead to a larger effect of the disperse phase on the properties of the matrix. This effect has been known for centuries; however, new materials and processing techniques in recent years have enabled the possibility of reducing the size of the disperse phase to the nanometer scale and in that case the experimental results do not follow the models. This change in the behavior of composites when the nanometer range is reached (the so-called nanoeffect) is what makes nanotechnology and nanomaterials very interesting. Actually, nowadays the scientific world is undergoing a nanorevolution in almost all the fields of science and engineering, and food packaging is not an exception [1,2].

Although there are examples of natural occurring nanocomposites (e.g., bones or wood) and nanocomposite materials obtained long time ago [3], research in nanocomposites is relatively new: The first nanocomposite was obtained by Toyota in 1986 [4] reporting that mechanical, thermal, and barrier properties of nylon–clay composite materials improved to a significant extent by reinforcement with less than 5 wt% of nanoclay. Since then the amount of scientific papers published every year has increased exponentially, revealing the great interest aroused in the scientific community (see Figure 10.1). The reason why nanocomposites were not discovered until 25 years ago is the lack of techniques for both obtaining and characterizing nanomaterials.

The reason behind this interest is that below the micrometer range, the bulk properties of the disperse phase (e.g., elastic modulus, yield strength, thermal conductivity, or electrical resistivity) cease to rule the final behavior of the resulting composite, and other properties (mainly surface properties) come into play (e.g., chemical affinity, the isoelectric point, or polarity). Therefore, the traditional models do not apply for these types of systems, and this enables the achievement of new interesting combinations of properties. Among the most interesting things that the use of nanoadditives can provide is the achievement

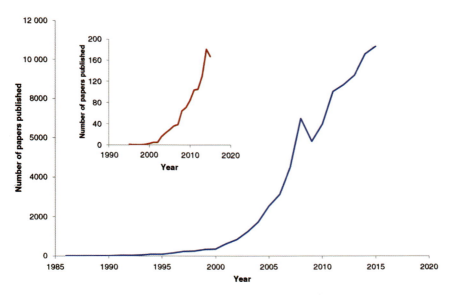

Figure 10.1 The number of papers published per year on nanocomposites. *Inset:* The number of papers published per year on nanocomposites and packaging. (*Source:* Scopus.)

of the effect of the addition of high loading of microadditives with much lower amount of additives; for instance, Giannelis first reported an increase in the gas barrier properties of a polymer with nanoparticles compared to a conventional microfiller, showing that in order to achieve an equivalent improvement with microfiller, 10 times more amount of filler was needed [5]. Another relevant innovation of the nanoadditives with respect to the conventional composites is that under determined circumstances, the use of nanoparticles instead of microparticles does not result in the undesirable decrease of other properties or, at least, not to the same extent. In this sense, Okada *et al.* [6] reported, for the first time, a reinforcing effect of montmorillonite nanolayers in a nylon without a decrease in the impact resistance, which would be impossible to reach with conventional polymer reinforcement. Therefore, these new nanoscale structures displaying a high surface-to-volume ratio may become ideal for applications that involve composite materials, chemical reactions, drug delivery, controlled and immediate release of substances in active and functional packaging technologies, and energy storage, for instance, in intelligent packaging [7].

A commonly accepted definition of nanocomposites could be a composite material in which at least one dimension of the dispersed phase is in the nanometer range [8]. However, there is a consensus in the scientific community that the nanoeffects take place for dimensions below 100 nm. Thus, one can distinguish three types of nanocomposites, depending on how many dimensions of the dispersed particles are in the nanometer range (below 100 nm). When all three dimensions are below 100 nm, the nanoparticles are called isodimensional (e.g., spherical silica nanoparticles, semiconductor nanoclusters, or fullerenes). When there are two dimensions below 100 nm and the third is above, we talk about nanotubes or nanowhiskers. The nanoparticles with just one dimension in the nanometer range, being the other two much larger, are known as layered. In the field of packaging, examples of the three kinds of nanocomposites can be found (e.g., nanoparticles of metals and oxides, carbon nanotubes, and more frequently nanocellulose). However, the most common and interesting are nanocomposites comprising a thermoplastic matrix and a layered nanoparticle. Different layered nanoparticles have been used in the recent years as nanofillers in polymer-layered nanocomposites (graphene and graphene oxide, ceramic nanodisks, layered double hydroxides, metal chalcogenides, etc.); nevertheless, the most widely studied and technologically relevant polymer nanocomposites are those using a clay mineral (so-called nanoclays) as the nanoparticle, due to their availability, low cost, significant enhancements, and relatively simple processability [9].

Nanoclay is the common name used for the clay materials used as nanoadditives to obtain nanocomposites or nanofluids. However, the term nanoclay or even the term clay can be a fairly vague concept, thus a definition is necessary. The Clay Minerals Society (CMS), together with the Association Internationale pour l'Etude des Argiles (AIPEA), define a clay as follows: "The term 'clay' refers to a naturally occurring material composed primarily of fine-grained minerals, which is generally plastic at appropriate water contents and will harden when

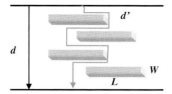

Figure 10.2 Tortuosity of a diffusion path.

dried or fired. Although clay usually contains phyllosilicates, it may contain other materials that impart plasticity and harden when dried or fired. Associated phases in clay may include materials that do not impart plasticity and organic matter" [10]. There are many kinds of laminar nanoclays that have been used as nanofillers in packaging applications; however, phyllosilicates and more specifically montmorillonite are the most widely used [11].

There is extensive scientific literature about polymer-layered nanocomposites, both from a general point of view and for the specific field of packaging. Some good reviews on the field have been published in the last years [7,9,12–20]. There is a common agreement that probably the most interesting advantage of the polymer-layered composites with respect to the pristine polymers is the considerable increase in the barrier properties of the resin when the nanoparticles are incorporated. The incorporation of high-aspect-ratio nanoparticles in a polymer matrix results in an increase in the tortuosity in the path that the permeants have to cover in order to diffuse across the polymer layer (see Figure 10.2). There are several models in order to predict the barrier properties of polymer-layered nanocomposites [21–23], but all of them consider this tortuosity factor as a key parameter and highly dependent on the degree of dispersion of the nanolayers within the polymer matrix.

Hence, the main issue in processing of polymer laminar nanocomposites is to achieve the sufficient interaction between the nanofiller and the polymer so as to achieve a favorable morphology. In this sense, the morphology of the nanocomposites can be classified into three different types, depending on the structure of the nanoclays and the interaction with the polymer chains: aggregated, intercalated, and exfoliated.

The aggregated morphology occurs when the polymer chains are not able to break apart the stacking of the clay and, therefore, small aggregates are found in the composite. If these aggregates are above the nanometer range, the composite is no longer a nanocomposite, but a microcomposite. When the size of the aggregates is within the nanometer range and the aspect ratio (length-to-thickness ratio) is still very high, they can act almost as a single nanofiller, being thus a nanocomposite.

In the intercalated structures, one or several polymer chains are inserted in the interlayer gallery, thus keeping some kind of parallel order of the clay layers but with a high degree of interaction with the polymer. Sometimes, this intercalated structure can be found in combination with aggregated structures of small amount of layers forming structures with a high aspect ratio, which are frequently called tactoids.

The exfoliated state takes place when the dispersion of the sheets of the clay is fully achieved and, therefore, no interaction between sheets occurs. The ideal exfoliated state is that in which the clay platelets are homogeneously distributed throughout the polymer matrix.

Generally speaking, layered nanocomposites never exhibit a pure structure; on the contrary, their morphology is a combination of all three possibilities with a dominant one.

10.3
Nanostructured Layers

Depending on the final intended use, the requirements on barrier properties for food packaging may include protection against oxygen, aromas, water vapor, and/or grease/oils. From the industrial point of view, barrier coatings are typically based on synthetic polymers due to their low cost and availability. For instance, ethylene vinyl alcohol (EVOH) possesses excellent barrier properties to oxygen, oils, and aroma in dry conditions, but due to its hygroscopic nature, this polymer absorbs water under high relative humidity conditions and so it loses its excellent barrier properties. Therefore, in the current market, this layer is usually coated with an additional layer of polyolefins to provide a water barrier. However, in light of the environmental concern on generated waste when using conventional polymers, biodegradable and compostable materials are being extensively investigated as a green alternative for synthetic polymers.

In general, biodegradable polymers present inferior barrier and thermal properties than their petroleum-based counterparts and thus, there is an industrial interest to enhance their barrier properties while maintaining their inherently good properties of transparency and biodegradability. Biodegradable materials such as polysaccharides (i.e., starch, chitosan), proteins (i.e., gluten), lipids (i.e., waxes and fatty acids), and polyesters (i.e., polyhydroxyalkanoates (PHAs) or polylactic acid (PLA)) are being used as packaging constituents, although they are mostly blended in combination with other biopolymers to form biodegradable composite layers with improved oxygen, aroma, and water vapor barrier properties.

It is worth mentioning that the main challenge in the development of biodegradable nanolayers for food packaging materials is to develop biodegradable interlayer bioadhesives that favor the adhesion between layers; this can be achieved by means of nanotechnology. For instance, Fabra *et al.* [24] developed high oxygen barrier structures, even at high relative humidity conditions, by combining layers of biopolyesters with bioadhesive interlayers based on proteins and polysaccharides. To this end, bioadhesives were developed by electrospinning. Specifically, Fabra *et al.* [24] demonstrated that biopolyester-based multilayers prepared with electrospun zein interlayers significantly reduced oxygen permeability values (by up to 70%) of PLA and PHA biopolymers, although the water vapor barrier properties depended on the biopolymers used as outer layers. As opposed to the PHA materials performance, PLA did not improve the water vapor permeability values of these films. Furthermore, Fabra et al. [25] pointed out that fibrillar structures obtained for zein and pullulan electrospun interlayers significantly contributed to improve the barrier performance of biopolyester-based multilayer systems, whereas beaded microstructures obtained by electrospraying whey protein isolate did not improve the barrier performance.

For hydrophilic compounds such as polysaccharides (i.e., thermoplastic starch) and proteins (i.e., wheat gluten), nanostructured coatings can provide certain advantages compared to polymer blends [26,27]. For instance, moisture sensitivity of these hydrophilic polymers is not completely protected in blends due to the fact that the phase distribution is close to the surface. However, the electrospinning process can also be used to develop continuous nanostructured layers to be used as coatings to avoid the moisture uptake of hydrophilic-based proteins and polysaccharide thermoplastic materials. It has been proved that this processing technology improved the adhesion between layers, avoiding partial or even a complete delamination between the layers, which has been reported to occur when they are assembled without synthetic adhesives [24,25,28]. Besides, barrier properties of packaging materials can be even improved by encapsulating nano-reinforcing agents (i.e., bacterial cellulose nanowhiskers (BCNW)) into electrospun coatings. In this regard, taking advantage of the preincorporation method recently developed by Martínez-Sanz et al. [28] in which enhanced barrier and mechanical properties were obtained when BCNW was incorporated into biodegradable matrices by means of the electrospinning process, Fabra et al. [29] showed that the incorporation of BCNW in one of the layers of a multilayer system led to a decrease in the oxygen permeability values. Therefore, nanocomposite coatings based on biodegradable matrices in combination with nanocellulose fillers and even other nanofillers (i.e., nanoclays) can be used to develop coating barrier layers. From the results reported in the literature, it can be concluded that the electrospinning process can be useful to tailor the barrier properties of food packaging materials according to the final intended use. This type of technology to produce nanostructured interlayers and coatings has been recently scaled up by a company called Bioinicia S.L., Paterna, Spain (www.bioinicia.com).

Another important tool to produce nanoscopically structured layers is the sequential layer-by-layer adsorption of polymers on solid substrates [30,31]. By means of this technique, several biopolymers with opposite charges can be used to produce nanolayered coatings assembled layer-by-layer. For instance, two polysaccharides with opposite charges, chitosan and sodium alginate, were deposited onto aminolyzed/charged PET (A/C PET), and the resulting multilayer nanofilm showed lower water permeability than what could be expected for the materials with which the film was built [32]. These satisfactory results were explained based on different interactions that are established between adjacent alginate and chitosan layers, which may increase the tortuosity of the material, thus decreasing its effective permeability to water molecules. Pinheiro et al. [30] also showed that both the water and oxygen permeabilities of the nanolayered structures prepared by alternating κ-carrageenan and chitosan nanolayers were apparently lower than those from the control A/C PET material and were even estimated to be better than those of films made from the individual materials. Thus, these results are encouraging and suggest the potential of nanolayered biopolymer combinations for improving food quality and extending shelf life.

10.4
Conclusion and Future Prospects

Nanotechnology is currently offering unique possibilities that have been scaled up to design more efficient passive high barrier technologies, which take tremendous advantages in the high surface-to-volume ratios of nanomaterials such as nanoclays, nanocellulose, nanostructured layers, and others in composite, interlayer, or coating forms. From a food safety viewpoint, a complete control and understanding over the potential migration of nanomaterials and/or of their modifiers needs be assessed and proven to legislators to design safe high barrier packaging.

References

1 Chaudhry, Q., Scotter, M., Blackburn, J., Ross, B., Boxall, A., Castle, L., and Watkins, R. (2008) Applications and implications of nanotechnologies for the food sector. *Food Addit. Contam.*, **25** (3), 241–258.

2 Duncan, T.V. (2011) Applications of nanotechnology in food packaging and food safety: barrier materials, antimicrobials and sensors. *J. Colloid Interface Sci.*, **363** (1), 1–24.

3 Twardowski, T.E. (2007) *Introduction to Nanocomposite Materials: Properties, Processing, Characterization*, DEStech Publications, Inc.

4 Okada, A. and Usuki, A. (2006) Twenty years of polymer–clay nanocomposites. *Macromol. Mater. Eng.*, **291** (12), 1449–1476.

5 Giannelis, E.P. (1996) Polymer layered silicate nanocomposites. *Adv. Mater.*, **8** (1), 29–35.

6 Okada, A., Kawasumi, M., Kurauchi, T., and Kamigaito, O. (1987) Synthesis and characterization of a nylon 6–clay hybrid. *Polymer Prepr.*, **28**, 447–448.

7 Sanchez-Garcia, M.D. and Lagarón, J.M. (2009) Nanocomposite packaging material, in *The Wiley Encyclopedia of Packaging Technology*, John Wiley & Sons, Inc., New York, pp. 807–813.

8 Alexandre, M. and Dubois, P. (2000) Polymer-layered silicate nanocomposites: preparation, properties and uses of a new class of materials. *Mater. Sci. Eng. R Rep.*, **28** (1–2), 1–63.

9 de Azeredo, H.M.C. (2009). Nanocomposites for food packaging applications. *Food Res. Int.*, **42** (9), 1240–1253.

10 Guggenheim, S. and Ave, S.O. (1995) Report Definition of Clay and Clay Mineral: Joint Report of the AIPEA Nomenclature and CMS Nomenclature Committees. Clay Miner., **43** (2), 255–256.

11 Lagarón, J.M. and Cabedo, L. (2015) Polylactide/clay nano-biocomposites, in *Poly(Lactic Acid) Science and Technology: Processing, Properties, Additives and Applications* (A. Jiménez, M. Peltzer, and R. Ruseckaite), RSC Polymer Chemistry Series, Royal Society of Chemistry, pp. 215–224.

12 Akbari, Z., Ghomashchi, T., and Moghadam, S. (2007) Improvement in food packaging industry with biobased

nanocomposites. *Int. J. Food Eng.*, **3** (4), 1556–3758.
13 Arora, A. and Padua, G.W. (2010) Review: nanocomposites in food packaging. *J. Food Sci.*, **75** (1), R43–R49.
14 de Azeredo, H.M.C. (2009). Nanocomposites for food packaging applications. *Food Res. Int.*, **42** (9), 1240–1253.
15 Farhoodi, M. (2015) Nanocomposite materials for food packaging applications: characterization and safety evaluation. *Food Eng. Rev.*, **8** (1), 35–51.
16 Lagarón, J.M., Cabedo, L., Cava, D., Feijoo, J.L., Gavara, R., and Gimenez, E. (2005) Improving packaged food quality and safety. Part 2: nanocomposites. *Food Addit. Contam.*, **22** (10), 994–998.
17 Ray, S., Quek, S.Y., Easteal, A., and Chen, X.D. (2006) The potential use of polymer–clay nanocomposites in food packaging. *Int. J. Food Eng.*, **2** (4). doi: 10.2202/1556-3758.1149.
18 Rhim, J.-W., Park, H.-M., and Ha, C.-S. (2013) Bio-nanocomposites for food packaging applications. *Prog. Polym. Sci.*, **38** (10–11), 1629–1652.
19 Sorrentino, A., Gorrasi, G., and Vittoria, V. (2007) Potential perspectives of bio-nanocomposites for food packaging applications. *Trends Food Sci. Technol.*, **18** (2), 84–95.
20 Youssef, A.M. (2013) Polymer nanocomposites as a new trend for packaging applications. *Polym Plast. Technol.*, **52** (7), 635–660.
21 Choudalakis, G. and Gotsis, A.D. (2009) Permeability of polymer/clay nanocomposites: a review. *Eur. Polym. J.*, **45** (4), 967–984.
22 Cui, Y., Kumar, S., Rao Kona, B., and van Houcke, D. (2015) Gas barrier properties of polymer/clay nanocomposites. *RSC Adv.*, **5** (78), 63669–63690.
23 Minelli, M., Baschetti, M.G., and Doghieri, F. (2011) A comprehensive model for mass transport properties in nanocomposites. *J. Membr. Sci.*, **381** (1–2), 10–20.
24 Fabra, M.J., Lopez-Rubio, A., and Lagaron, J.M. (2014) Nanostructured interlayers of zein to improve the barrier properties of high barrier polyhydroxyalkanoates and other polyesters. *J. Food Eng.*, **127**, 1–9.
25 Fabra, M.J., López-Rubio, A., and Lagaron, J.M. (2014). On the use of different hydrocolloids as electrospun adhesive interlayers to enhance the barrier properties of polyhydroxyalkanoates of interest in fully renewable food packaging concepts. *Food Hydrocolloid.*, **39**, 77–84.
26 Bélard, L., Dole, P., and Avérous, L. (2009) Study of pseudo-multilayer structures based on starch–polycaprolactone extruded blends. *Polym. Eng. Sci.*, **49** (6), 1177–1186.
27 Fabra, M.J., López-Rubio, A., and Lagaron, J.M. (2013) High barrier polyhydroxyalcanoate food packaging film by means of nanostructured electrospun interlayers of zein. *Food Hydrocolloid.*, **32**, 106–114.
28 Martínez-Sanz, M., Lopez-Rubio, A., and Lagaron, J.M. (2012) Optimization of the dispersion of unmodified bacterial cellulose nanowhiskers into polylactide via melt compounding to significantly enhance barrier and mechanical properties. *Biomacromolecules*, **13**, 3887–3899.
29 Fabra, M.J., López-Rubio, A., Ambrosio-Martin, J., and Lagaron, J.M. (2016) Improving the barrier properties of thermoplastic corn starch-based films containing bacterial cellulose nanowhiskers by means of PHA electrospun coatings. *Food Hydrocolloid.*, **61**, 261–268.
30 Pinheiro, A.C., Bourbon, A.I., Medeiros, B.G.S., da Silva, L.H.M., da Silva, M.C.H., Carneiro-da-Cunha, M.G., Coimbra, M.A., and Vicente, A.A. (2012) Interactions between κ-carrageenan and chitosan in nanolayered coatings: structural and transport properties. *Carbohydr. Polym.*, **87**, 1081–1090.
31 Weiss, J., Takhistov, P., and McClements, D.J. (2006) Functional materials in food nanotechnology. *J. Food Sci.*, **71**, r107–r116.
32 Carneiro-da-Cunha, M., Cerqueira, M.A., Souza, B.W.S., Carvalho, S., Quintas, M.A.C., Teixeira, J.A., and Vicente, A.A. (2010) Physical and thermal properties of a chitosan/alginate nanolayered PET film. *Carbohydr. Polym.*, **82**, 153–159.

11
Nanotechnologies for Active and Intelligent Food Packaging: Opportunities and Risks

Nathalie Gontard,[1] Stéphane Peyron,[2] Jose M. Lagaron,[3] Yolanda Echegoyen,[3,4] and Carole Guillaume[2]

[1]INRA, UMR IATE, Place P. Viala, 34060 Montpellier, France
[2]Montpellier University, UMR IATE, Place P. Viala, 34060 Montpellier, France
[3]CSIC, Institute of Agrochemistry and Food Technology, Avda. Agustin Escardino, 7, 46980 Paterna (Valencia), Spain
[4]Universitat de València, Science Education Department, Facultat de Magisteri, Avda. dels Tarongers, 4, 46022 València, Spain

11.1
Introduction and Definitions

Food packaging is a key player in the smorgasbord of actions to prevent food losses and wastes from production up to consumption [1]. Packaging defines around the food a headspace atmosphere whose composition controls food degradation rate and is controlled by the mass transfer through the packaging (transfer of gases, water vapor, etc.). Preservative-modified atmosphere packaging can be passive or active (e.g., by using gas or vapors absorbing packaging materials) [2]. Packaging materials can also be active in contact with food (antimicrobial surface acting by contact). Moreover, the so called "intelligent" packaging are able to monitor the product quality, trace the critical points, and give more complex information throughout the supply chain such as storage conditions. All these new functionalities of packaging are contributing to improving sustainable food consumption.

In the extensive field of nanoscience research and development, food packaging is, with no doubt, the most active, promising, and advanced area in terms of research and development as well as commercial applications with a compound annual growth rate of 12.7% [3]. The development strategy of food packaging engineered nanomaterials (ENM) aims either to enhance the mechanical and barrier properties of conventional packaging materials and/or to provide innovative active and intelligent functionalities, with a very special expectation for sustainable materials such as bio-based plastics. It has been demonstrated in different studies that, as for any other new technology, the public is more willing

Nanotechnology in Agriculture and Food Science, First Edition. Edited by Monique A.V. Axelos and Marcel Van de Voorde.
© 2017 Wiley-VCH Verlag GmbH & Co. KGaA. Published 2017 by Wiley-VCH Verlag GmbH & Co. KGaA.

to accept nanotechnology in "out of food" applications than those where nanoparticles are directly added to the food [4,5]. Food packaging applications seem more likely to manage with consumer acceptance and regulatory attention in the short term than "inside" food applications. A wide variety of nanoparticles such as organic/mineral nanospheres, -tube, or -sheet is the key aspect in the development of innovative packaging materials for food.

This paper focuses on providing a comprehensive review on the current development of nanotechnology in the field of active and intelligent packaging for food. As defined in the food contact material framework regulation 1935/2004, active and intelligent packaging deliberately incorporates active or intelligent components intended to release or to absorb substances into, onto, or from the packaged food or the environment surrounding the food, or to provide the intended information, with their conditions of use. Covered topics include how nanotechnology is, or is intended to be, used for creating innovative and breakthrough functionalities in both active and intelligent packaging areas, and which potential detrimental safety issues related to migration from food contact materials has to be potentially feared.

11.2
Nanomaterials in Active Packaging for Food Preservation

Active packaging relates to the incorporation of additives to the packaging systems with the purpose of maintaining or extending the shelf life and product quality. The active additives can be incorporated directly into the packaging matrix, attached to the interior of the packaging material or introduced inside the package in separate containers such as sachets [6].

Active packaging does not only provide an inert barrier from external conditions but also provides dynamic, rather than the conventional passive, protection to the food inside, and has an active role in food preservation [7]. The additives absorb or release different substances into or from the food and the surrounding environment [8], thus promoting food preservation. Most used active packaging materials include substances that absorb ethylene, oxygen, carbon dioxide, moisture, flavors or odors, and other materials that release antioxidants, carbon dioxide, antimicrobial agents, or flavors [9].

One alternative is the use of new active packaging systems based on nanoparticles that have attracted much attention in the food industry with the aim of replacing conventional food preservation systems [10], in which levels of additives can be in excess of what is required for an efficient active material [11] and can lead to adverse organoleptic properties.

11.2.1
Nanocomposites with Antioxidant Properties

Oxygen (O_2) is directly or indirectly responsible for the degradation of many foods. Direct oxidation reactions are responsible, for example, of rancidity of vegetable oils and browning of fruits. Food deterioration can also be produced by indirect action of O_2 due to food spoilage by aerobic microorganisms. Very low O_2 levels inside the packaging can be maintained by incorporating O_2 scavengers, which is useful for several applications.

Xiao-e *et al.* [12] described the photocatalytic activity of nanocrystalline titania: UV illumination of nanocrystalline TiO_2/polymer films in the presence of excess organic hole scavengers resulted in the oxygen removal in a closed environment. According to the authors, it can be used for packaging many oxygen-sensitive products, but a major flaw is the UVA light requirement.

In another approach, different thermoplastic polymers (PET, PP, FEP, LLDPE, and nylon) have been infused with metal and metal oxide nanoparticles and it has been discovered that films containing less than 1 wt% Pd and Pt nanoparticles were active as oxygen scavengers and reduced the oxygen flux by two orders of magnitude [13]. Iron nanoparticles with average diameter of 110 nm have also been studied as oxygen scavengers to prevent lipid oxidation [14], and tested with roasted sunflower seeds and walnuts. They found that the scavenger capacity of nano iron was almost 1.4 times over that of conventional iron powders. Khalaj *et al.* [15] studied the effect of the incorporation of iron nanoparticles modifying the nanoclay on the physical and mechanical properties of PP nanocomposites. The addition of the nanoparticles had a reverse effect on the intercalation and exfoliation of the clay to some extent and compensated the effect of clay on the change in crystallization temperature and crystallinity. No significant changes were observed in the tensile strength and elongation after the addition of the iron nanoparticles, but there was a substantial enhancement of the yield strength and in the rigidity of the nanocomposite.

Other studied approach was to incorporate iron nanoparticles in cellulosic matrix films in the form of films inside PE/nylon packages [16]. It was found that the use of these films inside the packages was effective on growth inhibition of *Aspergillus flavus* because of the oxygen absorber properties of the iron nanoparticles.

11.2.2
Nanocomposites with Antimicrobial Properties

Antimicrobial food packaging has been developed with different kinds of nanoparticles, but nanosilver is the most widespread [17], even with commercialized items already in the market. Already in ancient times wine and water were stored in silver vessels due to the known antimicrobial properties of silver. It has many advantages over other antimicrobial agents: silver has a broad spectrum and is toxic to numerous strains of fungi, bacteria, algae, and some viruses, with varying degrees of toxicity. In its elemental form it is shelf stable for long periods of time. Silver can also be easily incorporated into many different materials such as plastics and textiles, making it especially useful for food packaging applications.

The mechanism of activity of silver is still unknown, despite the known antimicrobial properties of bulk silver. There is also some controversy about the manner in which silver nanoparticles (AgNPs) are toxic to bacterial cells: some postulate that the activity comes from the Ag^+ ions detached from the surfaces of AgNPs which will act by the same mechanisms as conventional silver antimicrobials [18], but some research [19] shows that silver nanoparticles are more toxic than the equivalent amount of dissociated silver ion. It has also been demonstrated that particles of different shapes, sizes or other characteristics may behave differently, even in the same conditions [20]. Morones *et al.* [21] showed that the amount of silver ions released from silver nanoparticles in their experiment was too low to account completely for their toxicity.

Inorganic nanoparticles can be easily incorporated into the polymers to create antimicrobial nanocomposites [22] and that is one of the biggest advantages over molecular antimicrobials. AgNPs can be engineered to remain potent antimicrobial agents for long periods of time [23] due to their controlled release properties. And this makes nanosilver/polymer composites very attractive materials for food packaging. The antimicrobial activity of the silver nanocomposites is dependent on different factors that affect the Ag^+ release rate like the degree of polymer crystallinity, filler type (silver zeolites or silver nanoparticles), hydrophobicity of the matrix, and particle size.

Nanocomposites with many different polymers have been developed. Among them the most used are conventional polymers. In fact, there are many commercial products, especially in the Asian and American markets, based on polypropylene and polyethylene therephtalate, of low-density polyethylene nanocomposites with silver nanoparticles as BlueMoonGoods™ fresh box silver nanoparticle food storage containers (BlueMoonGoods, LLC, USA), Fresherlonger™ miracle food storage containers and plastic storage bags (Sharper Image®, USA), Oso fresh food storage containers and nanosilver NS-315 water bottle (A-DO Global, Korea), nanosilver storage box (Baoxianhe) (Quan Zhou Hu Zheng Nano Technology Co., Ltd.®, China), nanosilver baby mug cup (Baby Dream® Co., Ltd., Korea), and nanosilver salad bowl (Changmin Chemicals, Korea). In a research stage are other conventional polymer nanocomposites like polystyrene impregnated with silver nanoparticles, in this case, synthesized in an ecofriendly manner using marine bacterial isolate culture filtrate [24] and low-density polyethylene films created by extrusion, and the silver nanoparticles were embedded in distinct carriers (silica and titanium dioxide) [25], linking the silver nanoparticles to the LDPE surface by peptides [26], or sometimes even combined with modified atmosphere [27].

As there are concerns about the use of nonbiodegradable packaging materials over exhausting natural resources, other environmental burdens, and food safety, there is an increasing demand for biodegradable packaging materials (biopolymers) created from renewable sources as an alternative to conventional plastics, especially for disposable applications and use in short-term packaging [28–30]. For this reason, biopolymers have also been used to fabricate antimicrobial packaging, like coatings of biocomposites of polylactic acid (PLA) obtained by solvent casting containing a silver-based antimicrobial layered silicate additive that were found to have strong antimicrobial activity against Gram-negative *Salmonella spp* [31] or renewable and biodegradable nanocomposites with poly (hydroxybutyrate-co-valerate) (PHBV) [32,33] have been shown to have a strong and sustained (for seven months) antibacterial activity against *Salmonella enterica* and *Listeria monocytogenes*, with a surprising oxygen permeability drop of about 56% compared to the neat polymer.

Other nanomaterials have also been found to have antimicrobial properties. Hamal *et al.* [34] demonstrated that titanium dioxide codoped with silver, carbon, and sulfur can serve as a multifunctional generic biocide as well as a visible light-activated photocatalyst. NanoMgO was found to have better bactericidal efficacy than TiO_2 in experiments with *Bacillus subtilis var. niger* and *Staphylococcus aureus* [26] and the results showed that the bactericidal efficacy of nano-MgO increased with decreasing particle size. Ramyadevi *et al.* [35] demonstrated that Cu nanoparticles synthesized by a polyol method showed antimicrobial activity against five bacterial strains and three fungal strains, although they showed more inhibitory activity in bacteria than fungi, and Kruk *et al.* [36] also confirmed that monodisperse copper nanoparticles synthesized by the chemical reduction method had high activity against bacteria and fungi, with values similar to those of silver nanoparticles but at a much lower cost. Chitosan nanoparticles have been used to treat silk filaments and fabric [37] and the bacterial reduction against *S. aureus* was maintained after 20 launderings. In addition, the chitosan nanoparticle treatment also improved other characteristics of the fabric, like the breaking strength and the wrinkle resistance. ZnO nanoparticles also presented excellent antibacterial action against *Escherichia Coli* on chitosan films [38] and in this case there was no appreciable antibacterial activity of the chitosan itself, probably due to the high molecular weight chitosan used in this study. Xie *et al.* [39] also investigated the antibacterial activity of zinc oxide nanoparticles, in this case against *Campylobacter jejuni*, and they concluded that the antibacterial mechanism was most likely due to disruption of the cell membrane and oxidative stress. Azam *et al.* [40] studied the antimicrobial activity of metal oxide nanoparticles (ZnO, CuO, and Fe_2O_3) against different Gram-positive and Gram-negative bacteria and found that in all cases the antimicrobial activity increased with a decrease in the particle size, where the ZnO nanoparticles whee the ones with the highest bactericidal activity and the iron oxide nanoparticles had the least bactericidal activity. Also, their results indicate that the nanomaterials were more effective against Gram-positive bacterial strains compared to Gram-negative bacterial strains. Other morphologies of nanomaterials have also been proved to be effective against bacteria, like carbon nanotubes (both SWNTs and MWNTs) or colloidal graphite [41].

11.3
Nanotechnology for Intelligent Packaging as Food Freshness and Safety Monitoring Solution

11.3.1
Stakes and Challenges of Nano-Enabled Intelligent Packaging

Food date labels (including "best before," "sell by," and "display until") are key tools for preventing household food waste and foodborn disease. Food date labels provide an approximate date to food edibility expiration, based on a set of idealized assumptions about: (i) initial food quality and contamination, (ii) average expected storage and transport conditions (especially the temperature history and the composition of the atmosphere surrounding the food such as oxygen and moisture content), and (iii) the limiting factor for consumer safety (e.g., maximum level of potential pathogen contamination) and/or consumer acceptability (quality parameters). However, all these assumptions are not always corresponding to real conditions. For example, deviations from the expected/recommended storage and transport conditions could lead to the premature deterioration of the food and cause harm to the consumer due to the high content of toxins or pathogens. Due to the margin of safety applied for food date label determination, safe and good quality food may be discarded when initial food quality and food storage and transport conditions are better than anticipated. Moreover, the distinction between the different types of existing labels is difficult for many consumers. Globally, it is estimated that inaccuracies in, or misunderstanding of, food date labels cause over 20% of the avoidable disposal of still-edible food (according to the preparatory study on food waste across EU 27, EC report in 2010). In order to limit/avoid the safety and wastes issues resulting from the margin of safety and from the misunderstanding of conventional food date labels, a new generation of intelligent packaging is developing to provide accurate and easy to read information on food safety and quality. Intelligent packaging is growing at a faster rate if compared to the active packaging, with a compound annual growth rate of 12.7% during 2014–2015 according to "Global market study on nano-enabled packaging for food and beverages: intelligent packaging to witness highest growth by 2020 (from persistent market research)." The unique optical, chemical, biochemical, and electrical properties of nanoscale particles offer breakthrough innovation routes to develop intelligent packaging solutions able to solve such problems.

The speed and accuracy of small molecules detection has been drastically changed by noninvasive nanotechnology-based sensors [4]. Nanotechnology potentiality is not only about developing miniaturized version of classical machine [42] but also about creating nanodevices with new functionalities stemming from the small size and unusually sensitive optical, electrical, magnetic properties of nanoparticles. Nanosensors are used to convey information about nanoparticles to the macroscopic world [43]. Because nanotechnologies can be easily embedded into the packaging materials, their potentiality increases exponentially [44]. Nanosensors can be incorporated into the packaging matrix itself or in a label or a coating, to add an intelligent function to food packaging. These so-called "intelligent packaging" are able to provide to the user, appropriate in-time information about the quality and the safety of the packed food either indirectly by monitoring storage conditions such as time–temperature history or in-package gas concentration responsible for food degradation rate, or directly by monitoring the presence of chemical and microbial markers of food quality and safety.

11.3.2
Main Principles of Involved Nano-Enabled Sensing

Sensing food quality and safety is based on detecting the changes of physical–chemical properties of nanoparticles induced by target markers (Figure 11.1). The constituents and the structures are tailored (organic, inorganic constituents, pores, core, coating, multilayers, etc.) to reach the requested sensitivity. For example, surface of nanoparticles is typically modified with reacting substances that have strong affinity with the target markers. When such a reactive surface comes in contact with the target molecules, the optical, fluorescence, conductivity, magnetic, and so on, properties of the nanoparticles change. The response of the nanosensing systems is tailored by adjusting the size, the structure, and the surface functionalization of the nanoparticles. Thereby a variety of sensing nanoplatforms have been demonstrated. For example, numerous multifunctional nanoparticles have been designed to perform complex tasks simultaneously using plasmonic, upconversion, semiconductor, or magnetic nanoparticles, which are electromagnetically active and used as scaffolds upon which a diverse range of multifunctionality can be built [45]. The potentialities range from very sensitive and expensive complex up to low-cost and user-friendly systems. Systems devoted to intelligent food packaging require easy-to-read (especially when it is intended to be read by the consumer), cheap (to match food cost, including fresh foods), robust (reliable in targeted conditions), and safe (no migration expected toward the food) systems that are able to provide on time information.

Optical and visual read out of nanoparticles that relates molecular events to colorimetric and fluorimetric signals is possible by naked eyes or using simple standard digital camera. Beyond these traditional colorimetric indicators,

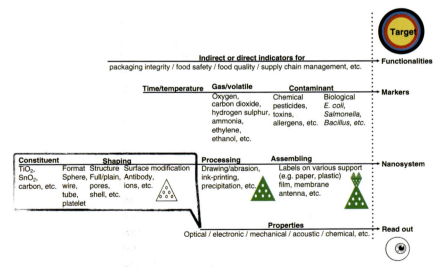

Figure 11.1 Overview of nano-enabled intelligent food packaging systems.

electrochemical properties of nanoscale particles offer innovative routes of wireless tracking of food quality and safety. An important advance is the integration of sensors of food spoilage markers, to RFID (radio frequency identification) tags allowing, during the whole food chain, *in situ*, continuous and wireless tracking of quality and safety of packed foods [46–49]. Passive RFID tags that do not contain onboard power sources are low-cost, long operational life, small-sized, and light-weighted devices. These interesting properties make them promising and realistic candidates for being applicable as intelligent food packaging. Among the large portfolio of existing and studied sensing materials, nanosensors constitute a new and exciting field of research, supported by recent advances in nanotechnology such as nanowires, nanotubes, and nanopores [44,50].

11.3.3
Indirect Nano-Enabled Indicators of Food Quality and Safety

The simplest and widely studied and used nanoindicator systems are based on the color change of metallic (plasmonic) nanoparticles in the presence of the targeted markers [51]. Zhang *et al.* [52] developed a time–temperature indicator based on cylindrical nanoparticles structured with a gold core and a silver surface. Silver shell growth rate with temperature induced a blue shift in the surface plasmon band and a color shift from red to green that can be visually observed. Nanoparticles structure is adjusted in order to make the impact of temperature on food spoilage micro-organisms growth (*E. Coli* was tested as reference microorganism in this study), well correlated with the impact of temperature on silver shell growth rate. As industrial applications example, Timestrip® has created the iStrip, which is a packaging indicator that starts out red and changes color if the temperature drops below freezing [53]. The gold nanoparticles change color based on the size of their aggregates with a temperature-dependent aggregation behavior. The technology uses gold nanoparticles that irreversibly aggregate together at temperatures below freezing. This time–temperature indicator type works without any contact with the food, and therefore is expected to be of no safety concern for migration toward the food.

Oxygen is a key element of food degradation such as oxidation of vitamin, unsaturated fatty acids or pigments, enzymatic reactions or microbial growth. Modified atmosphere by gas flush or gas absorbing and/or releasing systems, such as the wide spread oxygen scavengers, are largely used to improve food preservation and monitoring oxygen concentration in food packaging headspace remains a key challenge of food quality and safety control. Numerous noninvasive nanoenabled gas-sensing systems have been developed. A photoactivated indicator based on TiO_2 or SnO_2 nanoparticles combined with a redox dyes indicator (methylene blue or indigotetrasulfonate) has been developed [54,55]. UV light activation leads to the accumulation on the nanoparticles of photogenerated electrons that reduce the redox dye and bleached its color which is restored gradually with small amount of oxygen. The applicability of a similar UV-activated indicator (ion-pairing of methylene blue with nanoTiO_2 as sensitizer) as intelligent packaging was further refined by printing it on different hydrophobic polymer film [56]. This system could be easily used to noninvasively detect the unexpected presence of oxygen due to compromised oxygen absorption or seal integrity, during the whole shelf-life of the packed food. It does not require any contact with food and could be simply included in an oxygen permeable compartment within the closed packaging.

CO_2 is also an important gas to monitor, either for controlling food preservation conditions in modified atmosphere packaging or as food spoilage indicators. A noninvasive CO_2 indicator applicable to modified atmosphere packaging has been developed based on luminescent dyes standardized by fluorophore-encapsulated polymer nanobeads with 0.8–100% detection range and 1% resolution. Cross-sensitivity with oxygen is minimized by immobilizing the reference luminophore in a polymer nanobeads [57].

Moisture uptake of dry food and moisture loss of high moisture food are accountable for quality and safety food deterioration. A simple concept and low-cost optical humidity detection based on a hybrid metal/polymer film has been developed by Luechinger et al. [58]. Carbon coated copper nanoparticles of 20–50 nm are dispersed in a polymer film. When water vapor swells the polymer increasing its thickness, the distance between nanoparticles increases, which leads to a change in optical behavior. These color changes were assigned to surface plasmon resonance effects at low relative humidity (RH) and to a thin film interference at high humidity. This moisture indicator system does not require any contact with the food, of which it can be separated by a moisture permeable film.

Optical nanochemosensors for oxygen, temperature, ammonia, or carbon dioxide were demonstrated to be producible by precipitation as simple and versatile method [59]. Precipitation enables entrapping lipophilic indicators (such as iridium coumarin complex, ruthenium polypyridyl complexes, or metalloporphyrins for probing oxygen sensivity) into nanobeads based on various polymers such as poly(methyl methacrylate), polystyrene, polyurethanes, ethylcellulose, and other polymers. Sensing properties of the beads can be tuned by choosing the appropriate indicator. These beads can be used on their own, combined with others beads for multianalyte sensing or dispersed in a polymer matrix.

Others strategies of moisture and gas indicators based on nano-enabled electrochemical detection have also been investigated. For example, as humidity sensor based on $Na_2Ti_3O_7$ nanotubes were hydrothermally synthesized and coated on Al_2O_3 ceramic substrate to fabricate humidity sensors using Ag–Pd as interdigitated electrodes [60]. The impedance of this high performance humidity nanosensor changes about five orders of magnitude within humidity range from 11 to 95% RH. Gold-poly(vinyl pyrrolidone) core–shell nanocomposites were incorporated into a capacitive-type humidity sensor as a dielectric [61]. The capacitance sensitivity of the nanosensor was of about −136 Hz/% as the RH increased from 11.3 to 93%, with good reproducibility and stability, fast response, and low hysteresis. Silicon nanowires were reported as water vapor and oxygen sensing [62,63] based on mechanisms of contact resistance across two nanowires and the surface resistance along the individual nanowire. A high oxygen sensing ability was reported for SnO_2 nanobelt functionalized with Pd catalyst particles, due to the enhanced catalytic dissociation of the molecular oxygen on the Pd nanoparticles surface and subsequent diffusion to the oxide surface [64]. Nanoparticles such as cerium (IV) oxide or nanostructures such as nanoporous anodic aluminum oxide were reported to be very sensitive sensors to humidity [65–67] to be interfaced to RFID tags. Nanopores increase considerably the sensitivity of the layer by enabling sorption of large amount of water vapor, leading to a subsequent large increase in permittivity, and thus sensitivity, of the nanoporous layer. The feasibility and large potentiality of combining RFID tags with humidity sensors have already been demonstrated [68–71].

11.3.4
Direct Nano-Enabled Indicators of Food Quality and Safety

Currently, one of the most promising area of intelligent packaging research and development is about direct indicators. A direct indicator aims to monitor *in situ* the quality and the safety of the packed food by reacting in a tailored manner with target markers of food chemical and microbial contamination or degradation, such as ethanol, ammonia, biogenic amines, volatile nitrogen compounds, carbon dioxide, pesticides and toxins or pathogens.

Optical read out strategies based on gold and/or iron nanoparticles (e.g., oligonucleotides surface functionalized or not) agglomeration, which is induced by target molecules and results on visually observable color shift, is used for mercury or melamine contamination but also milk proteins content evaluation or even foodborn pathogens in a complex matrix [72–75] Fluorescence change of nanoparticles in the presence of a target molecule is also an existing strategy. For example, the organic fluorophore at the free end of nanoparticles (so-called quantum dot) surface peptides is cut by botulinum neurotoxins, which leads to the organic fluorophore to diffuse and stop the fluorescence resonance energy transfer (FRET) mechanism. The customer can directly observe the resulting change in fluorescence by holding the packaging foil against illumination or read it using the camera of a smartphone [76]. Combination of optical and fluorescence read out is also possible as demonstrated, for example, for pesticides such as organophosporus or carbamates [77]. These pesticides inhibit the activity of acetylcholinesterase, responsible for the fluorescence (due to rhodamine B release) and the agglomeration (because differently covered) of gold nanoparticles covered by fluorescent rhodamine B. Nanoparticles enable a convenient and sensitive read out either visually or by using simple devices such as smartphone's camera. These last systems require a direct contact with food (containing the target marker) and therefore require technical adjustment before being safely applicable to food packaging.

The fluorescence quenching of nanofibrils of perylen-based fluorophore or the conductance change of microrods of $ZnO-TiO_2$ nanocomposites enable sensitive detection of gaseous amine [78,79] which are indicators of fish and meat spoilage. By functionalizing multiwalled carbon nanotubes with nanocrystal of silver, a selective, room temperature ammonia gas-sensor was recently developed [80]. Silver hollow sites enable the orientation of the adsorbed ammonia molecules leading to a net charge transfer from ammonia toward the hybrid carbon nanotubes whose conductance is subsequently changed. Similarly, a green ammonia sensor working at ambient temperature was developed [81] from guar gum and silver nanoparticles, whose interface creates an active site for interaction with ammonia, leading to the conductance change of a thin film appropriate for food packaging applications.

A highly selective and sensitive ethylene nanosensor was developed to monitor ripening and senescence process of different fresh fruits [82]. This nanosensor relies on the conductivity change of single-walled carbon nanotubes combined with copper complexes in polystyrene particles, inspired by nature, to specifically recognized the target ethylene.

Many electrochemical nanosensor for food microbial spoilage markers are based on monitoring the changing conductivity of conductive nanoparticles (e.g., single-walled carbon nanotubes) on the surface of which selective antibodies are bound. This strategy was used for detection of the toxin of cyanobacteria, the presence of aflatoxins in milk, ochratoxin which is a foodborn fungal contaminant, of Staphyloccccoccal enterotoxin B or bacteria such as *Listeria monocytogenes*, Bacillus, Salmonella and Escherichia, as well as viruses [4].

In a general way, single-walled carbon nanotubes are useable for the selective detection of many gas, vapor, or microorganisms after chemical modifications of their surface [83,84] However, they require expensive and bulky equipment, complicated solubilization and stabilization process, and so on. For overcoming such drawbacks, a simple, versatile, and solvent free mechanical abrasion of compressed powders of sensing materials on the surface of cellulosic papers (like drawing with a pencil on a paper) was developed [85]. Even though the performance of the device, that is, the change of conductance of the paper, was evaluated on gaseous ammonia, the technology is applicable for processing many other gaseous, volatile, and microbial sensors. Similar processing effort were conducted by depositing such single-walled carbon nanotubes on paper or other substrates, and using binding solution-based process such as drop-casting, dip-coating, and ink-jet printing [85,86] A paper recently discussed the evolution toward the first RFID-enabled wireless sensor network infrastructure using ultra-high frequency/radio frequency RFID-enabled sensor nodes and inkjet-printed electronics technologies on flexible and paper substrates for the first time [86].

11.4
Potential Safety Issues and Current Legislation

As any material intended to be in contact with food (FCM), active and intelligent materials (AIM) must comply with the provisions of the framework Regulation (EC) 1935/2004 [87] that defines three requirements to ensure safe and quality food: FCM shall not transfer their components into the food in quantities that could endanger human health; FCM shall not change the composition of the food in an unacceptable way; and FCM shall not deteriorate the taste, odor or texture of the food. While respecting these requirements, which define the principle of inertia, AIM are intended to detect internal or external environmental change and to respond by changing their own properties or attributes and hence the internal package environment. As a consequence, some aspects of Regulation (EC) 1935/2004 appeared in contradiction with the function of AIM that precisely interact with the packaged food product and, in some cases, promote intentional mass transfer. Instead, a Commission Regulation (EC 450/2009) [88] has been laid down terming the most important aspects of AIM application framework. According to this Regulation, AIM must be "suitable and effective" and its components must comply with some requirements, namely, (i) they must be included in a positive list or (ii) they must comply with the provisions applicable to food (e.g., with food additives regulations) or (iii) they are separated from the food by a functional barrier, that should completely avoid the potential migration into the food product. About this last point, the Regulation sets two exceptions: mutagenic, carcinogenic, or toxic to reproduction substances and engineered nanomaterials (ENM).

The actual definition of "nanomaterial" may still be considered to be controversial. In the United States, the Food and Drug Administration (FDA) has not established regulatory definitions of "nanomaterial," "nanotechnology," "nanoscale," or other related terms admitting that these terms are commonly used in relation to the engineering (i.e., deliberate manipulation, manufacture, or selection) of materials that have at least one dimension in the size range of approximately 1–100 nm [89]. FDA considers that evaluations of safety, effectiveness, public health impact, or regulatory status of nanotechnology products should consider any unique properties and behaviors that the application of nanotechnology may impart. The various draft guidance documents the Agency has issued to date on the topic reiterate FDAs view that nanotechnology represents an "evolving state of the science," and that a single definition of nanotechnology and its related terms is not appropriate for regulatory purposes until more can be learned about the interaction of nanomaterials with biological system.

For its part, the European Commission defines a nanomaterial as "a natural, incidental, or manufactured material containing particles, in an unbound state or as an aggregate and where, for 50% or more of the particles in the number size distribution, one or more external dimensions is in the size range 1–100 nm" [90]. Actually, little is known about the risk that could be linked to the use of ENM in FCM and the strategic bias of the European Food Safety Authority is to propose a suitable risk assessment approach taking into account the "specific properties of the ENM in addition to those common to the equivalent nonnanoforms." On this basis, the EU had already edited statutory texts and technical guidance that mention nanomaterials. In particular, the EU Plastics Regulation (EC) 10/2011 [91] meeting the terms of the aforementioned AIM Regulation EC 450/2009 specifically states that new technologies that engineer substances in particle size that exhibit chemical and physical properties that significantly differ from those at a larger scale, for example, nanoparticles, should be assessed on a case-by-case basis as regards their risk until more information is known about such new technology. On this basis and taking into account the lack of knowledge about their potential toxicity (the oral exposure to ENM having received less attention than the dermal or inhalation pathways), the concept of functional barrier used to prevent migration of contaminants, which are not evaluated by health authorities, cannot be applied in the specific case of nanocomposite packaging including nanodimensional substances. This statement notably differentiates nanoform and nonnanoform active agents, the last ones being authorized to be used behind a functional barrier, provided that they fulfill certain criteria and their migration remains below a given detection limit. In the case of AIM, the safety issue is depending on the type of applied strategy and the degree of food-packaging interaction. The active packaging including antimicrobial agents, generally used as surface-active biocides is intended to actively interact with the food. In this case, the risk is directly related to the toxicological concern of the substance which involves hazard identification and dose response analysis. Up till today, very few substances have been evaluated and titanium nitride is the unique named "nano" substance registered in the positive list of the Regulation (EC) 10/2011 and authorized to be used in PET bottles without any specific migration limit. Recently, zinc oxide nanoparticles was approved for use in food contact materials in a scientific opinion published by EFSA [92] with a specific migration limit of 0.05 mg/kg which correspond to a value established on soluble ionic zinc as it was established that zinc oxide does not migrate in nanoform.

In the case of selective barrier or intelligent/smart packaging (atmosphere control/temperature monitor), the active substance can be embedded in the polymeric matrix or enclosed in a pouch or sachet, which is placed inside a sealed package. In this case, the risk assessment additionally includes the exposure quantification which is related both to the consumption pattern of the affected population and to the migration extent of NP resulting from the food contact usage.

In absence of any harmonized procedure, the difficulty surrounding evaluation of the release of migrating ENMs from food contact materials into complex food matrix remains a key point for assessing the human exposure. In this way, the case-by-case approach involves a disparity of criteria and a high degree of uncertainty until more data are available and the assessment becomes standardized. This consideration reflects the early review conducted in this way as soon as 2010 at the conference titled "Nanotechnologies in the Food and Agriculture Sectors: Potential Food Safety Implications" [93].

Ongoing research efforts remain thus necessary to explore the quantitative dependence of migration on the various ENM characteristics including particle size, shape, surface properties, and aggregation state. New analytical methods and equipment are yet available to measure/detect infime fraction of migrating inorganics nanoparticles [4,94]. Lots of investigations were guided on nanosilver, which is currently the nanoparticle most employed in commercial active products [3] and on this issue, the up-to-date scientific opinion of EFSA [95] recognizes that the relevance of the available toxicological studies to the safety evaluation could not be established regarding the lack of specifications of particles migrating in the nanoscale. In the same way, a scientific opinion dealing with the safety evaluation of the active substance iron(II) modified bentonite intended to be incorporated in packages as oxygen absorber concluded that such active substances does not raise a safety concern for the consumer if the particle size distribution is in the micron range. Caution appears to be exercised in case the substance is incorporated with compatibilizer at levels up to 15% that should induce the formation of nanoparticles due to the clay exfoliation [96]. The precautionary principle applies to material including nanoclays in absence of reliable information on their migration extent into the food. Therefore, the transport mechanism of nanoparticles remains an open question considering that the occurrence of nanoparticles in food could distinctly result from the release of the whole particles or from the postmigration assembly of ionic forms.

11.5
Conclusions and Perspectives

Nanotechnology is currently offering unique possibilities to design more efficient active technologies, which take tremendous advantages in the high surface to volume ratios of nanomaterials either in coating or in a composite form. Nanotechnology not only provides more efficient activity of the functional component at lower dosing but also easy incorporation into the matrix without adverse impact on the selected good properties of the matrix such as transparency, mechanical, and barrier performance. It is often even the case where the active nanomaterials also promote increased passive properties such as barrier and other physical properties. It is also clear that from a food safety view point, a complete control and understanding over the nonintended migration and, in the case where certain specific migration is allowed under which form the active species are released, needs be assessed and proven to legislators to design functional but safe packaging. Nanotechnology and new processing technology advances are also clearly providing breakthrough foundations for designing cost efficient intelligent labels applicable to food packaging. These indirect or direct indicators of food quality and safety are able to integrate multiple gas and volatile sensors useable on flexible substrate such as packaging materials, with a special recent attention to wireless sensors that could potentially set the foundation for the truly convergent wireless sensor *ad hoc* networks of the future. The lack of regulatory support providing standardized risk assessments of ENM through integration of hazard and exposure evaluations might hinder their market success and/or consumer acceptance. The lack of regulatory support providing standardized risk assessments of ENM through integration of hazard and exposure evaluations might hinder their market success and/or consumer acceptance. It is important to note that numerous nano-enabled active and intelligent packaging are "safe-by-design" as nanoparticles are integrated in systems that prevent direct contact with food and potential migration toward the food. In the case of nanoparticles potentially migrating toward the food, the precautionary principle currently recommended in Europe promotes a case-by-case evaluation of the ENM. International and collaborative nanosafety research effort is now necessary to progress on the toxicological profiles and then propose a hazard classification of nanosubstances constituting the basis of a universal risk management.

References

1 Angellier-Coussy, H., Guillard, V., Guillaume, C., and Gontard, N. (2013) Role of packaging in the smorgasbord of action for sustainable food consumption. *Agro Food Ind. Hi Tech.*, **24** (3), 15–19.

2 Guillaume, C., Chalier, P., and Gontard, N. (2008) Modified atmosphere packaging using environmentally compatible and active food packaging materials in *Environmentally Compatible Food Packaging*, CRC Press, pp. 396–418.

3 Bumbudsanpharoke, N. and Ko, S. (2015) Nano-food packaging: an overview of market, migration research, and safety regulations. *J. Food. Sci.*, **80** (5), R910–R923.

4 Duncan, T.V. (2011) Applications of nanotechnology in food packaging and food safety: barrier materials, antimicrobials and sensors. *J. Colloid Interf. Sci.*, **363** (1), 1–24.

5 Siegrist, M., Stampfli, N., Kastenholz, H., and Keller, C. (2008) Perceived risks and perceived benefits of different nanotechnology foods and nanotechnology food packaging. *Appetite*, **51** (2), 283–290.

6 Restuccia, D., Spizzirri, U.G., Parisi, O.I., Cirillo, G., Curcio, M., Iemma, F., Puoci, F., Vinci, G., and Picci, N. (2010) New EU regulation aspects and global market of active and intelligent packaging for food industry applications. *Food Control*, **21** (11), 1425–1435.

7 Lim, L. (2011) Active and intelligent packaging materials. *Compr. Biotechnol.*, **1**, 629–644.

8 Brody, A.L. (2001) What's active in active packaging. *Food Technol.*, **55**, 104–106.

9 Vermeiren, L., Devlieghere, F., Van Beest, M., De Kruijf, N., and Debevere, J. (1999) Developments in the active packaging of foods. *Trends Food Sci. Technol*, **10** (03), 77–86.

10 Cho, J.Il., Lee, S.H., Lim, J.S., Koh, Y.J., Kwak, H.S., and Hwang, I.G. (2011) Detection and distribution of food-borne bacteria in ready-to-eat foods in Korea. *Food Sci. Biotechnol.*, **20** (2), 525–529.

11 Kanmani, P. and Rhim, J.W. (2014) Physicochemical properties of gelatin/silver nanoparticle antimicrobial composite films. *Food Chem.*, **148**, 162–169.

12 Xiao-e, L., Green, A.N.M., Haque, S.A., Mills, A., and Durrant, J.R. (2004) Light-driven oxygen scavenging by titania/polymer nanocomposite films. *J. Photochem. Photobiol., A*, **162**, 253–259.

13 Yu, J., Liu, R.Y.F., Poon, B., Nazarenko, S., Koloski, T., Vargo, T., Hiltner, A., and Baer, E. (2004) Polymers with palladium nanoparticles as active membrane materials. *J. Appl. Polym. Sci.*, **92** (2), 749–756.

14 Mu, H., Gao, H., Chen, H., Tao, F., Fang, X., and Ge, L. (2013) A nanosised oxygen scavenger: preparation and antioxidant application to roasted sunflower seeds and walnuts. *Food Chem.*, **136** (1), 245–250.

15 Khalaj, M.-J., Ahmadi, H., Lesankhosh, R., and Khalaj, G. (2016) Study of physical and mechanical properties of polypropylene nanocomposites for food packaging application: nano-clay modified with iron nanoparticles. *Trends Food Sci. Technol.*, **51**, 41–48.

16 Botrel, D.A., Soares, N.D.F.F., Fernandes, R.V.D.B., and De Melo, N.R. (2011) Iron nanoparticles film on *Aspergillus flavus* inhibition. *Ital. J. Food Sci.*, **23**, 103–106.

17 Rai, M., Yadav, A., and Gade, A. (2009) Silver nanoparticles as a new generation of antimicrobials. *Biotechnol. Adv*, **27** (1), 76–83.

18 Lok, C.-N., Ho, C.-M., Chen, R., He, Q.-Y., Yu, W.-Y., Sun, H., Tam, P.K.-H., Chiu, J.-F., and Che, C.-M. (2007) Silver nanoparticles: partial oxidation and antibacterial activities. *J. Biol. Inorg. Chem.*, **12** (4), 527–534.

19 Limbach, L.K., Wick, P., Manser, P., Grass, R.N., Bruinink, A., and Stark, W.J. (2007) Exposure of engineered nanoparticles to human lung epithelial cells: influence of chemical composition and catalytic activity on oxidative stress. *Environ. Sci. Technol.*, **41** (11), 4158–4163.

20 Eby, D.M., Schaeublin, N.M., Farrington, K.E., Hussain, S.M., and Johnson, G.R. (2009) Lysozyme catalyzes the formation of antimicrobial silver nanoparticles. *ACS Nano*, **3** (4), 984–994.

21 Morones, J.R., Elechiguerra, J.L., Camacho, A., Holt, K., Kouri, J.B., Ramírez, J.T., and Yacaman, M.J. (2005) The bactericidal effect of silver nanoparticles. *Nanotechnology*, **16** (10), 2346–2353.

22 Althues, H., Henle, J., and Kaskel, S. (2007) Functional inorganic nanofillers for transparent polymers. *Chem. Soc. Rev.*, **36** (9), 1454–1465.

23 Roe, D., Karandikar, B., Bonn-Savage, N., Gibbins, B., and Roullet, J.B. (2008) Antimicrobial surface functionalization of plastic catheters by silver nanoparticles. *J. Antimicrob. Chemother.*, **61** (4), 869–876.

24 Youssef, A.M. and Abdel-Aziz, M.S. (2013) Preparation of polystyrene nanocomposites based on silver nanoparticles using marine bacterial for packaging. *Polym. Plast. Technol.*, **52** (6), 607–613.

25 Becaro, A.A., Puti, F.C., Correa, D.S., Paris, E.C., Marconcini, J.M., and Ferreira, M.D. (2015) Polyethylene films containing silver nanoparticles for applications in food packaging: characterization of physico-chemical and anti-microbial properties. *J. Nanosci. Nanotech.*, **15** (3), 2148–2156.

26 Huang, L., Li, D.Q., Lin, Y.J., Wei, M., Evans, D.G., and Duan, X. (2005) Controllable preparation of Nano-MgO and investigation of its bactericidal properties. *J. Inorg. Biochem.*, **99** (5), 986–993.

27 Azlin-Hasim, S., Cruz-Romero, M.C., Morris, M.A., Cummins, E., and Kerry, J.P. (2015) Effects of a combination of antimicrobial silver low density polyethylene nanocomposite films and modified atmosphere packaging on the shelf life of chicken breast fillets. *Food Pack. Shelf Life*, **4**, 26–35.

28 Weber, C.J., Haugaard, V., Festersen, R., and Bertelsen, G. (2002) Production and applications of biobased packaging materials for the food industry. *Food Addit. Contam.*, **19**, 172–177.

29 Siracusa, V., Rocculi, P., Romani, S., and Rosa, M.D. (2008) Biodegradable polymers for food packaging: a review. *Trends Food Sci. Technol*, **19** (12), 634–643.

30 Lagaron, J.M., Ocio, M.J., and Lopez-Rubio, A. (eds) (2011) *Antimicrobial Polymers. Antimicrobial Polymers*, 1st edn, John Wiley & Sons, Ltd, Hoboken.

31 Busolo, M.A., Fernandez, P., Ocio, M.J., and Lagaron, J.M. (2010) Novel silver-based nanoclay as an antimicrobial in polylactic acid food packaging coatings. *Food Addit. Contam. A*, **27** (11), 1617–1626.

32 Castro-Mayorga, J.L., Martinez-Abad, A., Fabra, M.J., Olivera, C., Reis, M., and Lagaron, J.M. (2014) Stabilization of antimicrobial silver nanoparticles by a polyhydroxyalkanoate obtained from mixed bacterial culture. *Int. J. Biol. Macromol.*, **71**, 103–110.

33 Castro-Mayorga, J.L., Fabra, M.J., and Lagaron, J.M. (2016) Stabilized nanosilver based antimicrobial poly(3-hydroxybutyrate-co-3-hydroxyvalerate) nanocomposites of interest in active food packaging. *Innov. Food Sci. Emerg. Technol.*, **33**, 524–533.

34 Hamal, D.B., Haggstrom, J.A., Marchin, G.L., Ikenberry, M.A., Hohn, K., and Klabunde, K.J. (2010) A multifunctional biocide/sporocide and photocatalyst based on titanium dioxide (TiO_2) codoped with

silver, carbon, and sulfur. *Langmuir*, **26** (4), 2805–2810.

35 Ramyadevi, J., Jeyasubramanian, K., Marikani, A., Rajakumar, G., and Rahuman, A.A. (2012) Synthesis and antimicrobial activity of copper nanoparticles. *Mater. Lett.*, **71**, 114–116.

36 Kruk, T., Szczepanowicz, K., Stefańska, J., Socha, R.P., and Warszyński, P. (2015) Synthesis and antimicrobial activity of monodisperse copper nanoparticles. *Colloids Surf. B*, **128**, 17–22.

37 Lu, Y.-H., Chen, Y.-Y., Lin, H., Wang, C., and Yang, Z.-D. (2010) Preparation of chitosan nanoparticles and their application to *Antheraea pernyi* silk. *J. Appl. Polym. Sci.*, **117** (6), 3362–3369.

38 Bajpai, S.K., Chand, N., and Chaurasia, V. (2010) Investigation of water vapor permeability and antimicrobial property of Zinc Oxide nanoparticles-loaded chitosan-based edible film. *J. Appl. Polym. Sci.*, **115** (2), 674–683.

39 Xie, Y., He, Y., Irwin, P.L., Jin, T., and Shi, X. (2011) Antibacterial activity and mechanism of action of zinc oxide nanoparticles against Campylobacter jejuni. *Appl. Environ. Microbiol.*, **77** (7), 2325–2331.

40 Azam, A., Ahmed, A.S., Oves, M., Khan, M.S., Habib, S.S., and Memic, A. (2012) Antimicrobial activity of metal oxide nanoparticles against Gram-positive and Gram-negative bacteria: a comparative study. *Int. J. Nanomed.*, **7**, 6003–6009.

41 Kang, S., Mauter, M.S., and Elimelech, M. (2009) Microbial cytotoxicity of carbon-based nanomaterials: implications for river water and wastewater effluent. *Environ. Sci. Technol.*, **43** (7), 2648–2653.

42 Kautt, M., Walsh, S.T., and Bittner, K. (2007) Global distribution of micro-nano technology and fabrication centers: a portfolio analysis approach. *Technol. Forecast Soc. Change*, **74** (9), 1697–1717.

43 Loveridge, D. and Saritas, O. (2009) Reducing the democratic deficit in institutional foresight programmes: a case for critical systems thinking in nanotechnology. *Technol. Forecast Soc. Change*, **76** (9), 1208–1221.

44 Bowles, M. and Lu, J. (2014) Removing the blinders: a literature review on the potential of nanoscale technologies for the management of supply chains. *Technol. Forecast Soc. Change*, **82** (1), 190–198.

45 Del Pino, P. (2014) Tailoring the interplay between electromagnetic fields and nanomaterials toward applications in life sciences: a review. *J. Biomed. Opt.*, **19** (10), 101507.

46 Realini, C.E. and Marcos, B. (2014) Active and intelligent packaging systems for a modern society. *Meat. Sci.*, **98** (3), 404–419.

47 Martinez-Olmos, A., Fernandez-Salmeron, J., Lopez-Ruiz, N., Rivadeneyra Torres, A., Capitan-Vallvey, L.F., and Palma, A.J. (2013) Screen printed flexible radiofrequency identification tag for oxygen monitoring. *Anal. Chem.*, **85** (22), 11098–11105.

48 Zampolli, S., Elmi, I., Cozzani, E., Cardinali, G.C., Scorzoni, A., Cicioni, M. . . . , and Becker, T. (2008) Ultra-low-power components for an RFID Tag with physical and chemical sensors. *Microsyst. Technol.*, **14**, 581–588.

49 Abad, E., Zampolli, S., Marco, S., Scorzoni, A., Mazzolai, B., Juarros, A. . . . , and Sayhan, I. (2007) Flexible tag microlab development: gas sensors integration in RFID flexible tags for food logistic. *Sens. Actuators B Chem.*, **127** (1), 2–7.

50 Yonzon, C.R., Stuart, D.A., Zhang, X., McFarland, A.D., Haynes, C.L., and Van Duyne, R.P. (2005) Towards advanced chemical and biological nanosensors – an overview. *Talanta*, **67** (3), 438–448.

51 Kelly, K.L., Coronado, E., Zhao, L.L., and Schatz, G.C. (2003) The optical properties of metal nanoparticles: the influence of size, shape, and dielectric environment. *J. Phys. Chem. B*, **107** (3), 668–677.

52 Zhang, C., Yin, A.X., Jiang, R., Rong, J., Dong, L., Zhao, T., Sun, L.D., Wang, J., Chen, X., and Yan, C.H. (2013) Time-temperature indicator for perishable products based on kinetically programmable Ag overgrowth on Au nanorods. *ACS Nano*, **7** (5), 4561–4568.

53 Ranjan, S., Dasgupta, N., Chakraborty, A.R., Melvin Samuel, S., Ramalingam, C., Shanker, R., and Kumar, A. (2014) Nanoscience and nanotechnologies in food industries: Opportunities and

research trends. *J. Nanopart. Res*, **16** (6), 2464.

54 Mills, A. and Hazafy, D. (2009) Nanocrystalline SnO2-based, UVB-activated, colourimetric oxygen indicator. *Sens. Actuators B Chem.*, **136** (2), 344–349.

55 Lee, S.K., Sheridan, M., and Mills, A. (2005) Novel UV-activated colorimetric oxygen indicator. *Chem. Mater.*, **17** (10), 2744–2751.

56 Suman, Gaur, V., Kumar, P., and Jain, V.K. (2012) Nanomaterial-based opto-electrical oxygen sensor for detecting air leakage in packed items and storage plants. *J. Exp. Nanosci.*, **7** (6), 608–615.

57 von Bultzingslöwen, C., McEvoy, A.K., McDonagh, C., MacCraith, B.D., Klimant, I., Krause, C., and Wolfbeis, O.S. (2002) Sol-gel based optical carbon dioxide sensor employing dual luminophore referencing for application in food packaging technology. *Analyst*, **127** (11), 1478–1483.

58 Luechinger, N.A., Loher, S., Athanassiou, E.K., Grass, R.N., and Stark, W.J. (2007) Highly sensitive optical detection of humidity on polymer/metal nanoparticle hybrid films. *Langmuir*, **23** (6), 3473–3477.

59 Borisov, S.M., Mayr, T., Mistlberger, G., Waich, K., Koren, K., Chojnacki, P., and Klimant, I. (2009) Precipitation as a simple and versatile method for preparation of optical nanochemosensors. *Talanta*, **79** (5), 1322–1330.

60 Zhang, Y.Y., Wu, J., Zhang, Y.Y., Guo, W., Ruan, S., Yonzon, C.R., Stuart, D.A., Zhang, X., McFarland, A.D., Haynes, C.L., and Van Duyne, R.P. (2005) Characterization and humidity sensing properties of the sensor based on $Na_2Ti_3O_7$ nanotubes. *J. Nanosci. Nanotech.*, **14** (6), 4303–4307.

61 Chen, X.J., Zhang, J., Ma, D.F., Hui, S.C., Liu, Y.L., and Yao, W. (2011) Preparation of gold–poly(vinyl pyrrolidone) core–shell nanocomposites and their humidity-sensing properties. *J. Appl. Polym. Sci.*, **121** (3), 1685–1690.

62 Zhou, X.T., Hu, J.Q., Li, C.P., Ma, D.D.D., Lee, C.S., and Lee, S.T. (2003) Silicon nanowires as chemical sensors. *Chem. Phys. Lett*, **369** (1–2), 220–224.

63 Elibol, O.H., Morisette, D., Akin, D., Denton, J.P., and Bashir, R. (2003) Integrated nanoscale silicon sensors using top-down fabrication. *Appl. Phys. Lett.*, **83** (22), 4613–4615.

64 Kolmakov, A., Klenov, D.O., Lilach, Y., Stemmer, S., and Moskovitst, M. (2005) Enhanced gas sensing by individual SnO_2 nanowires and nanobelts functionalized with Pd catalyst particles. *Nano. Lett.*, **5** (4), 667–673.

65 Kim, Y., Jung, B., Lee, H., Kim, H., Lee, K., and Park, H. (2009) Capacitive humidity sensor design based on anodic aluminum oxide. *Sens. Actuators B Chem.*, **141** (2), 441–446.

66 Balde, M., Vena, A., and Sorli, B. (2015) Fabrication of porous anodic aluminium oxide layers on paper for humidity sensors. *Sens. Actuators B Chem.*, **220**, 829–839.

67 Lacquet, B.M. and Swart, P.L. (1993) A new electrical circuit model for porous dielectric humidity sensors. *Sens. Actuators B Chem.*, **17** (1), 41–46.

68 Jedermann, R., Behrens, C., Westphal, D., and Lang, W. (2006) Applying autonomous sensor systems in logistics-combining sensor networks, RFIDs and software agents. *Sens. Actuators. A Phys.*, **132**, 370–375.

69 Potyrailo, R.A., Mouquin, H., and Morris, W.G. (2008) Position-independent chemical quantitation with passive 13.56-MHz radio frequency identification (RFID) sensors. *Talanta*, **75** (3), 624–628.

70 Ruhanen, A., Hanhikorpi, M., Bertuccelli, F., Colonna, A., Malik, W., Ranasinghe, D., Sanchez Lopez, T., Yan, N., and Tavilampi, M. (2008) Sensor-enabled RFID tag handbook. Available at http://www.bridge-project.eu/.

71 Fiddes, L.K. and Yan, N. (2013) RFID tags for wireless electrochemical detection of volatile chemicals. *Sens. Actuators B Chem.*, **186**, 817–823.

72 Liu, D., Qu, W., Chen, W., Zhang, W., Wang, Z., and Jiang, X. (2010) Highly sensitive, colorimetric detection of mercury(II) in aqueous media by quaternary ammonium group-capped gold nanoparticles at room temperature. *Anal. Chem.*, **82** (23), 9606–9610.

73 Zhu, K., Zhang, Y., He, S., Chen, W., Shen, J., Wang, Z., and Jiang, X. (2012) Quantification of proteins by functionalized gold nanoparticles using click chemistry. *Anal. Chem.*, **84** (10), 4267–4270.

74 Song, J., Wu, F., Wan, Y., and Ma, L. (2015) Colorimetric detection of melamine in pretreated milk using silver nanoparticles functionalized with sulfanilic acid. *Food Control*, **50**, 356–361.

75 Guo, L., Xu, Y., Ferhan, A.R., Chen, G., and Kim, D.H. (2013) Oriented gold nanoparticle aggregation for colorimetric sensors with surprisingly high analytical figures of merit. *J. Am. Chem. Soc.*, **135** (33), 12338–12345.

76 Petryayeva, E. and Algar, W.R. (2013) Proteolytic assays on quantum-dot-modified paper substrates using simple optical readout platforms. *Anal. Chem.*, **85** (18), 8817–8825.

77 Liu, D., Chen, W., Wei, J., Li, X., Wang, Z., and Jiang, X. (2012) A highly sensitive, dual-readout assay based on gold nanoparticles for organophosphorus and carbamate pesticides. *Anal. Chem.*, **84** (9), 4185–4191.

78 Che, Y., Yang, X., Loser, S., and Zang, L. (2008) Expedient vapor probing of organic amines using fluorescent nanofibers fabricated from an *n*-type organic semiconductor. *Nano. Lett.*, **8** (8), 2219–2223.

79 Che, Y. and Zang, L. (2009) Enhanced fluorescence sensing of amine vapor based on ultrathin nanofibers. *Chem. Commun* (34), 5106–5108.

80 Cui, S., Pu, H., Lu, G., Wen, Z., Mattson, E.C., Hirschmugl, C. . . . , and Chen, J. (2012) Fast and selective room-temperature ammonia sensors using silver nanocrystal-functionalized carbon nanotubes. *ACS Appl. Mater. Interfaces*, **4** (9), 4898–4904.

81 Pandey, S., Goswami, G.K., and Nanda, K.K. (2013) Nanocomposite based flexible ultrasensitive resistive gas sensor for chemical reactions studies. *Sci. Rep.*, **3**, 2082.

82 Esser, B., Schnorr, J.M., and Swager, T.M. (2012) Selective detection of ethylene gas using carbon nanotube-based devices: utility in determination of fruit ripeness. *Angew. Chem., Int. Ed.*, **51** (23), 5752–5756.

83 Su, H.C., Zhang, M., Bosze, W., Lim, J.-H., and Myung, N.V. (2013) Metal nanoparticles and DNA co-functionalized single-walled carbon nanotube gas sensors. *Nanotechnology*, **24** (50), 505502.

84 Kauffman, D.R. and Star, A. (2008) Carbon nanotube gas and vapor sensors. *Angew. Chem., Int. Ed*, **47** (35), 6550–6570.

85 Mirica, K.A., Weis, J.G., Schnorr, J.M., Esser, B., and Swager, T.M. (2012) Mechanical drawing of gas sensors on paper. *Angew. Chem., Int. Ed.*, **51** (43), 10740–10745.

86 Lakafosis, V., Rida, A., Vyas, R., Yang, L., Nikolaou, S., and Tentzeris, M.M. (2010) Progress towards the first wireless sensor networks consisting of inkjet-printed, paper-based RFID-enabled sensor tags. *P. IEEE*, **98** (9), 1601–1609.

87 European Commission (2004) Regulation (EC) No 1935/2004 of the European Parliament and of the Council of 27 October 2004 on materials and articles intended to come into contact with food. *Official J. Eur. Union*, **47** (L338), 4–17.

88 European Commission (2009) Commission Regulation (EU) No 450/2009 of 29 May 2009 on active and intelligent materials and articles intended to come into contact with food. *Official J. Eur. Union*, **52** (L135), 3–11.

89 U.S., Food and Drug Administration (2014) Food and Drug Administration – guidance for industry considering whether an FDA-regulated product involves the application of nanotechnology. *Biotechnol. Law Rep.*, **30** (5), 613–616.

90 European Union Executive (2011) Commission recommendation of 18 October 2011 on the definition of nanomaterial (2011/696/EU). *Official J. Eur. Union*, **L275**, 38–40.

91 European Commission (2011) Commission Regulation (EU) No 10/2011 of 14 January 2011 on Plastic materials and articles intended to come into contact with food. *Official J. Eur. Union*, (L12), 1–89.

92 European Food Safety Authority (2015) Scientific opinion on the safety evaluation

of the substance zincoxide, nanoparticles, uncoated and coated with [3(methacryloxy)propyl] trimethoxysilane, for use in food contact materials. *EFSA J.*, **13** (4), 4063–4071.

93 Food and Agriculture Organization/World Health Organization (2010) FAO/WHO Expert Meeting on the Application of Nanotechnologies in the Food and Agriculture Sectors: Potential Food Safety Implications: Meeting Report. Joint FAO/WHO activities on nanotechnologies.

94 Noonan, G.O., Whelton, A.J., Carlander, D., and Duncan, T.V. (2014) Measurement methods to evaluate engineered nanomaterial release from food contact materials. *Comp. Rev. Food Sci. Food Saf.*, **13** (4), 679–692.

95 European Food Safety Authority (2016) Scientific opinion on the re-evaluation of silver (E 174) as food additive. *EFSA J.*, **14** (1), 1–64.

96 European Food Safety Authority (2012) Scientific Opinion on the safety evaluation of the active substance iron (II) modified bentonite as oxygen absorber for use in active food contact materials. *EFSA J.*, **10** (10), 2906–2917.

12
Overview of Inorganic Nanoparticles for Food Science Applications

Xavier Le Guével

BIONAND, The Andalusian Centre for Nanomedicine and Biotechnology, Severo Ochoa, 35, Parque Tecnológico de Andalucía, Málaga 29590, Spain

12.1 Introduction

Inorganic nanoparticles (NPs) including metal and metal oxide-based NPs have emerged as potential nanomaterial in many fields: optoelectronics [1], catalysis [2], and in biomedical applications for diagnostic [3] and therapy [4,5]. The intrinsic properties of inorganic material at the nanoscale level such as mechanical, thermal, optical, antimicrobial properties find obvious advantages in food sciences to improve long term and sterility of food packaging and to design new composites for food additives. Detection of food contamination (chemical, pathogen) at different stages from the processing to the storage is reckoned as another crucial aspect in food science. It can benefit from the progress of inorganic NPs-based sensors and actuators over the past 25 years [6,7] to engineer the next generation of smart nanocomposites to act as platform for food analysis. We present here, a brief overview of the use of inorganic NPs for packaging, food additives, and food analysis to illustrate the benefit of engineered metal-based NPs in food science.

NP migration and toxicity will be only briefly mentioned in this chapter, as those aspects will be further extensively developed in Chapters 18 and 19, respectively.

12.2 Food Packaging, Processing, and Storage

Food packaging is considered to be the most prominent market on the use of metal and metal oxide NPs in food science with, for example, the high success of silver (Ag) NPs as antibacterial agents present in many manufactured products [8]. Inorganic NPs are usually incorporated in polymer to form nanocomposites with improved packaging properties to prevent antimicrobial activities, to enhance food storage, or to detect contamination [9].

12.2.1
Antimicrobial Activities

Metal (Ag, Cu) and metal oxide (ZnO, CuO, TiO_2) NPs-based composites are extensively investigated over the past 15 years to prevent the attachment, the growth of bacteria on packaging, and their migration to the food [9–11]. Silver is reckoned to have the greatest antimicrobial activities compared to other materials against a large family of Gram-positive and Gram-negative microorganisms. Recent studies have demonstrated the cytotoxic effect of Ag NPs at low concentration in presence of *Staphylococcus Aureus*, which is very encouraging considering the high resistance of this bacteria against various antibiotics [12,13]. Many research groups have extensively investigated this behavior to elucidate the mechanism of the antimicrobial activity of Ag NPs whether if it is originated from their physicochemical properties (size, surface, charge, nature of the ligand) or from the silver ion leakage. It appears that antibacterial activity of silver is mainly associated to silver ion leakage from the Ag NP, which interact by various mechanisms to the membrane and to the cell components such as proteins and nucleic acids leading to severe cell damages [14]. Ag NPs of different sizes from 10 to 200 nm have been incorporated directly or grown *in situ* by chemical approach in various types of matrices such as ceramic, zeolites, clays, hydrogels, or thermoplastic polymers [15]. For example, Mohan and his colleagues prepared highly monodispersed 200 nm Ag NPs by reduction of silver nitrate within an hydrogel matrix made of N-isopropylacrylamide (NIPAM) and sodium acrylate (SA) [16]. Those nanocomposites were tested against a broad range of bacteria at different concentrations and conditions (temperature, humidity, and exposure). Galeano *et al.* demonstrated that Ag NPs loaded in zeolites and coated on stainless steel surface could inactivate three Bacillus species [17]. Several studies indicated a strong antibacterial activity from Ag NPs by a slow release of silver ions without significant NP migration from the packaging. This observation was confirmed by measuring Ag NP migration from a number of commercially available food storage products using highly sensitive physicochemical characterizations [18].

Zinc oxide (ZnO) NPs are probably the second most promising inorganic NPs with silver for their antimicrobial properties bringing also additional enhanced mechanical and thermal resistance in packaging development [19]. ZnO NPs could be produced by physical, mechanical, and chemical processes [20], and were tested against various types of bacteria notably by modifying their size and surface chemistry [21]. The antibacterial activity of ZnO NPs is attributed to the leakage of toxic ions from the particle surface, to the interaction of ZnO NPs with microorganisms, and to the production of reactive oxygen species after light irradiation [19]. However, like for all the inorganic NPs present in food packaging, studies on ZnO NP migration through packaging and possible toxicity are scarce and are highly needed [8].

Despite a lower antibacterial activity than Ag NPs, copper and copper oxide NPs are also potential nanomaterials that could kill *E. Coli* and *S. Aureus* [22] mainly by releasing copper ions causing multiple toxic effects such as generation of reactive oxygen species, lipid peroxidation, protein oxidation, and DNA degradation [23].

TiO_2 NPs exhibit antibacterial properties especially with an enhanced efficacy upon UV irradiation. This photocatalytic effect to kill *E-Coli* has been demonstrated using TiO_2 coated on glass [24] and on propylene films [25].

12.2.2
Physical Barrier

On of the first application of inorganic NPs for food science was based on the high mechanical and thermal properties of those nanomaterials that could improve the mechanical strength and thermal resistance of packaging. Metal oxide NPs such as silica [26,27], titania [28], or zinc oxide [29] could drastically increase the young modulus of polymers such as polypropylene, polyvinylalcohol, starch, or cellulose that are usually used to produce food packaging. The "Nanoclays" family is also extensively investigated to improve the physical barrier of packaging but it will be discussed in another chapter of this book.

An important feature of packaging is based on keeping the freshness through transport and storage. Thus, efficient physical barriers that prevent gas and moistures to be transported from outside are high desirable. In this context, TiO_2 NPs have been included in polymer matrices and could oxidize ethylene under UV light irradiation allowing to regulate his concentration in packaging and therefore improve the shelf-life of fresh products [30]. Using this approach, nanocomposites containing TiO_2 NPs have shown remarkable results as light-driven oxygen scavenger and are present in commercial products for packaging to keep food freshness for a long period.

Other NPs such as iron oxide [31], platinum [32], and paladium [33] are active oxygen scavengers to eliminate residual O_2, even if they were used on the first place to enhance mechanical resistance of packaging. Those NPs could oxidize organic molecules such as methanol or ethylene and increase the delay of oxygen transport by filling the pores inside the polymer matrices developed for packaging. Different companies in Japan and in the United States have manufactured packaging (sachets, bottles) containing iron oxide that enable to reduce oxygen levels to 0.01%.

12.3
Supplements/Additives

Additives and supplements are extensively developed to preserve the flavor, texture, and appearance (colorants), to improve taste and nutritional value, and to reduce the use of fat, salt, sugar for safety and freshness reasons.

Two types of metal oxide NPs are present as food additives in large variety of food products: silicon dioxide (SiO_2), also known as synthetic amorphous silica or E551 and titanium oxide (TiO_2) which corresponds to 17–35% of titanium oxide food grade (E171) [34,35]. It should be highlighted that presence of the metal oxide intentionally or nonintentionally at the nanosize dimension is highly dependent on the production and extraction techniques used to obtain those food grades. TiO_2 (E171) can be found in many food products to provide packaging (tablets, capsules), cross-linking, and opacity (cottage and mozzarella cheeses, horseradish cream and sauces, and lemon curd), also in sweets where it is often used to provide a barrier between different colors. Even if this additive is considered safe, many studies still pursue to evaluate the toxicity of titanium dioxide especially for long-term exposure [36]. The food additive E551 containing SiO_2 NPs finds his interest as anticaking agents that keep powders or granulated materials such as milk powder, powdered sugar, tea, and coffee powders used in vending machines, table salt, and so on, flowing freely. In a similar perspective than TiO_2 NPs, recent studies have demonstrated significant cytotoxicity of E551 at high doses inducing strong reactive oxygen species human lung fibroblast cells [37].

Other inorganic nanoparticles are used as supplement in food made of calcium carbonate (E170), magnesium salts (E504, E511, and 553), selenium [38] for specific nutrional purpose, and iron oxide and hydroxide NPs (E172), which acts as coloring agent in food and drink (salmon pastes, shrimp pastes, meat pastes, cake and dessert packets, and soups).

12.4
Food Analysis

We have to differentiate two aspects of NPs in food analysis to prevent any confusion: (i) the detection of inorganic nanoparticles in food using spectroscopic, optical, and microscopy techniques and, (ii) biosensors based on inorganic nanoparticles to detect contamination (pathogens, toxins, pesticides, chemical compounds, etc.) with high sensitivity and selectivity.

12.4.1
NP Detection in Food

Because inorganic NPs are present as additive in food and in the manufacturing of several types of packages, their dosage in food are needed to answer to the regulatory aspect implemented in the legislation. A range of analytical technique such as (1) high-performance liquid chromatography (HPLC), (2) ultra-performance liquid chromatography (UPLC), (3) field flow fractionation (FFF), and (4) capillary electrophoresis are available to provide information on NP concentration and properties in food. Linsinger *et al.* reported an extensive review on the detection and quantification of engineered NPs in food from the analytical challenges to the validity approaches notably for silver and metal oxide NPs [39]. Using spectroscopy and spectrometry techniques Servin *et al.* showed in a recent study the transfer of TiO_2 NPs from soil to food chain, in cucumber in this case [40].

12.4.2
Nanoparticle-Based Sensors

The development of inorganic NPs-based sensors in food science benefit from the intense researches pursed over the past 25 years in the multidisciplinary fields of NP synthesis and their biofunctionalization, optical and electric detection, signal amplification, and engineering [41]. Thus, several hybrid nanosystems coupling biomolecules (enzymes, DNA, proteins, and antibodies) to inorganic molecules such as metal (Au, Ag, Cu) NPs, Quantum Dots (QDs), and metal oxide (SiO_2) NPs have been designed for food analysis [42–45]. Those biosensors are generally divided in two distinct categories related to their mode of transduction: optical and electrochemical. Furthermore, magnetic NPs bring another useful feature for food analysis by providing a fast separation between the complexes mixture and the NP-analyte system under external magnetic field. This allows rapid, simple, highly selective, and sensitive measurements when integrated to optical and electrochemical techniques.

12.4.2.1 Optical Detection
Noble metal NPs such as silver and gold exhibit strong interaction with light and are highly sensitive to their environment, thanks to their optical properties [46]. The ability to produce those NPs with high monodispersity and to decorate them easily with a large library of natural or synthetic molecules that specifically recognize chemical/biological agents offer an ideal optical platform for biosensing [47]. Hence, those remarkable features have been exploited to design metal NPs-based nanobiosensors to detect pathogens such as bacteria (*Salmonella*, *Escherichia coli* O157, *S. Aureus*) [48–52], pesticides (DDT, organophosphorus, and carbamate pesticides), [53,54] and mycotoxins [55] at ppm and ppb levels based on the change of the optical signal in absorption (surface plasmon resonance) or in scattering (surface-enhanced raman scattering). We can cite among the many studies on Au NP-based biosensors, the work of Chang *et al.*, which developed a rapid, ultrasensitive, low cost, and nonpolymerase chain reaction (PCR)-based method using aptamer-conjugated Au NPs. This nanoprobe could detect single *S. Aureus* cell within 1.5 h [56].

Semiconductors quantum dots (QDs) are rather small NPs (<15 nm), which exhibit remarkable optical properties such as high quantum yield, tunable and narrow fluorescence emission from visible to infrared region, high photostability, and possibility of multiplexing. Their surface could be modified with organic moiety or biomolecules and therefore those functionalized QDs provide a useful tool to perform multiplexing assays to determine contamination in food [57]. For example, this has been applied to measure simultaneously three food-borne pathogenic bacteria: *Salmonella typhimurium, Shigella flexneri,* and *E. coli O157: H7* in a food matrix at 10^{-3} cfu/ml using antibodies-conjugated QDs [58]. In another nanosystem, the lectin wheat germ agglutinin was grafted on QDs surface to specifically target sialic acid and N-acetylglucosaminyl residues present on bacterial cell walls [59].

Chemistry of silica enables to produce NPs with high control of their morphology (shape, size, and pores), loading, and surface chemistry [60]. Different type of hybrid silica-based on silica core/metal (Au, Ag) shells or silica NPs doped with metal NPs, QDs, or organic dyes were used as biolabels to detect foodbornes, pesticide residues, and other toxic contamination present in food [61–63].

Multifunctional NPs associating magnetic NPs to metal (Au, Ag) NPs have shown excellent performance as optical nanoprobes in terms sensitivity and selectivity to detect bacteria in food [64]. For instance, Wang *et al.* reported the use of NPs composed of gold and iron oxide NPs and functionalized with antibodies which specifically recognize two types of bacteria *S. enterica serovar typhimurium* and *S. aureus*. Those pathogens were separated by magnetic field and detected in spinach solution at 10^3 CFU/ml by SERS [65]. Using the same approach, fluorescent compounds (QDs, lanthanides complexes, and organic dyes) were combined to magnetic NPs in all-in-one particle to design optical multiplex pathogen detection assays [66–68].

12.4.2.2 Electrochemical Sensing

Au NPs display also remarkable conductivity and catalytic properties that allow analyte detection with high sensitivity by using electrochemical and electrocatalytic sensing. Rotello's group reported in an extensive review new strategies using Au NPs to detect bacteria and toxic chemical compounds [47].

12.5
Conclusion and Perspective

In summary, inorganic NPs have found many applications in the different branches of food science to (i) design active packaging, (ii) develop new food additives/encapsulation and, (iii) detect contamination with high sensititivity and selectivity in devices integrated or not in the packaging. The synthesis of those inorganic NPs and their inclusion in polymer matrices to produce smart nanocomposites are already in their mature stage enabling the development of a myriad of new formulations with various products that are already on the market.

This next generation of nanocomposites containing inorganic NPs has shown significant added values to the previous materials with antimicrobial properties, oxygen scavenging, enhanced thermal and resistance strength, and coloring.

The need to detect accurately (high sensitivity, selectivity) chemical and pathogens in food in real time production could benefit from the advances of inorganic NPs for sensing application. This young field of food science is reckoned to have great application in the future.

Despite those encouraging results, assessment of NP migration through packaging and long-term exposure to inorganic NPs require standardization and more *in vivo* studies. This issue is absolutely pivotal considering the number of those nanocomposites present in the market and the poor biodegradility/biocompatibility of such nanomaterials, which could lead to high risks for human and the environment.

Acknowledgment

This work was supported by the Instituto de Salud Carlos III (ISCII) on the project No CP12/03 310 cofinanced by European Regional Development Fund (ERDF).

References

1 Nie, Z., Petukhova, A., and Kumacheva, E. (2010) Properties and emerging applications of self-assembled structures made from inorganic nanoparticles. *Nat. Nanotechnol.*, **5**, 15–25.

2 Osterloh, F.E. (2008) Inorganic materials as catalysts for photochemical splitting of water. *Chem. Mater.*, **20**, 35–54.

3 Ladj, R., Bitar, A., Eissa, M., Mugnier, Y., Le Dantec, R., Fessi, H., and Elaissari, A. (2013) Individual inorganic nanoparticles: preparation, functionalization and *in vitro* biomedical diagnostic applications. *J. Mater. Chem. B*, **1**, 1381–1396.

4 Nam, J., Won, N., Bang, J., Jin, H., Park, J., Jung, S., Jung, S., Park, Y., and Kim, S. (2013) Surface engineering of inorganic nanoparticles for imaging and therapy. *Adv. Drug Deliv. Rev.*, **65**, 622–648.

5 Lohse, S.E. and Murphy, C.J. (2012) Applications of colloidal inorganic nanoparticles: from medicine to energy. *J. Am. Chem. Soc.*, **134**, 15607–15620.

6 Michalet, X., Pinaud, F.F., Bentolila, L.A., Tsay, J.M., Doose, S., Li, J.J., Sundaresan, G., Wu, A.M., Gambhir, S.S., and Weiss, S. (2005) Quantum dots for live cells, *in vivo* imaging, and diagnostics. *Supramol. Sci.*, **307**, 538–544.

7 Liu, J. and Lu, Y. (2003) A colorimetric lead biosensor using DNAzyme-directed assembly of gold nanoparticles. *J. Am. Chem. Soc.*, **125**, 6642–6643.

8 Hannon, J.C., Kerry, J., Cruz-Romero, M., Morris, M., and Cummins, E. (2015) Advances and challenges for the use of engineered nanoparticles in food contact materials. *Trends Food Sci. Technol.*, **43**, 43–62.

9 Llorens, A., Lloret, E., Picouet, P.A., Trbojevich, R., and Fernandez, A. (2012) Metallic-based micro and nanocomposites in food contact materials and active food packaging. *Trends Food Sci. Technol.*, **24**, 19–29.

10 Duncan, T.V. (2011) Applications of nanotechnology in food packaging and food safety: Barrier materials, antimicrobials and sensors. *J. Colloid Interface Sci.*, **363**, 1–24.

11 Bodaghi, H., Mostofi, Y., Oromiehie, A., Zamani, Z., Ghanbarzadeh, B., Costa, C., Conte, A., and Del Nobile, M.A. (2013) Evaluation of the photocatalytic antimicrobial effects of a TiO_2 nanocomposite food packaging film by *in vitro* and *in vivo* tests. *LWT Food Sci. Technol.*, **50**, 702–706.

12 Li, W.R., Xie, X.B., Shi, Q.S., Duan, S.S., Ouyang, Y.S., and Chen, Y.B. (2011)

Antibacterial effect of silver nanoparticles on *Staphylococcus aureus*. *Biometals*, **24**, 135–141.

13 Shahverdi, A.R., Fakhimi, A., Shahverdi, H.R., and Minaian, S. (2007) Synthesis and effect of silver nanoparticles on the antibacterial activity of different antibiotics against *Staphylococcus aureus* and *Escherichia coli*. *Nanomedicine*, **3**, 168–171.

14 Holt, K.B. and Bard, A.J. (2005) Interaction of silver(I) ions with the respiratory chain of *Escherichia coli*: an electrochemical and scanning electrochemical microscopy study of the antimicrobial mechanism of micromolar Ag. *Biosci. Biotechnol. Biochem.*, **44**, 13214–13223.

15 Azeredo, H.M.C.d. (2009) Nanocomposites for food packaging applications. *Food Res. Int.*, **42**, 1240–1253.

16 Murali Mohan, Y., Lee, K., Premkumar, T., and Geckeler, K.E. (2007) Hydrogel networks as nanoreactors: a novel approach to silver nanoparticles for antibacterial applications. *Polymer*, **48**, 158–164.

17 Galeano, B., Korff, E., and Nicholson, W.L. (2003) Inactivation of vegetative cells, but not spores, of *Bacillus anthracis*, *B. cereus*, and *B. subtilis* on stainless steel surfaces coated with an antimicrobial silver- and zinc-containing zeolite formulation. *Appl. Environ. Microbiol.*, **69**, 4329–4331.

18 Addo Ntim, S., Thomas, T.A., Begley, T.H., and Noonan, G.O. (2015) Characterisation and potential migration of silver nanoparticles from commercially available polymeric food contact materials. *Food Addit. Contam. Part A Chem. Anal. Control Expo. Risk Assess.*, **32**, 1003–1011.

19 Espitia, P.J.P., Soares, N.F.F., Coimbra, J.S.R., de Andrade, N.J., Cruz, R.S., and Medeiros, E.A.A. (2012) Zinc oxide nanoparticles: synthesis, antimicrobial activity and food packaging applications. *Food Bioproc. Technol.*, **5**, 1447–1464.

20 Wang, Z.L. (2004) Zinc oxide nanostructures: growth, properties and applications. *J. Phys. Condens Matter.*, **16**, R829–R858.

21 Padmavathy, N. and Vijayaraghavan, R. (2008) Enhanced bioactivity of ZnO nanoparticles – an antimicrobial study. *Sci. Technol. Adv. Mater.*, **9**, 035004.

22 Ruparelia, J.P., Chatterjee, A.K., Duttagupta, S.P., and Mukherji, S. (2008) Strain specificity in antimicrobial activity of silver and copper nanoparticles. *Acta Biomater.*, **4**, 707–716.

23 Chatterjee, A.K., Chakraborty, R., and Basu, T. (2014) Mechanism of antibacterial activity of copper nanoparticles. *Nanotechnology*, **25**, 135101.

24 Kim, Y., Choi, Y., Kim, S., Park, J., Chung, M., Song, K.B., Hwang, I., Kwon, K., and Park, J. (2009) Disinfection of iceberg lettuce by titanium dioxide-UV photocatalytic reaction. *J. Food Prot.*, **72**, 1916–1922.

25 Chawengkijwanich, C. and Hayata, Y. (2008) Development of TiO_2 powder-coated food packaging film and its ability to inactivate *Escherichia coli in vitro* and in actual tests. *Int. J. Food Microbiol.*, **123**, 288–292.

26 Dorigato, A., Sebastiani, M., Pegoretti, A., and Fambri, L. (2012) Effect of silica nanoparticles on the mechanical performances of Poly(lactic acid). *J. Polym. Environ.*, **20**, 713–725.

27 Wu, C.L., Zhang, M.Q., Rong, M.Z., and Friedrich, K. (2005) Silica nanoparticles filled polypropylene: effects of particle surface treatment, matrix ductility and particle species on mechanical performance of the composites. *Compos. Sci. Technol.*, **65**, 635–645.

28 Zhou, J.J., Wang, S.Y., and Gunasekaran, S. (2009) Preparation and characterization of whey protein film incorporated with TiO_2 nanoparticles. *J. Food Sci.*, **74**, N50–N56.

29 Díez-Pascual, A.M. and Díez-Vicente, A.L. (2014) Poly(3-hydroxybutyrate)/ZnO bionanocomposites with improved mechanical, barrier and antibacterial properties. *Int. J. Mol. Sci.*, **15**, 10950–10973.

30 Wang, K., Jin, P., Shang, H., Li, H., Xu, F., Hu, Q., and Zheng, Y. (2010) A combination of hot air treatment and nano-packing reduces fruit decay and maintains quality in postharvest Chinese

bayberries. *J. Sci. Food Agric.*, **90**, 2427–2432.

31 Chen, Z., Yin, J.J., Zhou, Y.T., Zhang, Y., Song, L., Song, M., Hu, S., and Gu, N. (2012) Dual enzyme-like activities of iron oxide nanoparticles and their implication for diminishing cytotoxicity. *ACS Nano*, **6**, 4001–4012.

32 Kajita, M., Hikosaka, K., Iitsuka, M., Kanayama, A., Toshima, N., and Miyamoto, Y. (2007) Platinum nanoparticle is a useful scavenger of superoxide anion and hydrogen peroxide. *Free Radic. Res.*, **41**, 615–626.

33 Yu, J., Liu, R.Y.F., Poon, B., Nazarenko, S., Koloski, T., Vargo, T., Hiltner, A., and Baer, E. (2004) Polymers with palladium nanoparticles as active membrane materials. *J. Appl. Polym. Sci.*, **92**, 749–756.

34 Yang, Y., Doudrick, K., Bi, X., Hristovski, K., Herckes, P., Westerhoff, P., and Kaegi, R. (2014) Characterization of food-grade titanium dioxide: the presence of nanosized particles. *Environ. Sci. Technol.*, **48**, 6391–6400.

35 Bouwmeester, H., Brandhoff, P., Marvin, H.J.P., Weigel, S., and Peters, R.J.B. (2014) State of the safety assessment and current use of nanomaterials in food and food production. *Trends Food Sci. Technol.*, **40**, 200–210.

36 Skocaj, M., Filipic, M., Petkovic, J., and Novak, S. (2011) Titanium dioxide in our everyday life; is it safe? *Radiol. Oncol.*, **45**, 227–247.

37 Athinarayanan, J., Periasamy, V.S., Alsaif, M.A., Al-Warthan, A.A., and Alshatwi, A.A. (2014) Presence of nanosilica (E551) in commercial food products: TNF-mediated oxidative stress and altered cell cycle progression in human lung fibroblast cells. *Cell Biol. Toxicol.*, **30**, 89–100.

38 Sarkar, B., Bhattacharjee, S., Daware, A., Tribedi, P., Krishnani, K.K., and Minhas, P.S. (2015) Selenium nanoparticles for stress-resilient fish and livestock. *Nanoscale Res. Lett.*, **10**, 371.

39 Linsinger, T.P.J., Chaudhry, Q., Dehalu, V., Delahaut, P., Dudkiewicz, A., Grombe, R., Von Der Kammer, F., Larsen, E.H., Legros, S., Loeschner, K., Peters, R., Ramsch, R., Roebben, G., Tiede, K., and Weigel, S.

(2013) Validation of methods for the detection and quantification of engineered nanoparticles in food. *Food Chem.*, **138**, 1959–1966.

40 Servin, A.D., Morales, M.I., Castillo-Michel, H., Hernandez-Viezcas, J.A., Munoz, B., Zhao, L., Nunez, J.E., Peralta-Videa, J.R., and Gardea-Torresdey, J.L. (2013) Synchrotron verification of TiO_2 accumulation in cucumber fruit: a possible pathway of TiO_2 nanoparticle transfer from soil into the food chain. *Environ. Sci. Technol.*, **47**, 11592–11598.

41 Gilmartin, N. and O'Kennedy, R. (2012) Nanobiotechnologies for the detection and reduction of pathogens. *Enzyme Microb. Technol.*, **50**, 87–95.

42 Shinde, S.B., Fernandes, C.B., and Patravale, V.B. (2012) Recent trends in *in vitro* nanodiagnostics for detection of pathogens. *J. Control. Release*, **159**, 164–180.

43 Neethirajan, S. and Jayas, D.S. (2011) Nanotechnology for the food and bioprocessing industries. *Food Bioproc. Technol.*, **4**, 39–47.

44 Pérez-López, B. and Merkoçi, A. (2011) Nanomaterials based biosensors for food analysis applications. *Trends Food Sci. Technol.*, **22**, 625–639.

45 Khot, L.R., Sankaran, S., Maja, J.M., Ehsani, R., and Schuster, E.W. (2012) Applications of nanomaterials in agricultural production and crop protection: a review. *Crop. Prot.*, **35**, 64–70.

46 Eustis, S. and El-Sayed, M.A. (2006) Why gold nanoparticles are more precious than pretty gold: noble metal surface plasmon resonance and its enhancement of the radiative and nonradiative properties of nanocrystals of different shapes. *Chem. Soc. Rev.*, **35**, 209–217.

47 Saha, K., Agasti, S.S., Kim, C., Li, X., and Rotello, V.M. (2012) Gold nanoparticles in chemical and biological sensing. *Chem. Rev.*, **112**, 2739–2779.

48 Miranda, O.R., Li, X., Garcia-Gonzalez, L., Zhu, Z.J., Yan, B., Bunz, U.H.F., and Rotello, V.M. (2011) Colorimetric bacteria sensing using a supramolecular enzyme-nanoparticle biosensor. *J. Am. Chem. Soc.*, **133**, 9650–9653.

49 Su, H., Ma, Q., Shang, K., Liu, T., Yin, H., and Ai, S. (2012) Gold nanoparticles as colorimetric sensor: a case study on *E. coli* O157:H7 as a model for Gram-negative bacteria. *Sens. Actuators B Chem.*, **161**, 298–303.

50 Khan, S.A., Singh, A.K., Senapati, D., Fan, Z., and Ray, P.C. (2011) Targeted highly sensitive detection of multi-drug resistant salmonella DT104 using gold nanoparticles. *Chem. Commun.*, **47**, 9444–9446.

51 Pissuwan, D., Cortie, C.H., Valenzuela, S.M., and Cortie, M.B. (2010) Functionalised gold nanoparticles for controlling pathogenic bacteria. *Trends Biotechnol.*, **28**, 207–213.

52 Phillips, R.L., Miranda, O.R., You, C.C., Rotello, V.M., and Bunz, U.H.F. (2008) Rapid and efficient identification of bacteria using gold-nanoparticle-poly(para-phenyleneethynylene) constructs. *Angew. Chem., Int. Ed.*, **47**, 2590–2594.

53 Lisa, M., Chouhan, R.S., Vinayaka, A.C., Manonmani, H.K., and Thakur, M.S. (2009) Gold nanoparticles based dipstick immunoassay for the rapid detection of dichlorodiphenyltrichloroethane: an organochlorine pesticide. *Biosens. Bioelectron.*, **25**, 224–227.

54 Liu, D., Chen, W., Wei, J., Li, X., Wang, Z., and Jiang, X. (2012) A highly sensitive, dual-readout assay based on gold nanoparticles for organophosphorus and carbamate pesticides. *Anal. Chem.*, **84**, 4185–4191.

55 Tothill, I.E. (2011) Biosensors and nanomaterials and their application for mycotoxin determination. *World Mycotoxin J.*, **4**, 361–374.

56 Chang, Y.C., Yang, C.Y., Sun, R.L., Cheng, Y.F., Kao, W.C., and Yang, P.C. (2013) Rapid single cell detection of Staphylococcus aureus by aptamer-conjugated gold nanoparticles. *Sci. Rep.*, **3**, 1863.

57 Vinayaka, A.C. and Thakur, M.S. (2010) Focus on quantum dots as potential fluorescent probes for monitoring food toxicants and foodborne pathogens. *Anal. Bioanal. Chem.*, **397**, 1445–1455.

58 Zhao, Y., Ye, M., Chao, Q., Jia, N., Ge, Y., and Shen, H. (2009) Simultaneous detection of multifood-borne pathogenic bacteria based on functionalized quantum dots coupled with immunomagnetic separation in food samples. *J. Agric. Food Chem.*, **57**, 517–524.

59 Kloepfer, J.A., Mielke, R.E., Wong, M.S., Nealson, K.H., Stucky, G., and Nadeau, J.L. (2003) Quantum dots as strain- and metabolism-specific microbiological labels. *Appl. Environ. Microbiol.*, **69**, 4205–4213.

60 Li, Z., Barnes, J.C., Bosoy, A., Stoddart, J.F., and Zink, J.I. (2012) Mesoporous silica nanoparticles in biomedical applications. *Chem. Soc. Rev.*, **41**, 2590–2605.

61 Kalele, S.A., Kundu, A.A., Gosavi, S.W., Deobagkar, D.N., Deobagkar, D.D., and Kulkarni, S.K. (2006) Rapid detection of *Escherichia coli* by using antibody-conjugated silver nanoshells. *Small*, **2**, 335–338.

62 Qi, G., Li, L., Yu, F., and Wang, H. (2013) Vancomycin-modified mesoporous silica nanoparticles for selective recognition and killing of pathogenic Gram-positive bacteria over macrophage-like cells. *ACS Appl. Mater. Interface*, **5**, 10874–10881.

63 Huang, X., Aguilar, Z.P., Li, H., Lai, W., Wei, H., Xu, H., and Xiong, Y. (2013) Fluorescent Ru(phen)3 2+-doped silica nanoparticles-based ICTS sensor for quantitative detection of enrofloxacin residues in chicken meat. *Anal. Chem.*, **85**, 5120–5128.

64 Sung, Y.J., Suk, H.J., Sung, H.Y., Li, T., Poo, H., and Kim, M.G. (2013) Novel antibody/gold nanoparticle/magnetic nanoparticle nanocomposites for immunomagnetic separation and rapid colorimetric detection of *Staphylococcus aureus* in milk. *Biosens. Bioelectron.*, **43**, 432–439.

65 Wang, Y., Ravindranath, S., and Irudayaraj, J. (2011) Separation and detection of multiple pathogens in a food matrix by magnetic SERS nanoprobes. *Anal. Bioanal. Chem.*, **399**, 1271–1278.

66 Wu, S., Duan, N., Shi, Z., Fang, C., and Wang, Z. (2014) Simultaneous aptasensor for multiplex pathogenic bacteria detection based on multicolor

upconversion nanoparticles labels. *Anal. Chem.*, **86**, 3100–3107.
67 Wan, Y., Sun, Y., Qi, P., Wang, P., and Zhang, D. (2014) Quaternized magnetic nanoparticles-fluorescent polymer system for detection and identification of bacteria. *Biosens. Bioelectron.*, **55**, 289–293.
68 Wang, H., Li, Y., Wang, A., and Slavik, M. (2011) Rapid, sensitive, and simultaneous detection of three foodborne pathogens using magnetic nanobead-based immunoseparation and quantum dot-based multiplex immunoassay. *J. Food Prot.*, **74**, 2039–2047.

13
Nanotechnology for Synthetic Biology: Crossroads Throughout Spatial Confinement

Denis Pompon, Luis F. Garcia-Alles, and Gilles Truan

LISBP, UMR 792, 135, avenue de Rangueil, 31077 Toulouse CEDEX 04, France

Synthetic biology and nanotechnology emerged together as complementary sciences and technologies addressing the engineering of functional assemblies. On one side, synthetic biology involves top-down methods (reverse engineering of rules governing biological functions with a strong emphasis on modularity) and bottom-up approaches (building original functions, structures, and properties with natural bioblocks). On the other side, nanotechnology covers a wide range of approaches aiming at exploiting the properties of mater at nanometer scales in order to build new macroscopic functions. The two fields initially developed independently, being reciprocally bound to the concepts of living and inert materials.

13.1
Convergence Between Nanotechnologies and Synthetic Biology

Synthetic biology and nanotechnology aim at designing novel functions with potential applications, particularly in the bioprocesses [1], medicine [2], and material [3,4] fields. The convergence between synthetic biology and nanotechnology can be inferred from their common exploitation of the spatial organization at the nano- to micrometer scales. Most natural biological processes rely on a delicate yet complicated organization of biological objects ranging from protein, nucleic acids, and diverse polymer structures giving rise to the supramolecular organization of living cells. Biological self-organization mechanisms are mostly encoded into their molecular components. In contrast, nanotechnology approaches frequently involve a large amount of externally encoded information, photolithographic approaches, for example. However, self-organization mechanisms now play an increasing role in nanotechnology and they constitute a major template for convergences with synthetic biology [5]. The synthetic biology concept of functional modules truly parallels the modular design of microelectronic

Nanotechnology in Agriculture and Food Science, First Edition. Edited by Monique A.V. Axelos and Marcel Van de Voorde.
© 2017 Wiley-VCH Verlag GmbH & Co. KGaA. Published 2017 by Wiley-VCH Verlag GmbH & Co. KGaA.

systems. One of the paradigm of synthetic biology is that one can design complex biological systems by combining standardized modules with predictable functions [6]. The underlying idea of our capacity to engineer complex functions by smart associations of functional elements is therefore inherent to both disciplines.

The interplay between synthetic biology and nanotechnology is now evident. A first example is the high throughput sequencing technology that exploits single molecule approaches and spatially confined enzymatic reactions into silicon-based nanostructures permitting signal processing [7]. A second one comes from artificial biological regulation networks that try to mimic the logic embedded in electronic digital devices [8]. Reciprocal examples also exist; biomimetic approaches drove the design of new super-hydrophobic or super-adhesive materials [9] or inspired fuzzy logic electronics with neural network capabilities [10]. Bio-based organic components can even be intimately interfaced with nonbiological structures resulting from standard nanofabrications [11]. Such hybrid devices are designed to implement information translation or transduction features (e.g., interconversion of chemical and electrical signals) and also bring novel properties to fabricated nano-objects like self-repair [12] or functional redundancies. They can translate in the development of low-cost, portable bio-based nanosensors for diagnostic in medicine [13], as *in vivo* reporters for early detection of health illness [14] or enhancement of specific human features and capabilities in neurosciences [15]. Crossroads between synthetic biology and nanotechnologies are too numerous and diverse to be covered in a single chapter. Among them, approaches taking profit of the nanoscale geometrical organization to create or encapsulate original biofunctions are considered as particularly promising routes for applications and will be the core subject of this chapter.

13.2
Spatially Constrained Functional Coupling in Biosystems

Natural biochemical reactions frequently occur within complex and compartmentalized architectures, which in turn play critical roles in coupling and/or modulating interactions between individual biocatalysts [16]. Cellular lipid-based organelles, proteinaceous compartments, multidomains proteins and scaffold-dependent complexes constitute the biological ways to confine enzyme reactions. Lipid-based organelles are life-critical structures that enable the compartmentation of cellular processes and reactions. However, their redesign remains extremely challenging due to their complexity and the large number of molecular components and transport mechanisms involved. Apart from these natural cellular compartments, the existence and roles of multifunctional proteins, multienzymatic complexes and metabolons was postulated years ago and further evidenced for many metabolic pathways or cellular processes.

In the case of coupled enzymatic reactions, confined/crowded environments do not only serve at increasing local concentrations of intermediates by restricting diffusion but also to limit unavoidable side reactions that could occur in free diffusing systems [17]. Spatial organization of cooperating catalysts also constitutes an important factor for controlling the elaboration and structure of biological polymers [18]. Two designs exist in natural and synthetic systems: (i) enzymes are linked or scaffolded, promoting a close proximity but the resulting macromolecule is still in contact with the cell medium and, (ii) enzymes are embedded in restricted compartments and protected from the cellular medium by a shell of biological macromolecules. In the first case, the resulting geometrical organization takes advantage of the concentration gradient of metabolic substrates or intermediates to facilitate their capture. A tighter geometrical fit between coupled enzymes, allowing direct intermediate transfers, might also be required when poorly- or not diffusible compounds (e.g., electrons) are involved. Alternatively, shell-restricted compartments can be viewed as specialized nanoreactors that confine metabolic intermediates without the need for highly constrained geometrical organization of the participating enzymes. Such shells can additionally control the selective trafficking of molecules/ions/electrons through the nanostructure. Recently, these compartments were also found to play a synchronizing role, constraining stochastic reaction networks and allowing the emergence of new properties like collective oscillations [19]. Both scaffolding and compartmentalization of biocatalysts can be designed and implemented using molecular biology approaches for *in vivo* applications, or using combinations of biological and chemical engineering when *in vitro* applications are targeted [20]. They will be analyzed in details in the following paragraphs.

13.3
Functional Coupling Through Scaffold-Independent Structures

Multidomain proteins represent the simplest natural mechanism to couple complementary functional modules [21]. They result from covalent polypeptide linkages encoded at the genetic level, making them easy to engineer by classical genetic approaches. Individual domains can be either catalytic or regulatory, thereby promoting a whole range of possible cascades of reactions and regulations. As the products of one enzymatic reaction are released into the cellular medium, diffusional dilution of intermediates can be counterbalanced by the physical proximity of the next enzyme in the metabolic pathway. In some cases, reaction intermediates are directly transferred from one catalytic site to the next one in the reaction chain without any release in the cellular medium, a phenomenon known as tunneling [22]. Modulation of multidomain protein composition or organization can be under control of natural or synthetic genetic mechanisms, for example, alternative splicing, or recombination events driving individual domain shuffling [23].

13.3.1
Functional Assembly Through Natural or Synthetic Fusions of Protein Domains

Protein fusions have been particularly developed for affinity purification, immunodetection, imaging, or fluorescent labeling. Synthetic multidomain enzymes were first described in 1987 involving either a fusion between β-galactosidase and galactokinase [24] or between a cytochrome P450 and its redox partner NADPH cytochrome P450 reductase [25]. But it was only a few years later that the beneficial role of catalyst proximity for the optimization of sequential reactions was recognized [26]. The spontaneous formation of naturally evolved fusions was then demonstrated following *in vivo* evolution of a recombinant *E. coli* strain carrying two synthetic genes for the production of glycerol and was at the origin of the yield improvement [27]. Other successful examples of artificially built multidomain enzymes involved in industrial processes like biomass degradation and utilization were further reported [28,29] and illustrated the optimization of stability and catalytic activities. Artificial fusions can also be used to create synthetic regulations. For example, a receptor domain recognizing external signal (e.g., light or calcium) can be artificially fused to a catalytic domain, thereby allowing the control of its function [30]. In genome engineering, domain fusions have been designed to associate a DNA specific affinity domain (zinc-fingers or TALE transcription activator domains) to a sequence editing catalytic domain (cleavage, recombination, or repair). The association between the recognition domain to a generic nuclease derived from the *FokI* enzyme have been used to generate targeted gene disruption, correction, or replacement [31,32]. As *FokI* is only active as a dimer, a very high specificity is achieved due to the requirement for simultaneous sequence recognition on both sides of the cutting site.

Nonetheless, although designing new multidomain architectures could seem straightforward, the successful design of efficient multidomain enzymes generally demands a combination of rational and random search through combinatorial approaches. Particularly, suitable length, composition, structure, and dynamics of the linker region are essential to achieve efficient synthetic fusions [33]. Such parameters are highly dependent on the structural features of individual domains and cannot always be rationally designed, raising practical limits. Computational tools are now offering useful help to design synthetic multidomain constructs [33], mainly by optimizing linker properties such as length, hydrophobicity, and secondary structure.

13.3.2
Functional Assembly Through Engineering of Natural or Synthetic Complexes

One of the first successful attempts to build synthetic supramolecular assembly was demonstrated by the design and construction of cyclic peptides able to self-assemble into long hollow cylinders. The resulting nanotubes could be inserted within membranes and ion transport across the phospholipid bilayer was fully reconstituted [34]. In the early 1990s, alternate approaches emerged with the two-hybrid systems [35], which aim at evidencing protein–protein interactions through a genetic test. The method takes profit of the activation of a reporter gene by a synthetic transcription factor. Two sets of protein fusions (Bait and Pray) are created between potentially interacting proteins and DNA recognition and a transcription promoting domain, reciprocally. Transcription of the reporter gene depends on the reconstitution of a functional transcription-activating complex through interaction between Bait and Prey mediated by interacting protein domains. This concept was extended to various couples of protein fusions dedicated to control formation of complexes bringing into spatial proximity catalytic partners that do not have natural tendency to interact. One advantage is that recognition domains can be of small sizes, limiting steric constraints. In addition, a wide range of recognition partners can be designed, including multisite complexes. However, building complicated interfaces still remains difficult, due to the fact that our basic scientific knowledge associated with the available computational tools are not always sufficiently performant to successfully design such interfaces [36].

13.4
Spatial Confinement Mediated by Natural and Synthetic Scaffolds

Scaffolding constitutes a more generic approach than protein fusions for the design of synthetic complexes. Scaffold structure encompasses a set of recognition domains able to specifically dock different functional partners and associate them into adapted geometrical configurations.

13.4.1
Protein-Based Scaffolds

A first example, directly inspired from the yeast mitogen-activated protein kinase pathway, involves a natural scaffold binding Ste5, Pbs2, and various signaling kinases [37]. Mutations in the scaffold specifically disrupt interactions with individual kinases and halt the associated transduction pathway. However, by engineering heterologous recruitment domains from mammals into mutated Ste5 scaffolding domains, the biological function was restored. This result evidenced that the same scaffolding domain can serve as template to promote the modular association of diverse biological functions via self-assembly. Other designs involved the same type of scaffold but different adapter domains, demonstrating flexibility in the recognition [38]. The advantage of scaffolding over protein fusions comes from the possibility of easy modulation of the geometry and the stoichiometry of recruited domains. Metabolic engineering improvement promoted by scaffolding enzymatic reactions catalyzing consecutive steps was described with the mevalonate metabolic pathway of *E. coli* [39]. The modularity of this system was controlled through the genetic regulation of scaffolds containing 1–3 copies of the binding domains allowing recruiting variable numbers of enzymatic units in the final assembly. The best configuration improved up to 77-fold the mevalonate production.

Scaffolds derived from natural cellulosomes are also widely used for engineering. Cellulosomes constitute large structures associating various cooperating glycolytic enzymes [40]. In these structures (see Figure 13.1d), a scaffold protein plays a central role in associating into the same polypeptide chain several affinity domains (cohesins). Each of these domains specifically interacts in turn with a target dockerin domain. Within the complex, the various enzymatic domains are individually fused with a different dockerin domain. This constitutes a highly modular and self-organized system that can be engineered to tune at will the

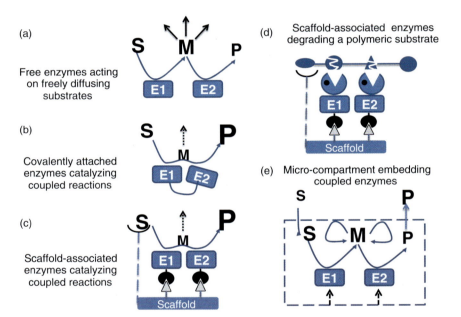

Figure 13.1 Scaffold and compartment-based confinement mechanisms for coupled enzymatic reactions. (a) Two enzymes (E1 and E2) free in solution convert a substrate S to a product P through intermediate M. (b) Same reaction with covalently-linked E1 and E2 enzymes limiting the diffusional escape of intermediate M and improving coupling by facilitating recapture of M. (c) Enzyme E1 and E2 are fused to affinity domains (black filled) specifically recognizing their respective scaffold-associated partners (triangles). (d) Enzymes are attached to a scaffold as in panel c but substrate is polymeric. Scaffold generates suitable geometry to direct enzymes E1 and E2 toward their respective targets on the structured polymer. In addition, scaffold can recruit polymer by recognizing some present motif (cellulosome like model). (e) Enzymes E1 and E2 are encapsulated into a microcompartment (dashed box). Selective shell permeability might concentrate substrate, restrict intermediate diffusion, and promote product release. Enzymes might be simply confined together, or further organized by direct or mediated interactions with the shell components (dashed arrows).

types and the structural organization of catalytic partners. Biosynthetic functions can be easily engineered by substituting catalytic domains in dockerin fusions without requiring modification of the scaffolding protein by itself [41]. Such modularity is even naturally found in different organisms where similar structures are involved in alternate functions [42]. Reciprocally, the natural cellulosome scaffold can be substituted by unrelated synthetic scaffolds, for example, derived from ankyrin assemblies, while keeping efficient cellulolytic activity [43].

13.4.2
Nucleic Acids-Based Scaffolds

Development of nucleic acids as structural bricks for the construction of rationally designed and highly organized materials is a typical orthogonal approach in synthetic biology. This technology takes advantage of predictive method using computer tools to build 1D- to 3D-scaffolds [44]. Both static assemblies and dynamic structures responding to changes in the external environment have been built including switches, walkers, and various devices exhibiting shape-triggered functions [45]. These scaffolds were used in turn to spatially organize functional nanostructures including biocatalysts and nanoparticles [46] or to build functional materials. In the case of enzymes, the main goal was to build enzymatic cascades in which spatial organization favors coupling [47,48]. In some instances, more sophisticated mechanisms were implemented involving, for example, a swinging arm to catalyze hydride transfer between dehydrogenases [49]. A major difficulty in designing DNA/RNA-scaffolds remains to control the specific associations between nucleic acid and protein components. Several strategies have been developed, involving either chemical- or enzyme-mediated covalent coupling between protein and nucleic acid tiles [50,51], self-recognition by protein of a specific DNA modification, protein recognition by single stranded region of the nucleic acid scaffold [52,53], fusion with DNA binding domains [54], or geometrical shape discrimination [55]. Assembly of DNA-based scaffolds followed by covalent coupling between components through chemical linkers was frequently used for *in vitro* demonstrations. However, applications requiring *in vivo* assemblies mostly involve noncovalent associations. Protein domains similar to the ones found in transcription factors or DNA-acting enzymes can be used to recognize nucleotide sequences on scaffold. Alternatively, 3D-folded nucleotide domains (aptamer) can be included in the scaffold to recognize specific protein components [56]. The use of RNA scaffolds appears particularly well adapted for *in vivo* engineering but RNA- or DNA-tiles also offer potential to assemble into large 3D-structures able to embed protein components [57]. An important application of RNA–protein hybrid scaffolds is the case of the CRISPR/Cas9 complexes. Natural complexes involved in bacterial defense systems were engineered to build-up synthetic nucleases able to be tuned as will to cut DNA sequence complementary to a given "guide RNA" [58]. These complexes use the RNA component as a sequence recognition module and the associate protein part as DNA cleaving catalyst.

While nucleic scaffold offers the advantage to be rationally designed using current computing tools, they suffer, with rare exceptions, from the limited availability of easily manageable tools to manage nucleic acid–peptide interactions. In addition, single-stranded DNA and RNA structures can have fairly short *in vivo* lifetimes or interfere with cell physiology.

13.5
Encapsulated Biosystems Involving Natural or Engineered Nanocompartments

Within organisms, compartmentalization plays general roles in isolating incompatible activities or biochemicals while maintaining necessary concentration gradients. Subcellular compartments like mitochondria, endoplasmic reticulum, or chloroplast not only provide intracellular surfaces for homing membrane-bound enzymes and complexes but also create spatially confined volumes allowing colocalization of metabolically related enzymes. As a consequence of the presence of physical barriers, specific transport mechanisms are evolved to negotiate exchanges with the surrounding environment. This section provides views on alternative implementations of compartmentalization, based on natural and derived assemblies that are currently exploited for a panoply of technological applications.

13.5.1
Lipid-Based Compartments

Lipid-based compartments are the most frequent structures found in living organisms. Their limiting layers are mostly made of amphipathic molecules that self-assemble, such structuration being mostly energized by the favorable packing of lipid hydrophobic tails together with concomitant exposition of polar heads to the aqueous phase. Lipid vesicles can be decorated on their surface with a large range of additional functional biocomponents, for example, proteins or oligosaccharides. They are relatively easy to prepare and their transport properties can be adapted to protect bioactive cargos or for controlled drug or effectors delivery [59]. Protein transport into natural lipid compartments can be engineered to implement or redirect synthetic metabolic pathways [60,61]. However, with few exceptions like the caveolin-driven formation of cytoplasmic vesicles in *E. coli* [62], their engineering usually involves *in vitro* approaches. Difficulties arise from the fact that resulting structures depend on the presence of a complex set of molecular effectors mediating for instance lipid biosynthesis, vesicle budding, or material exchanges [63].

13.5.2
Protein-Based Nanocompartments

Protein-based scaffolds offer the advantage of facilitated design using available genetic engineering tools. They are particularly adapted for the *in vivo* generation of a large variety of functionalized nanoscale structures. Generally speaking, natural protein scaffolds might be grouped in two classes, depending on whether structural and functional components are strongly integrated and inter-dependent or not. Larger structures belong to the second class, involving scaffolds that are more flexible in shape and/or size and encapsulate a wide range of biological or nonbiological cargos.

13.5.2.1 **Shell-Independent Nanocompartments**
The first type of structures corresponds to self-organized assemblies forming small internal cavities where their physiological roles are confined. Such compartments, found in chaperones, ferritin, proteasomes, lumazin synthase, and pyruvate dehydrogenase complexes, do not feature independent shell structure. Their principal characteristics and applications are compiled in Table 13.1 and some structures are shown in Figure 13.2. All these complexes have been considered for their bioengineering capacity. Ferritin was likely the most extensively exploited protein capsule. In nature, it ensures storage and transport of iron atoms as well as phosphates and hydroxide ions within its hollow interior. The biomineralizing properties of ferritin have been recurrently exploited to prepare nanoparticles of iron, metal oxides, carbon nanotubes, and other appealing materials for microelectronics and imaging [64]. The possibility to incorporate functional catalysts within the internal cavity was demonstrated through the realization of confined chemical reactions or by the preparation of polymers with narrow size distributions. Assemblies of similar size to ferritin are formed by chaperones and proteasomes to facilitate refolding/reparation and degradation of misfolded/damaged proteins, respectively. These structures organize as multimeric rings that pile on each other generating barrels with internal trans-piercing tunnels where substrates are bound [65,66]. Large pores allow access of target polypeptides to the interior volume, which is mostly surrounded by hydrophobic residues in chaperones or decorated with protease cleavage sites in proteasomes. Similarly to ferritin, chaperone organization has been investigated as mineralization vessel for iron and other transition metals and alloys [67], and for volume-constrained polymerization purposes. Small HSP cages with engineered cysteine residues were also used as vectors to convey and release drugs in a pH dependent manner (e.g., the antitumor agent doxorubicin) [68]. Contrarily, proteasomes, or the RNA-degrading counterpart called exosomes, have not yet being regarded as nano-technological tools.

The organization of pyruvate dehydrogenase (PDC) and lumazine/riboflavin synthase complexes (LRSC), which catalyze the conversion of pyruvate into acetyl-CoA or the last steps of riboflavin biosynthesis, respectively, have also inspired numerous applications. In both cases, the multienzymatic complex is built around a central core composed of dozens of subunits (E2 subunits in PDC and β-subunits of LRSC) [108,109]. Complementary catalytic activities are found either attached to the container outer surface (PDC) or inside (LRSC). Importantly for technological applications, the two surface-exposed and noncatalytic domains of the tridomain E2 subunit of PDC can be replaced by foreign peptides

Table 13.1 Structural properties and biotechnological applications of protein nanocompartments.

Family	Major function	Prototype example	Approximate size (nm)	Number of shell subunits	Largest pore (Å)	Emplacement of modification for applications	
						Outer surface	Interior
Ferritin	Iron storing	Human	13	24	<8	Targeting peptide [69] Vaccine eliciting [70]	Biomineralization [64] Confined chemical catalysis [71,72] Constrained polymerizations [71,72] Contrast agents [73] Drug delivery [69]
Proteasomes	Protein degradation	*Thermoplasma acidophilum*	11×11×15	28	13	Not existing	
Chaperones Chaperonins	Protein folding	GroEL/GroES *E. coli*	13×13×19	14/7	50	Targeting peptide [74]	Metal biomineralization [67] Catalysis [75]; Imaging [74] Drug release [68] Size-constrained polymerizations [76]
PDC	Central metabolism	*Azotobacter vinelandii*	13×13×13	24/60	35	Peptide display [77] Immune tolerance [78] Cell targeting [79]	Bioimaging [79] Drug delivery [79] Enhanced hydrophobicity [80]

Name	Function	Organism	Size (nm)			Applications	
LRSC	Vitamin B2 biosynthesis	*Aquifex aeolicus*	15	60	15	Vaccine eliciting [81]	Metal biomineralization [82], Cargo encapsulation [83,84], Bioimaging [85]
Encapsulins	Defense oxidative stress	*Thermotoga maritima*	22	60	<8	Peptide display [86], Cell targeting [86]	Drug delivery [86], Packaging luciferase [87], Bioimaging [86]
Viruses	Genetic material encapsulation	Cowpea chlorotic mottle virus	28[a]	180	18	Vaccine eliciting [88]	Drug delivery [89], Gene therapy [90], Bioimaging [91], Polymerization reactions [91], Biomineralization [92], Enzymatic catalysis [93–95]
Vaults	Not known	*Rattus norvegicus*	42 × 42 × 75	78/96[b]	30	Cell targeting [96]	Semiconducting polymer [97], Encapsulation of proteins [98,99], Vaccine formulation [96], Lysosomal release, drug delivery [100], Increase of hydrophobicity [101]
BMC	Enzyme encapsulation	carboxysomes *Synech. species*	110	≈ 4500[c]	15	Unknown	Enzyme encapsulation [102,103]

(*continued*)

Table 13.1 (Continued)

Family	Major function	Prototype example	Approximate size (nm)	Number of shell subunits	Largest pore (Å)	Emplacement of modification for applications	
						Outer surface	Interior
GV	Buoyancy	*Halobacterium salinarum*	100 × 700	Not known	6[d]	Epitope display [104] Vaccine eliciting [105] Enzyme display [106]	Not known
S-layers	Wrap cells	*Methanosarcina acetivorans*	Cell size	Not known	20–80	Basically exploited as scaffolding matrix [107] Stabilization of lipid membranes and liposomes [107]	

a) Viral capsids can be up to 200 nm diameter. However, up to 1 μm-large structure called gyruses are being characterized.
b) Controversial.
c) Estimated from dimensions of hexamer bricks and assuming viral capsid-like model.
d) Estimated from permeability tolerance to gases.

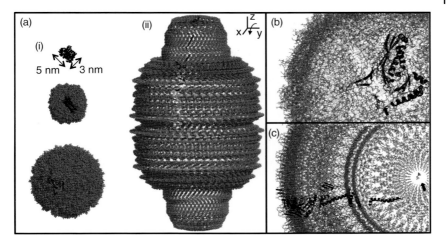

Figure 13.2 Comparison of the structural organization of some protein nano-compartments. (a) Shape and size comparison of biological assemblies formed by ferritin (PDB code 1FHA, middle object of (i)), encapsulin (3DKT, bottom object of (i)) and vaults (2QZV, (ii)). A stick representation is drawn, with a single monomer brick presented as black ribbons. For comparison, the top object of (i) is for the crystal structure of a green fluorescent protein (2WVR), with approximate dimensions indicated. (b) An enlarged view of encapsulin structure highlights a compact packing with presence of narrow interstices. (c) Similar but less enlarged representation of vaults after 90° rotation (as indicated in (a)) reveals the occurrence of 25–35 Å-wide pores. Molecules of ATP (indicated with arrows) are modeled above the nanocompartment surface for comparison in (b) and (c).

and even proteins without affecting the core assembly. Such simplified scaffolds, either as wild-type versions or after surface PEGylation elicited weak innate immune responses [78]. This and other engineered constructs with increased core hydrophobicity were exploited for drug delivery, bioimaging, or biodiagnosis applications (Table 13.1). In the case of LRSC, the internal α-subunits can be easily removed after dissociation and reassembling of the original icosahedric cage [108,110]. Resulting containers were used as template for mineralization of iron oxide. Protein engineering strategies were also applied to guide the electrostatic-driven encapsulation of (polycationic) decaarginine-tagged GFP [83]. The concept was further developed to identify coats with improved sequestering capabilities, which were obtained after directed evolution rounds in cells expressing R10-tagged HIV protease, a toxic agent when present free in the cytosol of the host [84]. Biomedical applications are also emerging (see Table 13.1).

13.5.2.2 Shell-Dependent Nanocompartments

These types of scaffolds can form large structures, making possible the encapsulation of many copies of biological components within a single vessel. Shell formation involves the noncovalent polymerization of a single or limited number of structural protein tiles. The biological function of the resulting assembly will be mostly defined by the embedded cargo, which in principle can be reprogramed to build new functions. Bacterial microcompartments (BMC), viral capsids, encapsulins, vaults, and gas vesicles fall in this category. *Surface (S)-layers*, which are self-assembled arrays of a (glyco)protein that wrap around many bacteria and archaea [107], might be also included here. However, although interesting as patterning elements and nanoscale building blocks, they seem hardly exploitable as encapsulating tools.

Encapsulins are icosahedral capsids built exclusively from pentamers that enclose important enzymes for cellular responses to oxidative-stress [111]. Many *viruses* are also icosahedric, yet coat shape and size can be very disparate. Diameters range between 20 and 200 nm, and can even extend up to the micrometer dimension in giant viruses [112]. Eukaryotic vaults, contrarily, are ellipsoidal capsules resulting from the lateral association of an elongated major vault protein (MVP) [113] (Figure 13.2), which naturally embed poly(ADP-ribose) polymerase, telomerase-associated protein and untranslated RNAs. Concerning gas vesicles (GV), these are large structures (30–250 nm wide, 50–1000 nm long) with species-characteristic shapes [114]. The gas-permeable 2 nm-thin layer, mostly made of protein A (GvpA) tiles associated to complementary protein components [115], is the key element that provides buoyancy properties to cyanobacteria and some archaea.

Exception made of GV, assembling–disassembling can often be triggered by simple means, such as pH shifts [116,117] or salt concentration changes [118] in the presence or absence of natural contents [119]. Besides, cargo loading can be programmed both *in vitro* and *in vivo* by varied means. Viruses package nucleic acids, their inner shell surfaces are therefore positively charged. This feature has been recurrently exploited to introduce nonviral cargoes, with applications in drug delivery, gene therapy, or bioimaging (Table 13.1). In the case of encapsulins, foreign enzymatic cargo can be targeted to the compartment interior via C-terminal signal sequences [87]. Similarly, an MVP-interacting domain can be exploited to direct foreign proteins into vaults lumen [98,99]. Anionic semiconducting polymers were also entrapped inside vaults, suggesting that the inner cavity surface is positively charged [97]. Alternatively, artificial contents can be directed into these compartments by means of covalent fusions with shell components. This permitted for instance to graft a short peptide on drug-containing encapsulins that efficiently targeted hepatocarcinoma cells [86]. Similar strategies are exploited in the case of vaults, since the final emplacement of foreign peptide/protein can be controlled by fusing at N- (inner side) or C-ter (outer surface) of MVP [113]. Regardless of such differences, all these compartments have inspired similar technological applications (summarized in Table 13.1). Apart from targeting medical uses in imaging or drug delivery, these capsules can be also exploited as molds for constraining polymerization of organic compounds or for biomineralization purposes [92,97,120]. Artifical nanoreactors encapsulating diverse enzymes, such as luciferase, lipases, peroxidases, or deaminases, were also described [87,93–95]. In another noteworthy work, vaults were engineered to display an internal lipophilic lumen capable of sequestering hydrophobic molecules [101]. The stability and weak immunogenicity of vaults have motivated intensive work aiming to exploit them as nanocarriers to elicit responses to encapsulated immunogenic peptides and proteins [113].

13.5.2.3 Bacterial Microcompartments: Framework for Enzymatic Nanoreactors

Despite of their stability, applications of above-mentioned nanocompartments in the field of synthetic biology are hampered by their small size and/or lack of suitable properties. As example, an encapsulin shell was estimated to provide enough volume to accommodate only one single copy of a 240 kDa multimeric peroxidase [111]. GV and some viral capsids can be much larger, but adapting shell permeability to specific confined reactivities appears challenging. Besides, mechanisms permitting to redirect GV contents are missing. Basically, technological applications are circumscribed to epitope display on surfaces for eliciting immune responses against viruses or pathogenic bacteria. In contrast, bacterial microcompartments (BMC) constitute serious candidates for the design of nanoreactors [121]. BMC are polyhedral organelles (roughly 100–200 nm in diameter) that encapsulate enzymes implicated in diverse metabolic processes. Large volumes and adapted shell permeability are therefore intrinsic BMC features. Recent bioinformatic searches in sequenced bacterial genomes permitted to identify 23 different BMC types [122]. Best characterized are the carboxysomes, which will be presented below, and the Pdu and Eut BMCs that mediate the metabolic utilization of 1,2-propanediol and ethanolamine, respectively. Other BMC encapsulate enzymes for catabolic action on varied compounds, such as ethanol (Etu BMC), L-fucose (PVM BMC), or L-rhamnose (PVM-like BMC). Catabolic processes occurring inside BMC generally involve toxic aldehyde intermediates [123,124], highlighting a major role of BMC shells in preserving cellular viability.

Functional BMC compartments naturally result from the products of 10–20 shell and cargo genes, which may be clustered in operons or segregated at different loci. Recombinant Pdu, Eut, and carboxysomes can be reconstituted following operon transfer to alternate hosts [103,125,126]. Loci coding for up to three functional different BMC were unveiled in 12 organisms, suggesting that strains with combinations of such modules should be feasible. Despite major differences in the nature of encapsulated enzymes, all BMC shells are structurally-related. Namely, they assemble from combinations of hexamers and pentamers that organize in 3D with BMC-specific folds. Pores of variable sizes (0–14 Å) observed in crystal structures of these shell bricks are assumed to function as gates for molecular exchange with the cytoplasm [121]. The common 3D folding suggests that bricks from different BMC types might be combined in artificial shells, either directly or after minor reengineering of interacting interfaces [127], a possibility that reinforces the avenues toward nanoreactors with shell permeability adapted to new encapsulated processes.

Carboxysomes constitute the only known anabolic BMC [121]. They are composed of 5–10 different shell proteins that confine RuBisCO and carbonic anhydrases (CA), necessary for CO_2 fixation in photosynthetic cyanobacteria and some chemoautotroph microorganisms. In addition, CcmM and CcmN proteins are found at the interface between the enzyme cargo and shell in β-carboxysomes. Depending on compartment size, 3000–5000 protein subunits can be assembled in the shells [128]. The physiological importance of carboxysome shell barrier is demonstrated by comparative studies with mutants lacking key shell components. Thus, cyanobacteria carrying defective carboxysomes need high concentrations of carbon dioxide to proliferate, contrasting with growth of wild-type strains in atmospheric concentrations [129,130]. Similar data suggest for other BMC a correlation between shell integrity and BMC-conferred growth advantages [121]. Encapsulation would prevent leaking of rapidly diffusing intermediates. The shell would permit to exert control on the exchange of substrates, products, ions, and enzyme cofactors between the carboxysome lumen and the cytoplasm. Shells might also limit entry of potential competitors and inhibitors. This might be especially important as strategy to reduce photorespiration [131], a side-reaction with O_2 (instead of CO_2) catalyzed also by RuBisCO. Colocalization of RuBisCO and carbonic anhydrase would also contribute in virtue of the increase of local CO_2 concentration around RuBisCO.

13.5.2.4 Engineering of Natural BMC

Engineering of BMC focuses intense investigations. The importance of compartmentalization is highlighted by theoretical simulations that indicate that crop yields might increase by over 30% if carboxysomes could be reconstituted in chloroplasts together with a few other carbon-concentrating cyanobacterial components [132]. As mentioned, an important step has been to show that BMC can be assembled in recombinant hosts after genetic loci transfer. This permitted to engineer an ethanol bioreactor by targeting pyruvate decarboxylase and alcohol dehydrogenase into Pdu compartments of *Citrobacter freundii* [102]. Transfer of engineered loci is a simple way for introducing new biochemical pathways of industrial interest in alternate microorganisms. Concerning the more challenging transfer to eukaryotes, a pioneering example is the formation of BMC structures after transient expression of multiple β-carboxysome proteins in chloroplasts of the plant *Nicotiana benthamiana* [133]. This study also demonstrated the possibility of targeting into such structures foreign cargo fused to the targeting CcmN peptide (see below).

Mastering mechanisms to control the embedding of functional biological components into BMC is critical for applications. Fan *et al.* reported that the N-terminal 10–14 amino acid long peptide of PduP of *Salmonella enterica* was responsible of directing its localization inside Pdu BMC [134]. Bioinformatics analysis mapped similar N-ter extensions in PduD and PduE subunits of the diol dehydratase as well as in EutC and EutG enzymes. Ulterior studies provided evidences for encapsulation being driven by the interaction between the N-ter targeting peptide and C-terminal helices of shell proteins PduA, PduJ, or PduK [102,135]. Functionally similar sequences were also identified in the C-terminus of CcmN of β-carboxysomes [136]. Importantly, inspection of all classified BMC gene clusters of different functionalities proved that homologs of the CcmN/PduP peptides are found at either the N- or C terminus, or even between domains, of at least one gene corresponding to a presumable encapsulated protein. These peptides are short (13–22 residues) with a portion predicted to adopt an amphipathic α-helical structure. Interestingly, tagged components usually bind with enzymes catalyzing the first or last step of the encapsulated metabolism, suggesting that such enzymes are harbored adjacent to the shell and that substrate and products flux is probably finely controlled.

Apart from controlling the integration of new cargo, exploiting BMC as nanoreactors for technological applications will require a good understanding of the molecular determinants of shell assembly and molecular transport across BMC. Our present knowledge on BMC shell biogenesis is mostly based on β-carboxysomes studies, which indicate that the assembly might occur in a stepwise manner. The presence of structurally organized seeds of RuBisCO/CA, with presumably interfacing CcmM and CcmN proteins, appears to precede the formation of the wrapping shell layer [137,138]. However, other studies indicate that less organized compartments form in the absence of such seeds [139,140]. In the case of α-carboxysomes, microscopy data suggest that cargo and shell proteins would assemble simultaneously [141], and several studies proved the possibility to reconstitute BMC in absence of RuBisCO [103,130,142]. These observations matched well with the absence of known cargo–shell interfacing proteins in α-carboxysomes. In contrast, the occurrence of large cargo polar bodies in mutant strains unable to form Pdu shells suggests that assembling might happen stepwise like in β-carboxysomes [143]. Another obscure area is to ascertain the relative importance to the overall BMC function of mechanisms of control of molecular permeability over effects related to enzyme scaffolding/colocalization. Necessarily, shells must allow transfer of substrates and products and permit regeneration or exchange of cofactors, ions, electrons, and so on. Protons are known to freely diffuse across the shells, according to data measured for α-carboxysomes incorporating a pH-sensitive GFP reporter [144]. However, the coat is expected to regulate the exchange with the cytosol of toxic intermediates and volatile molecules like CO_2 and O_2, suggesting that molecular sieving properties could be defined by the size and electrochemical properties of pores of multimeric shell bricks. Two major strategies are envisioned to reconfigure shell permeability, either by modifying the pore-flanking residues or by replacing a component by a shell unit from a different BMC, eventually engineered to interact with a novel shell environment [127].

13.6
Synthetically Designed Structures for Protein Coupling and Organization

Engineering of protein and nucleic acid building blocks are promising approaches to *de novo* design intracellular artificial nanocompartments. Preparing 3D-self-organized nucleic acid scaffolds is relatively straightforward, but suffers from drawbacks, for instance the difficulties to link shell components to compartment contents/decorations necessary for the precise nanotechnological applications. Protein-based compartments are in principle less subjected to such problems. However, engineering the complex network of inter-subunit interactions necessary to form stable protein-based nanocompartment still remains challenging.

Several approaches have been successful for the preparation of synthetic protein compartments. The first one is based on multidomain proteins comprising two self-assembling domains that naturally oligomerize with variable symmetries [145]. In this manner, Yeates and coworkers demonstrated the preparation of 15 nm-large cages constituted of 12 assembled fusion units incorporating dimerizing and trimerizing domains. More recently, cube-shaped cages and coil–coil dimerizing domains successfully led to the formation of nanometric cages, yet still composed by a mixture of different structures [146]. A second approach consists in redesigning new protein interfaces to guide the assembly toward geometrically defined architectures. Examples are the reconfiguration of the interface of a trimeric protein to drive association into 11- or 13-nm big protein cages made by association of 12- or 24-subunits [147], respectively, or the preparation of tetrahedral cages combining two different components with engineered contact surfaces [148]. These and other reports illustrate that designing protein nanocompartment architectures is nowadays feasible. These studies will be valuable for the future development of novel nanomaterials for cell targeting, drug delivery, immunotherapy, biomineralization, and so on. Contrarily, further improvements permitting to control cargo contents or cage permeability will need to be implemented to exploit them as reactor containers.

DNA-based containers offer an interesting alternative. Impressive demonstrations document the possibility to prepare dozens of 3D objects using a modular approaches exploiting self-assembly of DNA bricks [44]. Similar approaches led to the assembly of perfectly defined DNA cubes [149] or RNA polyhedrons [57]. Self-assembled DNA nanostructures with intricate curved surfaces in 3D were also prepared using DNA origami [150]. Encapsulation of proteins (cytochrome c) within a rigid tetrahedral DNA cage was also demonstrated [151], pointing out to the possibility of joining in the future the protein and nucleic acid engineering for complex artificial nanostructure design.

13.7
Future Directions

Forthcoming opportunities might emerge when nanostructured materials other than proteins or nucleic acids will be exploited to guide the structural and functional organization of biocomponents. Among building blocks, lipid assemblies, oligosaccharides, or other biopolymers (e.g., chitin, lignin) play critical roles in biomaterial structuration but to date remained mostly absent from synthetic biology approaches. These materials cannot be directly encoded at the genetic level and result from complex interplays involving both structures and cellular mechanisms. Indeed, technologies permitting to rationally control spatial organization at different scales of biomaterials that are not genetic remain to be designed. Research works on complex organisms like plants allow establishing relations of causality between the presence of specific genetically encoded materials and micro- or macroscopic organizations, but such knowledge remain widely insufficient to enable retroengineering. Due to the complexity of this natural net, it can be questioned whether the more efficient approach is to try to master such natural complexity or if completely orthogonal concepts have to be designed. However, more general routes, particularly implementable in living organisms remain to be designed. Convergences between synthetic biology and nanotechnologies are starting to have bodies and will certainly offer future codevelopment tools to generate synthetic supramolecular organizations with defined, complex, biological functions useful for industrial applications.

References

1 Khalil, A.S. and et Collins, J.J. (2010) Synthetic biology: applications come of age. *Nat. Rev. Genet.*, **11** (5), 367–379.
2 Mihail, C.R. (2003) Nanotechnology: convergence with modern biology and medicine. *Curr. Opin. Biotechnol.*, **14**, 337–346.
3 Rice, M.K. and et Ruder, W.C. (2014) Creating biological nanomaterials using synthetic biology. *Sci. Technol. Adv. Mater.*, **15**, 014401.
4 Bryksin, A.V., Brown, A.C., Baksh, M.M. et al. (2014) Learning from nature – novel synthetic biology approaches for biomaterial design. *Acta. Biomater.*, **10**, 1761–1769.
5 Ball, P. (2005) Synthetic biology for nanotechnology. *Nanotechnology*, **16** (1), R1–R8.
6 Purnick, P.E.M. and et Weiss, R. (2009) The second wave of synthetic biology: from modules to systems. *Nat. Rev. Mol. Cell Biol.*, **10** (6), 410–22.
7 Metzker, M.L. (2010) Sequencing technologies – the next generation. *Nat. Rev. Genet.*, **11** (1), 31–46.
8 Wang, B. and et Buck, M. (2014) Rapid engineering of versatile molecular logic gates using heterologous genetic transcriptional modules. *Chem. Commun. Camb.*, **50** (79), 11642–11644.
9 Liu, M., Zheng, Y., Zhai, J., and et Jiang, L. (2010) Bioinspired super-antiwetting interfaces with special liquid–solid adhesion. *Acc. Chem. Res.*, **43** (3), 368–377.
10 Fan, D., Sharad, M., Sengupta, A., and Roy, K. (2015) Hierarchical temporal memory based on spin-neurons and resistive memory for energy-efficient brain-inspired computing. *IEEE Trans. Neural. Netw. Learn. Syst.*, DOI: arXiv:1402.2902.
11 Collier, C.P. and et Simpson, M.L. (2011) Micro/nanofabricated environments for synthetic biology. *Curr. Opin. Biotechnol.*, **22**, 516–526.
12 Wong, T.-S., Kang, S.H., Tang, S.K.Y. et al. (2011) Bioinspired self-repairing slippery surfaces with pressure-stable omniphobicity. *Nature*, **477** (7365), 443–447.
13 Ye, H. and et Fussenegger, M. (2014) Synthetic therapeutic gene circuits in mammalian cells. *FEBS Lett.*, **588** (15), 2537–2544.
14 Eckert, M.A., Vu, P.Q., Zhang, K. et al. (2013) Novel molecular and nanosensors for *in vivo* sensing. *Theranostics*, **3** (8), 583–594.
15 Walsh, F., Balasubramaniam, S., Botvich, D., and et Donnelly, W. (2010) Synthetic protocols for nano sensor transmitting platforms using enzyme and DNA based computing. *Nano Commun. Netw.*, **1**, 50–62.
16 Conrado, R.J., Varner, J.D., and et DeLisa, M.P. (2008) Engineering the spatial

organization of metabolic enzymes: mimicking nature's synergy. *Curr. Opin. Biotechnol.*, **19**, 492–499.

17 Chen, A.H. and et Silver, P.A. (2012) Designing biological compartmentalization. *Trends Cell Biol.*, **22**, 662–670.

18 Vazana, Y., Barak, Y., Unger, T. *et al.* (2013) A synthetic biology approach for evaluating the functional contribution of designer cellulosome components to deconstruction of cellulosic substrates. *Biotechnol. Biofuels.*, **6** (1), 182.

19 Zieske, K. and et Schwille, P. (2015) Reconstituting geometry-modulated protein patterns in membrane compartments. *Methods Cell Biol.*, **128**, 149–163.

20 Schoffelen, S. and et van Hest, J.C.M. (2013) Chemical approaches for the construction of multi-enzyme reaction systems. *Curr. Opin. Struct. Biol.*, **23**, 613–621.

21 Apic, G., Gough, J., and et Teichmann, S.A. (2001) Domain combinations in archaeal, eubacterial and eukaryotic proteomes. *J. Mol. Biol.*, **310** (2), 311–325.

22 Miles, E.W. (2001) Tryptophan synthase: a multienzyme complex with an intramolecular tunnel. *Chem. Rec.*, **1** (2), 140–151.

23 Kelemen, O., Convertini, P., Zhang, Z. *et al.* (2013) Function of alternative splicing. *Gene*, **514** (1), 1–30.

24 Bülow, L. and et Mosbach, K. (1987) Preparation of bifunctional enzyme complexes by fusion of two genesa. *Ann. NY Acad. Sci.*, **501** (1), 44–49.

25 Murakami, H., Yabusaki, Y., Sakaki, T. *et al.* (1987) A genetically engineered P450 monooxygenase: construction of the functional fused enzyme between rat cytochrome P450c and NADPH-cytochrome P450 reductase. *DNA*, **6** (3), 189–197.

26 Bülow, L. and et Mosbach, K. (1991) Multienzyme systems obtained by gene fusion. *Trends Biotechnol.*, **9** (1), 226–231.

27 Meynial Salles, I., Forchhammer, N., Croux, C. *et al.* (2007) Evolution of a *Saccharomyces cerevisiae* metabolic pathway in *Escherichia coli*. *Metab. Eng.*, **9** (2), 152–159.

28 Wang, R., Xue, Y., Wu, X. *et al.* (2010) Enhancement of engineered trifunctional enzyme by optimizing linker peptides for degradation of agricultural by-products. *Enzyme Microb. Technol.*, **47** (5), 194–199.

29 Rizk, M., Antranikian, G., and et Elleuche, S. (2012) End-to-end gene fusions and their impact on the production of multifunctional biomass degrading enzymes. *Biochem. Biophys. Res. Commun.*, **428** (1), 1–5.

30 Volzing, K., Biliouris, K., and et Kaznessis, Y.N. (2011) proTeOn and proTeOff, new protein devices that inducibly activate bacterial gene expression. *ACS Chem. Biol.*, **6** (10), 1107–1116.

31 Miller, J.C., Tan, S., Qiao, G. *et al.* (2011) A TALE nuclease architecture for efficient genome editing. *Nat. Biotechnol.*, **29** (2), 143–148.

32 Kim, Y.G., Cha, J., and et Chandrasegaran, S. (1996) Hybrid restriction enzymes: zinc finger fusions to Fok I cleavage domain. *Proc. Natl. Acad. Sci. USA*, **93** (3), 1156–1160.

33 Yu, K., Liu, C., Kim, B.-G., and et Lee, D.-Y. (2015) Synthetic fusion protein design and applications. *Biotechnol. Adv.*, **33** (1), 155–164.

34 Ghadiri, M.R., Granja, J.R., and et Buehler, L.K. (1994) Artificial transmembrane ion channels from self-assembling peptide nanotubes. *Nature*, **369** (6478), 301–304.

35 Fields, S. and et Song, O. (1989) A novel genetic system to detect protein–protein interactions. *Nature*, **340** (6230), 245–246.

36 Stranges, P.B. and et Kuhlman, B. (2013) A comparison of successful and failed protein interface designs highlights the challenges of designing buried hydrogen bonds. *Protein Sci.*, **22** (1), 74–82.

37 Park, S.-H., Zarrinpar, A., and et Lim, W.A. (2003) Rewiring MAP kinase pathways using alternative scaffold assembly mechanisms. *Science*, **299** (5609), 1061–1064.

38 Moon, J. and et Park, S.-H. (2014) Reassembly of JIP1 scaffold complex in JNK MAP kinase pathway using heterologous protein interactions. *PLoS One*, **9** (5), e96797.
39 Dueber, J.E., Wu, G.C., Malmirchegini, G.R. *et al.* (2009) Synthetic protein scaffolds provide modular control over metabolic flux. *Nat. Biotechnol.*, **27** (8), 753–759.
40 Bayer, E.A., Belaich, J.-P., Shoham, Y., and et Lamed, R. (2004) The cellulosomes: multienzyme machines for degradation of plant cell wall polysaccharides. *Annu. Rev. Microbiol.*, **58**, 521–554.
41 Savakis, P.E., Angermayr, S.A., and et Hellingwerf, K.J. (2013) Synthesis of 2,3-butanediol by Synechocystis sp. PCC6803 via heterologous expression of a catabolic pathway from lactic acid and enterobacteria. *Metab. Eng.*, **20**, 121–30.
42 Peer, A., Smith, S.P., Bayer, E.A. *et al.* (2009) Noncellulosomal cohesin- and dockerin-like modules in the three domains of life. *FEMS Microbiol. Lett.*, **291** (1), 1–16.
43 Cunha, E.S., Hatem, C.L., and et Barrick, D. (2013) Insertion of endocellulase catalytic domains into thermostable consensus ankyrin scaffolds: effects on stability and cellulolytic activity. *Appl. Environ. Microbiol.*, **79** (21), 6684–6696.
44 Ke, Y., Ong, L.L., Shih, W.M., and et Yin, P. (2012) Three-dimensional structures self-assembled from DNA bricks. *Science*, **338**, 1177–83.
45 Zhang, D.Y. and et Seelig, G. (2011) Dynamic DNA nanotechnology using strand-displacement reactions. *Nat. Chem.*, **3** (2), 103–113.
46 Yang, Z., Liu, H., and et Liu, D. (2015) Spatial regulation of synthetic and biological nanoparticles by DNA nanotechnology. *NPG Asia Mater*, **7**, e161.
47 Wilner, O.I., Weizmann, Y., Gill, R. *et al.* (2009) Enzyme cascades activated on topologically programmed DNA scaffolds. *Nat. Nanotechnol.*, **4** (4), 249–254.
48 Delebecque, C.J., Lindner, A.B., Silver, P.A., and et Aldaye, F.A. (2011) Organization of intracellular reactions with rationally designed RNA assemblies. *Science*, **333** (6041), 470–474.
49 Fu, J., Yang, Y.R., Johnson-Buck, A. *et al.* (2014) Multi-enzyme complexes on DNA scaffolds capable of substrate channelling with an artificial swinging arm. *Nat. Nanotechnol.*, **9**, 531–6.
50 Laisne, A., Ewald, M., Ando, T. *et al.* (2011) Self-assembly properties and dynamics of synthetic proteo-nucleic building blocks in solution and on surfaces. *Bioconjug. Chem.*, **22** (9), 1824–1834.
51 Brglez, J., Ahmed, I., and et Niemeyer, C.M. (2015) Photocleavable ligands for protein decoration of DNA nanostructures. *Org. Biomol. Chem.*, **13** (18), 5102–5104.
52 Delebecque, C.J., Silver, P.A., and et Lindner, A.B. (2012) Designing and using RNA scaffolds to assemble proteins in vivo. *Nat. Protoc.*, **7**, 1797–807.
53 Weigand, J.E., Wittmann, A., and et Suess, B. (2012) RNA-based networks: using RNA aptamers and ribozymes as synthetic genetic devices. *Methods Mol. Biol. Clifton NJ*, **813**, 157–168.
54 Ngo, T.A., Nakata, E., Saimura, M. *et al.* (2014) A protein adaptor to locate a functional protein dimer on molecular switchboard. *Methods*, **67**, 142–50.
55 Yamazaki, T., Heddle, J.G., Kuzuya, A., and et Komiyama, M. (2014) Orthogonal enzyme arrays on a DNA origami scaffold bearing size-tunable wells. *Nanoscale*, **6** (15), 9122–9126.
56 Liu, Y., Zhu, Y., Ma, W. *et al.* (2014) Spatial modulation of key pathway enzymes by DNA-guided scaffold system and respiration chain engineering for improved N-acetylglucosamine production by *Bacillus subtilis*. *Metab. Eng.*, **24**, 61–9.
57 Severcan, I., Geary, C., Chworos, A. *et al.* (2010) A polyhedron made of tRNAs. *Nat. Chem.*, **2**, 772–9.
58 Gilbert, L.A., Larson, M.H., Morsut, L. *et al.* (2013) CRISPR-mediated modular RNA-guided regulation of transcription in eukaryotes. *Cell*, **154** (2), 442–451.
59 György, B., Hung, M.E., Breakefield, X.O., and Leonard, J.N. (2015) Therapeutic

applications of extracellular vesicles: clinical promise and open questions. *Annu. Rev. Pharmacol. Toxicol.*, **55**, 439–464.

60 Farhi, M., Marhevka, E., Masci, T. et al. (2011) Harnessing yeast subcellular compartments for the production of plant terpenoids. *Metab. Eng.*, **13** (5), 474–481.

61 Avalos, J.L., Fink, G.R., and et Stephanopoulos, G. (2013) Compartmentalization of metabolic pathways in yeast mitochondria improves the production of branched-chain alcohols. *Nat. Biotechnol.*, **31** (4), 335–341.

62 Walser, P.J., Ariotti, N., Howes, M. et al. (2012) Constitutive formation of caveolae in a bacterium. *Cell*, **150**, 752–63.

63 Bonifacino, J.S. and et Glick, B.S. (2004) The mechanisms of vesicle budding and fusion. *Cell*, **116** (2), 153–166.

64 Yamashita, I., Iwahori, K., and et Kumagai, S. (2010) Ferritin in the field of nanodevices. *Biochim. Biophys. Acta.*, **1800**, 846–57.

65 Hartl, F.U. and et Hayer-Hartl, M. (2002) Molecular chaperones in the cytosol: from nascent chain to folded protein. *Science*, **295** (5561), 1852–1858.

66 Groll, M., Ditzel, L., Löwe, J. et al. (1997) Structure of 20S proteasome from yeast at 2.4 A resolution. *Nature*, **386** (6624), 463–471.

67 McMillan, R.A., Howard, J., Zaluzec, N.J. et al. (2005) A self-assembling protein template for constrained synthesis and patterning of nanoparticle arrays. *J. Am. Chem. Soc.*, **127**, 2800–1.

68 Flenniken, M.L., Liepold, L.O., Crowley, B.E. et al. (2005) Selective attachment and release of a chemotherapeutic agent from the interior of a protein cage architecture. *Chem. Commun. Camb.*, 447–9.

69 Zhen, Z., Tang, W., Guo, C. et al. (2013) Ferritin nanocages to encapsulate and deliver photosensitizers for efficient photodynamic therapy against cancer. *ACS Nano*, **7** (8), 6988–6996.

70 Kanekiyo, M., Wei, C.-J., Yassine, H.M. et al. (2013) Self-assembling influenza nanoparticle vaccines elicit broadly neutralizing H1N1 antibodies. *Nature*, **499** (7456), 102–106.

71 Abe, S., Hirata, K., Ueno, T. et al. (2009) Polymerization of phenylacetylene by rhodium complexes within a discrete space of apo-ferritin. *J. Am. Chem. Soc.*, **131**, 6958–60.

72 Abe, S., Niemeyer, J., Abe, M. et al. (2008) Control of the coordination structure of organometallic palladium complexes in an apo-ferritin cage. *J. Am. Chem. Soc.*, **130**, 10512–4.

73 Wang, Q., Mercogliano, C.P., and et Löwe, J. (2011) A ferritin-based label for cellular electron cryotomography. *Structure*, **19** (2), 147–154.

74 Flenniken, M.L., Willits, D.A., Harmsen, A.L. et al. (2006) Melanoma and lymphocyte cell-specific targeting incorporated into a heat shock protein cage architecture. *Chem. Biol.*, **13** (2), 161–170.

75 Varpness, Z., Peters, J.W., Young, M., and et Douglas, T. (2005) Biomimetic synthesis of a H2 catalyst using a protein cage architecture. *Nano Lett.*, **5** (11), 2306–2309.

76 Abedin, M.J., Liepold, L., Suci, P. et al. (2009) Synthesis of a cross-linked branched polymer network in the interior of a protein cage. *J. Am. Chem. Soc.*, **131** (12), 4346–4354.

77 Domingo, G.J., Orru', S., and et Perham, R.N. (2001) Multiple display of peptides and proteins on a macromolecular scaffold derived from a multienzyme complex. *J. Mol. Biol.*, **305** (2), 259–267.

78 Molino, N.M., Bilotkach, K., Fraser, D.A. et al. (2012) Complement activation and cell uptake responses toward polymer-functionalized protein nanocapsules. *Biomacromolecules*, **13**, 974–81.

79 Ren, D., Kratz, F., and et Wang, S.-W. (2014) Engineered drug-protein nanoparticle complexes for folate receptor targeting. *Biochem. Eng. J.*, **89**, 33–41.

80 Ren, D., Dalmau, M., Randall, A. et al. (2012) Biomimetic design of protein nanomaterials for hydrophobic molecular transport. *Adv. Funct. Mater.*, **22** (15), 3170–3180.

81 Ra, J.-S., Shin, H.-H., Kang, S., and et Do, Y. (2014) Lumazine synthase protein cage nanoparticles as antigen delivery nanoplatforms for dendritic cell-based vaccine development. *Clin. Exp. Vaccine Res.*, **3** (2), 227–234.

82 Shenton, W., Mann, S., Cölfen, H. *et al.* (2001) Synthesis of nanophase iron oxide in lumazine synthase capsids. *Angew. Chem., Int. Ed.*, **40** (2), 442–445.

83 Seebeck, F.P., Woycechowsky, K.J., Zhuang, W. *et al.* (2006) A simple tagging system for protein encapsulation. *J. Am. Chem. Soc.*, **128**, 4516–7.

84 Worsdorfer, B., Woycechowsky, K.J., and et Hilvert, D. (2011) Directed evolution of a protein container. *Science*, **331** (6017), 589–592.

85 Song, Y., Kang, Y.J., Jung, H. *et al.* (2015) Lumazine synthase protein nanoparticle-Gd(III)-DOTA conjugate as a T1 contrast agent for high-field MRI. *Sci. Rep.*, **5**, 15656.

86 Moon, H., Lee, J., Min, J., and et Kang, S. (2014) Developing genetically engineered encapsulin protein cage nanoparticles as a targeted delivery nanoplatform. *Biomacromolecules*, **15**, 3794–801.

87 Tamura, A., Fukutani, Y., Takami, T. *et al.* (2015) Packaging guest proteins into the encapsulin nanocompartment from *Rhodococcus erythropolis* N771. *Biotechnol. Bioeng.*, **112** (1), 13–20.

88 Zeltins, A. (2013) Construction and characterization of virus-like particles: a review. *Mol. Biotechnol.*, **53** (1), 92–107.

89 Abbing, A., Blaschke, U.K., Grein, S. *et al.* (2004) Efficient intracellular delivery of a protein and a low molecular weight substance via recombinant polyomavirus-like particles. *J. Biol. Chem.*, **279** (26), 27410–27421.

90 Schaffer, D.V., Koerber, J.T., and et Lim, K. (2008) Molecular engineering of viral gene delivery vehicles. *Annu. Rev. Biomed. Eng.*, **10**, 169–194.

91 Lucon, J., Qazi, S., Uchida, M. *et al.* (2012) Use of the interior cavity of the P22 capsid for site-specific initiation of atom-transfer radical polymerization with high-density cargo loading. *Nat. Chem.*, **4**, 781–8.

92 Tseng, R.J., Tsai, C., Ma, L. *et al.* (2006) Digital memory device based on tobacco mosaic virus conjugated with nanoparticles. *Nat. Nanotechnol.*, **1** (1), 72–77.

93 Comellas-Aragonès, M., Engelkamp, H., Claessen, V.I. *et al.* (2007) A virus-based single-enzyme nanoreactor. *Nat. Nanotechnol.*, **2** (10), 635–639.

94 Inoue, T., Kawano, M., Takahashi, R. *et al.* (2008) Engineering of SV40-based nano-capsules for delivery of heterologous proteins as fusions with the minor capsid proteins VP2/3. *J. Biotechnol.*, **134**, 181–92.

95 Minten, I.J., Claessen, V.I., Blank, K. *et al.* (2011) Catalytic capsids: the art of confinement. *Chem. Sci.*, **2**, 358–362.

96 Champion, C.I., Kickhoefer, V.A., Liu, G. *et al.* (2009) A vault nanoparticle vaccine induces protective mucosal immunity. *PloS One*, **4** (4), e5409.

97 Ng, B.C., Yu, M., Gopal, A. *et al.* (2008) Encapsulation of semiconducting polymers in vault protein cages. *Nano Lett.*, **8** (10), 3503–3509.

98 Goldsmith, L.E., Pupols, M., Kickhoefer, V.A. *et al.* (2009) Utilization of a protein «shuttle» to load vault nanocapsules with gold probes and proteins. *ACS Nano*, **3**, 3175–83.

99 Kickhoefer, V.A., Garcia, Y., Mikyas, Y. *et al.* (2005) Engineering of vault nanocapsules with enzymatic and fluorescent properties. *Proc. Natl. Acad. Sci. USA*, **102** (12), 4348–4352.

100 Lai, C.-Y., Wiethoff, C.M., Kickhoefer, V.A. *et al.* (2009) Vault nanoparticles containing an adenovirus-derived membrane lytic protein facilitate toxin and gene transfer. *ACS Nano*, **3** (3), 691–699.

101 Buehler, D.C., Marsden, M.D., Shen, S. *et al.* (2014) Bioengineered vaults: self-assembling protein shell-lipophilic core nanoparticles for drug delivery. *ACS Nano*, **8**, 7723–32.

102 Lawrence, A.D., Frank, S., Newnham, S. *et al.* (2014) Solution structure of a bacterial microcompartment targeting peptide and its application in the construction of an ethanol bioreactor. *ACS Synth. Biol.*, **3** (7), 454–465.

103 Choudhary, S., Quin, M.B., Sanders, M.A. et al. (2012) Engineered protein nanocompartments for targeted enzyme localization. *PloS One*, **7** (3), e33342.

104 Sremac, M. and et Stuart, E.S. (2008) Recombinant gas vesicles from Halobacterium sp. displaying SIV peptides demonstrate biotechnology potential as a pathogen peptide delivery vehicle. *BMC Biotechnol.*, **8**, 9.

105 DasSarma, P., Negi, V.D., Balakrishnan, A. et al. (2014) Haloarchaeal gas vesicle nanoparticles displaying Salmonella SopB antigen reduce bacterial burden when administered with live attenuated bacteria. *Vaccine*, **32** (35), 4543–4549.

106 DasSarma, S., Karan, R., DasSarma, P. et al. (2013) An improved genetic system for bioengineering buoyant gas vesicle nanoparticles from Haloarchaea. *BMC Biotechnol.*, **13**, 112.

107 Sleytr, U.B., Huber, C., Ilk, N. et al. (2007) S-layers as a tool kit for nanobiotechnological applications. *FEMS Microbiol. Lett.*, **267** (2), 131–144.

108 Ritsert, K., Huber, R., Turk, D. et al. (1995) Studies on the lumazine synthase/riboflavin synthase complex of *Bacillus subtilis*: crystal structure analysis of reconstituted, icosahedral beta-subunit capsids with bound substrate analogue inhibitor at 2.4 A resolution. *J. Mol. Biol.*, **253** (1), 151–167.

109 Zhou, Z.H., McCarthy, D.B., O'Connor, C.M. et al. (2001) The remarkable structural and functional organization of the eukaryotic pyruvate dehydrogenase complexes. *Proc. Natl. Acad. Sci. USA*, **98** (26), 14802–14807.

110 Bacher, A., Ludwig, H.C., Schnepple, H., and et Ben-Shaul, Y. (1986) Heavy riboflavin synthase from *Bacillus subtilis*. Quaternary structure and reaggregation. *J. Mol. Biol.*, **187** (1), 75–86.

111 Sutter, M., Boehringer, D., Gutmann, S. et al. (2008) Structural basis of enzyme encapsulation into a bacterial nanocompartment. *Nat. Struct. Mol. Biol.*, **15** (9), 939–947.

112 Philippe, N., Legendre, M., Doutre, G. et al. (2013) Pandoraviruses: amoeba viruses with genomes up to 2.5 Mb reaching that of parasitic eukaryotes. *Science*, **341** (6143), 281–286.

113 Rome, L.H. and et Kickhoefer, V.A. (2013) Development of the vault particle as a platform technology. *ACS Nano*, **7** (2), 889–902.

114 Pfeifer, F. (2012) Distribution, formation and regulation of gas vesicles. *Nat. Rev. Microbiol.*, **10**, 705–15.

115 Beard, S.J., Hayes, P.K., Pfeifer, F., and et Walsby, A.E. (2002) The sequence of the major gas vesicle protein, GvpA, influences the width and strength of halobacterial gas vesicles. *FEMS Microbiol. Lett.*, **213** (2), 149–157.

116 Rahmanpour, R. and et Bugg, T.D.H. (2013) Assembly *in vitro* of *Rhodococcus jostii* RHA1 encapsulin and peroxidase DypB to form a nanocompartment. *FEBS J.*, **280**, 2097–104.

117 Goldsmith, L.E., Yu, M., Rome, L.H., and et Monbouquette, H.G. (2007) Vault nanocapsule dissociation into halves triggered at low pH. *Biochemistry (Mosc.)*, **46**, 2865–75.

118 Stray, S.J., Johnson, J.M., Kopek, B.G., and et Zlotnick, A. (2006) An *in vitro* fluorescence screen to identify antivirals that disrupt hepatitis B virus capsid assembly. *Nat. Biotechnol.*, **24** (3), 358–362.

119 Stephen, A.G., Raval-Fernandes, S., Huynh, T. et al. (2001) Assembly of vault-like particles in insect cells expressing only the major vault protein. *J. Biol. Chem.*, **276**, 23217–20.

120 Douglas, S.M., Dietz, H., Liedl, T. et al. (2009) Self-assembly of DNA into nanoscale three-dimensional shapes. *Nature*, **459** (7245), 414–418.

121 Chowdhury, C., Sinha, S., Chun, S. et al. (2014) Diverse bacterial microcompartment organelles. *Microbiol. Mol. Biol. Rev.*, **78**, 438–68.

122 Axen, S.D., Erbilgin, O., and et Kerfeld, C.A. (2014) A taxonomy of bacterial microcompartment loci constructed by a novel scoring method. *PLoS Comput. Biol.*, **10**, e1003898.

123 Chowdhury, C., Chun, S., Pang, A. et al. (2015) Selective molecular transport through the protein shell of a bacterial

microcompartment organelle. *Proc. Natl. Acad. Sci. USA*, **112**, 2990–5.
124 Penrod, J.T. and et Roth, J.R. (2006) Conserving a volatile metabolite: a role for carboxysome-like organelles in *Salmonella enterica. J. Bacteriol.*, **188** (8), 2865–2874.
125 Parsons, J.B., Dinesh, S.D., Deery, E. *et al.* (2008) Biochemical and structural insights into bacterial organelle form and biogenesis. *J. Biol. Chem.*, **283** (21), 14366–14375.
126 Bonacci, W., Teng, P.K., Afonso, B. *et al.* (2012) Modularity of a carbon-fixing protein organelle. *Proc. Natl. Acad. Sci. USA*, **109**, 478–83.
127 Cai, F., Sutter, M., Bernstein, S.L. *et al.* (2015) Engineering bacterial microcompartment shells: chimeric shell proteins and chimeric carboxysome shells. *ACS Synth. Biol.*, **4** (4), 444–453.
128 Tsai, Y., Sawaya, M.R., Cannon, G.C. *et al.* (2007) Structural analysis of CsoS1A and the protein shell of the *Halothiobacillus neapolitanus* carboxysome. *PLoS Biol.*, **5** (6), e144.
129 Rae, B.D., Long, B.M., Badger, M.R., and et Price, G.D. (2012) Structural determinants of the outer shell of β-carboxysomes in *Synechococcus elongatus* PCC 7942: roles for CcmK2, K3-K4, CcmO, and CcmL. *PLoS One*, **7** (8), e43871.
130 Menon, B.B., Dou, Z., Heinhorst, S. *et al.* (2008) *Halothiobacillus neapolitanus* carboxysomes sequester heterologous and chimeric RubisCO species. *PLoS One*, **3** (10), e3570.
131 Tcherkez, G.G.B., Farquhar, G.D., and et Andrews, T.J. (2006) Despite slow catalysis and confused substrate specificity, all ribulose bisphosphate carboxylases may be nearly perfectly optimized. *Proc. Natl. Acad. Sci. USA*, **103** (19), 7246–7251.
132 McGrath, J.M. and et Long, S.P. (2014) Can the cyanobacterial carbon-concentrating mechanism increase photosynthesis in crop species? A theoretical analysis. *Plant Physiol.*, **164** (4), 2247–2261.
133 Lin, M.T., Occhialini, A., Andralojc, P.J. *et al.* (2014) β-Carboxysomal proteins assemble into highly organized structures in Nicotiana chloroplasts. *Plant J. Cell Mol. Biol.*, **79** (1), 1–12.
134 Fan, C., Cheng, S., Liu, Y. *et al.* (2010) Short N-terminal sequences package proteins into bacterial microcompartments. *Proc. Natl. Acad. Sci. USA*, **107** (16), 7509–7514.
135 Fan, C., Cheng, S., Sinha, S., and et Bobik, T.A. (2012) Interactions between the termini of lumen enzymes and shell proteins mediate enzyme encapsulation into bacterial microcompartments. *Proc. Natl. Acad. Sci. USA*, **109** (37), 14995–15000.
136 Kinney, J.N., Salmeen, A., Cai, F., and et Kerfeld, C.A. (2012) Elucidating essential role of conserved carboxysomal protein CcmN reveals common feature of bacterial microcompartment assembly. *J. Biol. Chem.*, **287** (21), 17729–17736.
137 Chen, A.H., Robinson-Mosher, A., Savage, D.F. *et al.* (2013) The bacterial carbon-fixing organelle is formed by shell envelopment of preassembled cargo. *PLoS One*, **8** (9), e76127.
138 Cameron, J.C., Wilson, S.C., Bernstein, S.L., and et Kerfeld, C.A. (2013) Biogenesis of a bacterial organelle: the carboxysome assembly pathway. *Cell*, **155**, 1131–40.
139 Lin, M.T., Occhialini, A., Andralojc, P.J. *et al.* (2014) β-Carboxysomal proteins assemble into highly organized structures in Nicotiana chloroplasts. *Plant J. Cell Mol. Biol.*, **79** (1), 1–12.
140 Keeling, T.J., Samborska, B., Demers, R.W., and et Kimber, M.S. (2014) Interactions and structural variability of β-carboxysomal shell protein CcmL. *Photosynth. Res.*, **121** (2–3), 125–133.
141 Iancu, C.V., Morris, D.M., Dou, Z. *et al.* (2010) Organization, structure, and assembly of alpha-carboxysomes determined by electron cryotomography of intact cells. *J. Mol. Biol.*, **396** (1), 105–117.
142 Parsons, J.B., Frank, S., Bhella, D. *et al.* (2010) Synthesis of empty bacterial microcompartments, directed organelle protein incorporation, and evidence of filament-associated organelle movement. *Mol. Cell*, **38**, 305–15.

143 Havemann, G.D., Sampson, E.M., and et Bobik, T.A. (2002) PduA is a shell protein of polyhedral organelles involved in coenzyme B(12)-dependent degradation of 1,2-propanediol in *Salmonella enterica serovar typhimurium* LT2. *J. Bacteriol.*, **184** (5), 1253–1261.

144 Menon, B.B., Heinhorst, S., Shively, J.M., and et Cannon, G.C. (2010) The carboxysome shell is permeable to protons. *J. Bacteriol.*, **192** (22), 5881–5886.

145 Padilla, J.E., Colovos, C., and et Yeates, T.O. (2001) Nanohedra: using symmetry to design self assembling protein cages, layers, crystals, and filaments. *Proc. Natl. Acad. Sci. USA*, **98** (5), 2217–2221.

146 Patterson, D.P., Su, M., Franzmann, T.M. *et al.* (2014) Characterization of a highly flexible self-assembling protein system designed to form nanocages. *Protein Sci.*, **23**, 190–9.

147 King, N.P., Sheffler, W., Sawaya, M.R. *et al.* (2012) Computational design of self-assembling protein nanomaterials with atomic level accuracy. *Science*, **336** (6085), 1171–1174.

148 King, N.P., Bale, J.B., Sheffler, W. *et al.* (2014) Accurate design of co-assembling multi-component protein nanomaterials. *Nature*, **510** (7503), 103–108.

149 Chen, J.H. and et Seeman, N.C. (1991) Synthesis from DNA of a molecule with the connectivity of a cube. *Nature*, **350** (6319), 631–633.

150 Han, D., Pal, S., Nangreave, J. *et al.* (2011) DNA origami with complex curvatures in three-dimensional space. *Science*, **332** (6027), 342–346.

151 Erben, C.M., Goodman, R.P., and et Turberfield, A.J. (2006) Single-molecule protein encapsulation in a rigid DNA cage. *Angew. Chem., Int. Ed Engl.*, **45** (44), 7414–7417.

14
Modeling and Simulation of Bacterial Biofilm Treatment with Applications to Food Science

Jia Zhao,[1,4] *Tianyu Zhang,*[2] *and Qi Wang*[3,4]

[1]*University of North Carolina at Chapel Hill, Department of Mathematics, 3250 Phillips Hall, Chapel Hill, NC, 27599, USA*
[2]*Montana State University, Department of Mathematical Sciences, 211 Montana Hall, Bozeman, MT 59717, USA*
[3]*Beijing Computational Science Research Center, East Xibeiwang Road, Haidian District, Beijing 100931, China*
[4]*University of South Carolina, Department of Mathematics and NanoCenter, 1523 Greene Street, Columbia, SC 29208, USA*

14.1
Introduction

Nanotechnology has found its way in various aspects of food industry such as food processing, packaging, safety, and quality control. Nanotechnology can modify properties of packaging materials, increasing barrier properties, improving mechanical and heat resistance, developing active antimicrobial surfaces, and creating nanobiodegradable packaging materials [1]. Nanotechnology in food industry can be used to deliver nutrient to produce nanoformulated agrochemicals, enrich nutritional values, and generate novel products through bioactive encapsulation; it can also be used in the development of biosensors for detection of pathogens and chemical contaminants. Nanotechnology can also be further used to deliver antibacterial agents to targeted regions in food contaminated with biofilms to combat bacterial growth. This application combined with polymer–nanoparticle composites was initially devised to combat biofilm growth in medical treatment and industrial disinfection of biofilms [2,3].

Bacterial surface contamination – the adhesion, persistence, and colonization of surfaces by bacteria in the form of biofilms – is detrimental to ones' health and society. In food industry, it impacts food safety, arising from surface contamination or food deterioration. Biofilms consist of a complex mixture of bacteria, exopolysaccharides, DNA, and catalytic proteins, secreted by microorganisms after bacteria's adhesion onto surfaces. The development of biofilms consists of multistage processes, involving bacterial cells adhesion to the substrate excreting

Nanotechnology in Agriculture and Food Science, First Edition. Edited by Monique A.V. Axelos and Marcel Van de Voorde.
© 2017 Wiley-VCH Verlag GmbH & Co. KGaA. Published 2017 by Wiley-VCH Verlag GmbH & Co. KGaA.

exopolysaccharides (EPS) [4] to form an EPS matrix anchored on the surface and proliferation of bacterial cells. Biofilm growth is a dynamical process whose structure is largely influenced by environmental conditions, including the substrate properties and the surrounding media.

Biofilm-related problems cost the society literally billions of dollars every year, in terms of energy losses, equipment damage, product contamination, and medical cares. In food industry, in particular, biofilm prevention is one major concern for food safety. On the other hand, biofilms can also offer huge potential benefits in bioremediating hazardous waste sites, biofiltering municipal and industrial water and wastewater, forming biobarriers to protect soil and groundwater from contamination, and heap leaching in mining industry. It is important and theoretically challenging to unravel the complex dynamical properties of biofilms to prevent damages caused by unwanted biofilm growth and contamination and to utilize biofilms for good causes.

Mathematical modeling of biofilms is crucial to attaining a broader and deeper understanding of this complex microorganism. Mathematical models can be used not only to verify experimental findings but also to make qualitative and quantitative predictions that might well serve as guidelines for experimental design. There has been extensive research on mathematical modeling of biofilms in the last three decades. Some of the models were proposed and applied successfully in explaining and predicting biofilm dynamics in various conditions. Chronically, the development of biofilm models has undergone evolution from simple to complex, from low spatial dimensional models to multidimensional models, from single-species models to multispecies models, from steady-state models to time-dependent dynamical models, from pure growth models to models including biomass growth and biofilm–fluid interaction, from models describing a particular process to models incorporating many different processes occurring simultaneously, from qualitative models to quantitative models, from single-scale models to multiscale ones, and so on. With the innovative application of nanotechnology to treat and/or prevent biofilms, biofilm interaction between nanoparticles, bacteria-resistant substrata, and transport of nanoparticles through bacterial cell membranes has emerged as one of the important issues deserving a comprehensive theoretical investigation.

In this chapter, we will give a brief chronical review of some biofilm models developed in the past 30 years, most of which have been successfully applied to study biofilm growth. There have been several excellent review articles about biofilms from the biological, physiological, mechanical, and mathematical perspectives by some distinguished researchers in the field [5–8]. Following the brief review, we present three new models of biofilms on a bacteria-resistant substrate, nanoparticle-delivered antibacterial treatment, and cell–nanoparticle interaction, respectively. The nanoparticle–cell interaction in the context of biofilm treatment represents the state-of-the-art treatment of bacterial biofilms and substance delivered into the cell. They lay the foundation for studying nanoparticle delivery into the cell as well as onto the cell surface to achieve certain purposes in medical science and food science. This new modeling and simulation tools presented here have the potential to be applied to specific theoretical investigations into biofilm prevention and treatment as well as into biofilm prevention in food science and processing.

14.2 Review of Biofilm Models

Early efforts on mathematical modeling of biofilms can be traced back to the 1980s represented by Refs [9–14]. The authors of these works studied biofilms using continuum models primarily in one-space dimension. These studies centered on steady-state biofilm growth dynamics, including biofilm thickness and spatial distribution of microbial species and substrate concentration. For example, the multispecies biofilm model introduced by Wanner and Gujer [14] was based on a continuum description of the biofilm material and conservation principle; it predicts the evolution of the biofilm thickness, dynamics, and spatial distribution of microbial species and substrates in the biofilm. It also allows biomass detachment due to shear stress and sloughing.

Based on new experimental observations, an extended mixed-culture biofilms (MCB) model was introduced in Ref. [15] later. This model includes a more flexible description of transport of dissolved components in the biofilm and considers diffusive transport of particulate components in the biofilm solid matrix, changes in the biofilm liquid-phase volume fraction (porosity), and simultaneous detachment and attachment of cells and particles at the biofilm surface. This extended model can reproduce most of the new experimental data. This model is derived based on the general mass conservation principle and independent of any special setup except for limitations in one-space dimension. The model can be adjusted to any specific microbial kinetics or biomass detachment mechanism, and used to predict short- and long-term behavior of biofilm growth. Indeed, the model has been successfully applied to many studies of biofilms and helps enormously in understanding complex bulk interactions in multispecies biofilms. On the other hand, since it is a one-dimensional model, it significantly simplifies the interaction between the substrate and the biofilm and spatial dynamics of the biofilm. It however lacks the ability to characterize the multidimensional morphology of the biofilm. Various higher dimensional models with the biofilm–flow interaction were developed later.

In general, the growth mechanism of bacterial colonies is highly complex, in which the substrate concentration plays an important role. One way to model the growth pattern is through the class of models called the diffusion-limited aggregation (DLA) model. The DLA growth was first studied through computer simulations [16] and the first biological DLA example was a DLA colony of *Bacillus subtilis* given in Ref. [17]. More experimental and modeling work were done in Refs [18,19] later. The rule of the DLA model is as follows: One chooses a seed particle as the origin of a square lattice on a plane. Another particle is released far from the origin and is allowed to move at random. When it arrives at the nearest neighboring site to the origin, it sticks to the site. Then another particle is launched and it moves until it reaches the nearest neighboring site of a cluster made of the two particles. Through the repetition of these procedures, a cluster grows with an outwardly open and randomly branched structure. Its pattern is morphologically self-similar, implying that it is a fractal.

Another class of models for biofilms is called the discrete–continuum/cellular automaton (CA) model, originated from the cellular automaton model, for example, Conway's Game of Life [20]. It consists of a regular grid of cells, each in one of the finite numbers of states. The grid can be in any finite number of dimensions. Time is also discrete, and the state of a cell at time t is a function of the states of a finite number of cells (called its neighborhood) at time $t-1$. Every cell has the same rule for updating, based on the values in this neighborhood. Each time the rules are applied to the whole grid, a new generation is created. A clear advantage of CA model is that it can produce complex behavior despite of its simplicity, which makes it particularly attractive to simulate biological systems that often exhibit complex spatial structures [21–23].

CA models for biofilms incorporating the biological rules including bacteria reproduction and movement, cell-to-cell communication, and diffusion of nutrient were extensively studied in Refs [24–29]. These models can capture various biofilm growth patterns observed in experiments, and strongly suggest that biofilm structures are largely determined by the surrounding substrate concentration. In Ref. [29], the authors claimed that the three widely accepted conceptual models of biofilm structure – (i) penetrated water channel biofilms, (ii) heterogeneous mosaic biofilms, and (iii) dense confluent biofilms – can all be reproduced by CA models at very different substrate concentrations. Their apparent simplicity and great success in capturing the biofilm structure makes CA models very attractive. The model is a close approximation to molecular diffusion in a much coarser scale. The coarseness emphasizes the stochastic nature of the model, which is the main mechanism of producing heterogeneous biofilm morphology.

In Ref. [30], the authors developed a combined differential–discrete CA biofilm model. In this model, the substrate concentration was assumed to be continuous and governed by the conventional diffusion–reaction equation. The biomass density was computed by direct integration of the biomass balance equation, taking into account only the growth as a result of substrate consumption. The biomass spreading was modeled by the CA model. But instead of performing entirely random walks for the "walkers," newly formed biomass finds a place by "pushing" its neighbors to adjacent, unoccupied space. Numerical simulations in real time were carried out for both 2D and 3D, and the calculated values of global oxygen uptake rate, concentration profiles for oxygen, and biomass and colonies size were both qualitatively and quantitatively in agreement with experimental data in Refs [31–33].

In Ref. [34], the authors developed a three-dimensional multispecies biofilm model based on the CA approach and tested it with a two-species biofilm composed of sulfate-reducing bacteria and methanogens. The model predicts different biofilm structures in the absence of sulfate, when a syntrophic association between the two organisms develops, and in the presence of sulfate, when the two organisms compete for the available hydrogen. In Ref. [35], a 2D CA model was developed to study the effect of thickness of substrate concentration and hydrodynamic boundary layers in the biofilm structure, motivated by the experimental results of Refs [36,37]. The model predicts that when external mass transfer (for substrate) limitations are significant, biofilm develops more open structures; when the substrate concentration layer is reduced and external mass transfer is enhanced, biofilm develops denser and smooth structures. The simulation reproduced the "mushroom"-shaped biofilm described in Ref. [38].

A computer model called BacLAB based on the solute diffusion model coupled with the CA principle for bacterial growth was developed by Hunt *et al.* [39,40] to study the detachment mechanisms in biofilms. This computer model has been used to test various hypothetical detachment mechanisms in 3D simulations [41,42]. These studies show that detachment is a critical determinant of biofilm structures and of the dynamics of biofilm accumulation and loss.

For most of the biofilm models already described, the biofilm structure is primarily determined by the substrate concentration that is solely modeled by a diffusion process. However, heterogeneous structures, including nonuniform distribution of cells and polymers, variable biofilm thickness and surface shape, and variable density and porosity, have been observed and quantitatively measured by many researchers [43–45]). It is also found that hydrodynamics of the bulk fluid plays an important role in shaping the structure of biofilms through both convection of the nutrient and detachment of biomass by shear stress [46–48]. It is generally accepted that the formation of biofilm structure is the result of a combination of several simultaneously occurring biological, chemical, and physical processes, including cell transport to the substratum and attachment, biofilm generation by cellular growth and extracellular polymer production, biofilm detachment by shear stress, and substrate and product transport to and from the biofilm. Based on the previously established models and a better understanding of the biofilm properties through experiments, several biofilm models coupling the bulk fluid were developed.

14.2.1
Hybrid Discrete–Continuum Models

In Refs [49–51], a fully quantitative two- and three-dimensional biofilm model was developed that incorporates the flow over irregular biofilm surfaces, convective and diffusive mass transfer of substrate, bacterial growth, and biomass spreading. In this model, the biomass growth and spreading is modeled by the discrete CA model, the mass balance of the substrate is modeled by the continuity equation and convection–diffusion–reaction equations, respectively, and the flow field is governed by the momentum balance (Navier–Stokes) equation. The fluid flow affects the biofilm growth by regulating the substrate concentration at the biofilm–fluid interface, shearing the biofilm surface, and eroding the protuberances. The interaction is reciprocal since a new biofilm shape leads to a different boundary condition and thus different flow and substrate concentration.

In Ref. [51], a two-dimensional model was developed for detachment based on internal stress created by moving fluid past the biofilm. This model can model two biofilm detachment mechanisms – erosion (small-particle loss) and sloughing (large biomass particle removal) – in a unified way. The model predicts that erosion makes the biofilm surface smooth and sloughing leads to an increased biofilm surface roughness. Simulations also indicated an avalanche effect in biomass loss, and fast-growing biofilms have a faster detachment rate than slow-growing biofilm, and thus suffers instability in biofilm accumulation and abrupt biomass loss.

The hybrid discrete–continuum biofilm model was a fully quantitative model incorporating many physical, chemical, and biological processes affecting the biofilm development. The quantitative nature of the model not only produced simulation results in good agreement with experimental data but also predicted properties of biofilm and shed light on experimental studies. The model was widely accepted and successfully applied to several situations since its inception.

Another completely different discrete–continuum biofilm model was proposed in Refs [52,53] based on the immersed boundary method. In this microscale model, the microbes in the biofilm are treated as discrete entities (cells) that can swim through the action of flagella, the bulk fluid is treated as a continuum, and the formation of biofilm is achieved by cell–cell aggregation and cell–substratum adhesion that are modeled by generating appropriate binding forces between discrete representations of the organisms. The model includes the hydrodynamic interaction between the biofilm and the fluid, the substrate reaction, diffusion and convection, and the chemotactic responses of swimming microbes. Numerical simulations in 2D predicted that higher local cohesion forces and cell motility can greatly facilitate biofilm formation, cells may adhere preferentially to a saturated surface with chemotaxis, and changes of the channel geometry also have effects on the surface irregularities in biofilm formation.

14.2.2
Multidimensional Continuum Models

The hybrid discrete–continuum models of multiple steps described previously (CA models) can produce results in good agreement with experimental expectations. However, there are some drawbacks associated with the element of randomness in the step of the biomass redistribution. Namely, they are lattice dependent and not invariant to changes of coordinate system; multiple simulations with the same initial condition generally produce different outputs; the biomass, although a continuous variable according to growth kinetics, suffers discrete changes during splitting. This randomness can be viewed as an artifact from the details of the algorithm rather than the growth process modeled. Even though the random feature of these models determines the biofilm structure to some extent, sometimes a deterministic model is more desirable. Several deterministic continuum biofilm models were developed to address the need.

In Ref. [54], the authors developed a spatiotemporal continuum model under which the biomass spreading is described by a nonlinear density-dependent diffusion mechanism. Dockery and Klapper in Ref. [55] developed a continuum model of a single-substrate-limited biofilm growing into a static aqueous environment. Alpkvist and Klapper [56] extended the work in Ref. [55] and proposed a multidimensional continuum model for heterogeneous growth of biofilm systems with multiple species and multiple substrates while retaining the Darcy's law for the velocity. The model was tested for two systems. System 1 is the biofilm consisting of active and inert biomass with oxygen as substrate; system 2 is the biofilm consisting of heterotrophs, autotrophs, and inerts with limiting substrates of oxygen, acetate, and ammonium. Numerical simulations were performed in 2D and 3D. It predicted that for system 1, inactivation rate has an effect on the stability of the biofilm interface, namely, a higher inactivation rate will increase the depth of the active layer by locally decreasing the amount of active biomass. For system 2, the 2D simulation produced results in agreement with that in Ref. [14]; but the 3D simulation yielded the total averaged mass fractions of the biofilm components not adding to 1 due to the heterogeneity of the biofilm surface, a multidimensional phenomenon cannot be described by the 1D model in Ref. [14].

Cogan and Keener in Ref. [57] proposed a continuum model in which the biofilm is treated as a biological gel composed of extracellular polymeric substance (EPS) and water. The bacteria are enmeshed in the network and are the producers of the polymer. Gels absorb or expel solvent causing swelling or contraction due to osmotic pressure gradients, where the environmentally physical and chemical conditions affecting the biofilm morphology are built into the osmotic pressure. The model assumes that the primary forces that induce gel motion are applied to the fluid solvent and the EPS, and provides a mechanism of biomass redistribution through swelling of the EPS and the constitutive relationships. Here both the network and the solvent are treated as Newtonian fluids with a much larger viscosity for the network. This fundamental formulation of the model follows the two-fluid model [58]. However, the inertia is neglected so that the governing equation for the force balance is the Stokes equation for multifluids. This model lays out the framework of studying the effect of material and chemical properties (network viscosity, osmotic pressure) on biomass distribution, and produces results in qualitative agreement with previously developed models. The idea of using the network volume fraction θ_n to represent the biomass in the biofilm provides the possibility of modeling the whole biofilm system as a multiphase fluid.

Experimental results in Ref. [59] indicate that detachment of multicellular clumps is more significant than erosion of a single cell, possibly because clumps retain certain advantages of the biofilm. Klapper and Dockery in Ref. [60] proposed a more general multifluid model to explain this phenomenon by cohesion energy. Experiments [48,61] also indicate that biofilm responds elastically to mechanical stress in short timescales and it acts as a viscous fluid in long timescales. Early work of modeling a biofilm as a viscoelastic fluid also includes Ref. [62]. Researchers are now starting to build biofilm models using more sophisticated constitutive equations to capture the viscoelastic property of the biofilm.

Most of the models already described either track the biofilm–fluid interface explicitly [49–51,55] or assume that the solvent has zero velocity [54,57]. Motivated by Refs [55,57,60] and the work from other researchers, but in particular bothered by the lack of mathematical machinery to handle the inlet and outlet boundary conditions in the multifluid models, Zhang et al. [63,64] proposed a phase-field biofilm model. The model is based on the one-fluid two-component formulation in which the combination of extracellular polymeric substances (EPS) and the bacteria is effectively modeled as one-fluid component, while the collective ensemble of the substrate and the solvent is modeled as the other. The biofilm is assumed to be an incompressible continuum, in which the relative motion of the polymer network and the solvent relative to the average velocity is accounted for by binary mixing kinetics. This model can be shown to be a closure model of a more general kinetic theory for biofilm mixtures [65].

14.2.3
Individual-Based Modeling (IbM)

Kreft et al. [66,67] proposed the individual-based modeling (IbM) of biofilms. The model consists of two parts: one deals with growth and behavior of individual bacteria as autonomous agents; the other deals with the substrate and product diffusion and reaction. Bacterial cells are represented as hard spheres, with each cell having, besides variable volume and mass, a set of variable growth parameters. Each cell grows by consuming substrate and divides when a certain volume is reached. The pressure buildup due to the growth of biomass is released by maintenance of a minimum distance between neighboring cells. For each cell, the vector sum of all positive overlap radii with neighboring cells is calculated and then the position of the cell is shifted in the direction opposite to this vector. The substrate concentration is governed by reaction–diffusion equations, thus IbM is a hybrid discrete–continuous model. The biomass spreading mechanism in IbM allows cell movement with continuous distance and direction, which makes it a deterministic model and overcomes the drawbacks associated with the randomness in CA-based models. Numerical simulations in 2D showed that IbM model produced more confluent and rounded biofilm structure than the CA-based models, due to its deterministic and directionally unconstrained spreading of the biomass. Picioreanu et al. [68] extended the IbM model and proposed a particle-based multidimensional multispecies biofilm model. This model allows variable size biomass particles in order to model systems with large-scale heterogeneity. Numerical simulations predicted that if only the average flux of nutrients needs to be known, 2D and 1D models are very similar. However, the behavior of intermediates, which are produced and consumed in different locations within the biofilm, is better described in 2D and 3D models because of the multidirectional concentration gradients. The predictions of 2D or 3D models are also different from those of 1D models for slowly growing or minority species in the biofilm.

14.2.4
Other Models Related to Biofilm Properties

Most of the models already described concentrate on the biofilm structure resulting from biomass growth and spreading, substrate transport, and consumption. Experimental results also indicate that biofilm exhibits complicated physical, chemical, and biological properties during the interactions with the surrounding environment such as external fluid flow and antimicrobial agents. There are extensive studies on these biofilm properties and many models associated with that.

Bacterial cells inside biofilm generally have higher tolerance to chemical attack from antimicrobial agents compared to planktonic cells [69–71], and a widely accepted explanation that the biofilm matrix acts as a protective barrier against attack [72–74] proposed a theoretical framework to explain this phenomenon based on diffusion processes in biofilms. Besides the penetration barrier described above, another important hypothesis concerning the resistance mechanism to antimicrobial for bacterial biofilms is the existence of "persister" cells that are extremely tolerant of antimicrobials. Experimental results [75,76] demonstrate the existence of persisters and indicate that persisters are not genetic variants nor cells in a spore-like state nor cells caught coincidentally in a quiescent phase of cell division. Biofilm with persisters typically exhibits the biphasic disinfection curve. Cogan [77] proposed a phenotype-switching model between the persister and susceptible cells. The population of susceptible cells changes due to growth, death due to antibiotic action, loss due to transition to persister cells, and gain as the persistent cells revert back to susceptible cells. Klapper *et al.* [78] proposed a model that treats persister cells as senescent cells based on the idea that senescent cells are slow growing, and slow-growing cells are more tolerant to antimicrobials. It predicted the biphasic disinfection curve and the relative scarcity of persisters during growth phase as opposed to during stationary phase, providing a natural explanation for persistence-related phenomena.

Poplawski *et al.* [79] used the Glazier–Graner–Hogeweg (GGH) model coupled with a discrete–continuum model (PLH) developed by Picioreanu *et al.* [80] to study biofilm dynamics provided by the CompuCell3D modeling environment, a cell-oriented framework designed to simulate growth and pattern formation due to biological cells' behaviors. They simulated the growth of a single-species bacterial biofilm, and studied the roles of cell–cell and cell–flow field interactions in determining biofilm morphology. In their simulations they generalized the PLH model by treating cells as spatially extended, deformable bodies, differential adhesion between cells, and their competition for a substrate (nutrient), surface to produce a fingering instability that generates finger shapes of biofilms.

14.3
Biofilm Dynamics Near Antimicrobial Surfaces

In food packages, nanotechnology has been used to fabricate antimicrobial packaging materials for food storage and transportation. It is important to understand biofilm dynamics near the surface of the antimicrobial packaging material fabricated using nanotechnology [81]. Guided by such knowledge, one can then fabricate food contact surfaces to prevent bacteria from attachment and thereby formation of biofilms in the packaged food. For instance, in Ref. [82], authors showed that the alumina surface with nanoscale topography reduces attachment of biofilms formed by *Escherichia coli* and *Listeria* spp; in Ref. [83], authors reported that the bacterial biofilm growth pattern can be effectively controlled by nanostructural mechanics and geometry.

In contrast to the promising experimental findings, there have been very few theoretical investigations aiming at these issues. In this section, we present a phase field model for biofilms developed by the authors in the past [63] and study the interaction between biofilms and the boundary condition, which mimics the surface of the antimicrobial packaging material. The surface, also known as the substrate, is assumed to be coated with an antibacterial material presumably fabricated via nanotechnology. This study has a direct implication to the control of biofilms by an antimicrobial surface in food packaging industry.

Specifically, we model the biofilm as an incompressible binary mixture of effective biomass (EPS + bacteria) and solvent (water + nutrient), whose volume fractions are denoted by ϕ_n and ϕ_s, respectively. It implies that the incompressible condition reads $\phi_n + \phi_s = 1$. We use c to denote the concentration of nutrient. The extended Flory–Huggin's mixing free energy in the model is given by [63]

$$F = \int_\Omega d\mathbf{x} \left(\frac{\gamma_1}{2} |\nabla \phi_n|^2 + \gamma_2 \left[\frac{\phi_n}{N} \ln \phi_n + (1-\phi_n)\ln(1-\phi_n) + \chi \phi_n(1-\phi_n) \right] \right), \tag{14.1}$$

where γ_1 is a parameter that measures the strength of the conformation entropy, γ_2 controls the strength of the bulk mixing free energy, N is an extended polymerization index for the biomass, and χ is the mixing parameter [63].

The system of governing equations for the biofilm fluid mixture in the dimensionless variables is given by

$$\rho(\partial_t \mathbf{v} + \mathbf{v} \cdot \nabla \mathbf{v}) = -\nabla p + \nabla \cdot (2\eta \mathbf{D}) - \gamma_1 \nabla \cdot (\nabla \phi_n \otimes \nabla \phi_n),$$
$$\nabla \cdot \mathbf{v} = 0,$$
$$\frac{\partial \phi_n}{\partial t} + \nabla \cdot (\phi_n \mathbf{v}) = \nabla \cdot \left[\lambda \phi_n \nabla \frac{\delta F}{\delta \phi_n} \right] + C_1 \frac{c}{k_c + c} \phi_n \left(1 - \frac{\phi_n}{\phi_{max}} \right) - C_2 \phi_n d,$$
$$\frac{\partial \phi_s c}{\partial t} + \nabla \cdot (c\phi_s \mathbf{v}) = \nabla \cdot (D_s \phi_s \nabla c) - C_3 \frac{\phi_n c}{k_c + c}, \tag{14.2}$$

where d represents the disinfection effect of the antimicrobial substrate that is a prescribed function of space, and the chemical potential $\delta F/\delta \phi_n$ is calculated by

$$\frac{\delta F}{\delta \phi_n} = -\gamma_1 \Delta \phi_n + \gamma_2 \left(\frac{1}{N} \ln \phi_n + \frac{1-N}{N} - \ln(1-\phi_n) + \chi(1-2\phi_n) \right). \quad (14.3)$$

Here $\rho = \phi_n \rho_n + \phi_s \rho_s$ is the volume-averaged density, where ρ_n, ρ_s are the densities for biomass and solvent, respectively; $\eta = \phi_n \eta_n + \phi_s \eta_s$ is the volume-averaged viscosities, where η_n, η_s are the viscosities for biomass and solvent, respectively; and λ is the motility. C_1 is the growth rate, C_2 is the disinfection rate, C_3 is the nutrient consumption rate, k_c is the Monod constant, and Ds is the diffusion rate for nutrients.

The boundary condition for the model variables are given by

$$[c\mathbf{v}_s \phi_s - D_s \phi_s \nabla c] \cdot \mathbf{n}|_{y=0} = 0, \quad [c\mathbf{v}_s \phi_s - D_s \phi_s \nabla c] \cdot \mathbf{n}|_{x=0,L_x} = 0,$$
$$\nabla \phi_n \cdot \mathbf{n}|_{y=0,L_y} = 0, \quad \nabla \phi_n \cdot \mathbf{n}|_{x=0,L_x} = 0,$$
$$\left(v\phi_n - \lambda \phi_n \nabla \frac{\delta F}{\delta \phi_n} \right) \cdot \mathbf{n}|_{y=0,L_y} = 0, \quad \left(v\phi_n - \lambda \phi_n \nabla \frac{\delta F}{\delta \phi_n} \right) \cdot \mathbf{n}|_{x=0,L_x} = 0, \quad (14.4)$$
$$\mathbf{v} \cdot \mathbf{n}|_{y=0,L_y} = 0, \quad \mathbf{v} \cdot \mathbf{n}|_{x=0,L_x} = 0.$$

We also impose a nutrient feeding condition $c|_{y=L_y} = c^*$ in place of the zero-flux condition at the top.

Figure 14.1 depicts a comparative study of the biofilm growth in a rectangular domain with a regular boundary versus the one in the same domain with an antimicrobial boundary. It is apparent the biofilm growth is contained near the antimicrobial boundary compared with the case with an untreated boundary. It

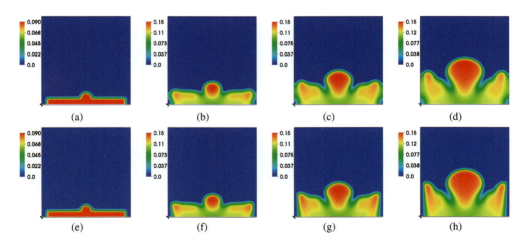

Figure 14.1 Impact of antimicrobial substrate on biofilm formation. (a–d) Snapshots of biofilm profiles at selected time slots in a domain with an untreated substrate. (e–h) Snapshots of biofilm profiles at four selected time slots in a domain with an antimicrobial substrate. This figure shows that antimicrobial substrates have disinfective impact on the nearby biofilm. It effectively disinfects attached bacteria such that biofilms maintain a low concentration near the substrate.

illustrates that an antimicrobial surface can effectively limit the growth of nearby biofilm, which is relevant to designing food packaging materials with antimicrobial properties.

14.4
Antimicrobial Treatment of Biofilms by Targeted Drug Release

Encapsulated nanoparticles have been used to deliver drugs to targeted places within a biofilm [84]. In Ref. [85], the authors developed a model to study drug and biofilm interaction for multispecies biofilms. Here, we employ the model to study antimicrobial treatment of biofilms by nanoparticle-delivered drug release. We model the biofilm together with its surrounding aqueous environment as a multiphasic complex fluid. The biofilm consists of the mixture of biomass and solvent, in which biomass is made up of bacteria of various phenotypes and their secreted by-product exopolysaccharide (EPS). Nutrient and antimicrobial agents are small-molecule substances dissolved in solvent. Their mass and volume fractions are negligibly small, which are therefore neglected in this model.

However, their chemical effects are retained. Let ϕ_{bs} be the volume fraction of the bacteria that are susceptible to antimicrobial agents, ϕ_{bp} the volume fraction of the bacteria that are persistent to the agent, ϕ_{bd} the volume fraction of the dead bacteria, and ϕ_p the volume fraction of EPS.

In addition, we denote the concentration of the nutrient and the antimicrobial agent as c and d, respectively. We denote ϕ_n the volume fraction of the biomass, consisting of all the volume fractions for the bacteria ϕ_b as well as EPS ϕ_p:

$$\phi_n = \phi_p + \phi_{bs} + \phi_{bp} + \phi_{bd}. \tag{14.5}$$

In addition to the volume fractions introduced above, the volume fraction of the solvent is denoted as ϕ_s. The incompressibility of the complex fluid mixture then implies

$$\phi_s + \phi_n = 1. \tag{14.6}$$

We assume the bacteria, regardless of whether they are live or dead, and the EPS mix with the solvent owing to the osmotic pressure. Then, we adopt the modified free energy introduced in Ref. [86] and denote it by F:

$$F = \int_\Omega d\mathbf{x} \left[\frac{\gamma_1}{2} k_B T |\nabla \phi_n|^2 + \gamma_2 k_B T \left(\frac{\phi_n}{N} \ln \phi_n + (1-\phi_n)\ln(1-\phi_n) + \chi \phi_n(1-\phi_n) \right) \right]. \tag{14.7}$$

This is the modified Flory–Huggins free energy with a conformational entropy, in which γ_1 and γ_2 parametrize the strength of the conformational entropy and the bulk mixing free energy, respectively, χ is the mixing parameter, N is the extended polymerization index for the biomass, k_B is the Boltzmann constant, and T is the absolute temperature.

The governing system of equations for the biofilm system is summarized as follows [85]:

$$\begin{cases} \rho\left(\dfrac{\partial \mathbf{v}}{\partial t}+\mathbf{v}\cdot\nabla\mathbf{v}\right) = \nabla\cdot(2\eta\mathbf{D})-\nabla p-\gamma_1\nabla\cdot(\nabla\phi_n\otimes\nabla\phi_n), \\ \nabla\cdot\mathbf{v} = 0, \\ \dfrac{\partial}{\partial t}\phi_i+\nabla\cdot(\phi_i\mathbf{v}) = \nabla\cdot\left(\lambda\phi_i\nabla\dfrac{\delta F}{\delta\phi_i}\right)+g_i, \quad i=\mathrm{bs,bp,bd,p}, \\ \dfrac{\partial \phi_s c}{\partial t}+\nabla\cdot(c\mathbf{v}\phi_s) = \nabla\cdot(D_c\phi_s\nabla c)+g_c, \\ \dfrac{\partial \phi_s d}{\partial t}+\nabla\cdot(d\mathbf{v}\phi_s) = \nabla\cdot(D_d\phi_s\nabla d)+g_d, \end{cases} \quad (14.8)$$

where $\eta = (\phi_b/Re_b)+(\phi_s/Re_s)+(\phi_p/Re_p)$, and the reactive terms for the component are given by

$$\begin{aligned} g_{\mathrm{bs}} &= \dfrac{C_2 c}{K_1+c}\left(1-\dfrac{\phi_{\mathrm{bs}}}{\phi_{\mathrm{bs}_0}}\right)\phi_{\mathrm{bs}}-b_{\mathrm{sp}}\phi_{\mathrm{bs}}+b_{\mathrm{ps}}\phi_{\mathrm{bp}}-\left(\dfrac{r_{\mathrm{bs}}K_{\mathrm{sd}}^2}{K_{\mathrm{sd}}^2+c}+\dfrac{C_3 d}{K_3+d}\right)\phi_{\mathrm{bs}}, \\ g_{\mathrm{bp}} &= b_{\mathrm{sp}}\phi_{\mathrm{bs}}-b_{\mathrm{ps}}\phi_{\mathrm{bp}}, \\ g_{\mathrm{bd}} &= \left(r_{\mathrm{bs}}\dfrac{K_{\mathrm{sd}}^2}{K_{\mathrm{sd}}^2+c^2}+\dfrac{C_3 d}{K_3+d}\right)\phi_{\mathrm{bs}}-r_{\mathrm{bd}}\phi_{\mathrm{bd}}, \\ g_p &= \left[\dfrac{C_5 c}{K_1+c}\phi_{\mathrm{bs}}+\dfrac{C_6 c}{K_1+c}\phi_{\mathrm{bp}}\right]\left(1-\dfrac{\phi_p}{\phi_{p_0}}\right), \\ g_c &= (\phi_{\mathrm{bs}}+\mu_2\phi_{\mathrm{bp}})\dfrac{C_7 c}{K_1+c}, \\ g_d &= -C_8\phi_n\dfrac{d}{K_3+d}, \end{aligned} \quad (14.9)$$

with the conversion rates

$$\begin{aligned} b_{\mathrm{sp}} &= \left(b_{\mathrm{sp}_1}\dfrac{k_{\mathrm{sp}_c}^2}{k_{\mathrm{sp}_c}^2+c^2}+b_{\mathrm{sp}_2}\dfrac{d^2}{k_{\mathrm{sp}_d}^2+d^2}\right)\left(1-\dfrac{\phi_{\mathrm{bp}}}{\phi_{\mathrm{bp}_0}}\right), \\ b_{\mathrm{ps}} &= b_{\mathrm{ps}_1}\dfrac{c^2}{k_{\mathrm{ps}_c}^2+c^2}\dfrac{k_{\mathrm{ps}_d}^2}{k_{\mathrm{ps}_d}^2+d^2}. \end{aligned} \quad (14.10)$$

All C's and K's are model parameters.

Figure 14.2 depicts a 3D simulation of antimicrobial treatment of a grown biofilm using nanoparticle-delivered antimicrobial agents. It simulates the pinpointed disinfectant effect by targeted delivery. It provides a mathematical framework and simulation toolbox for studying biofilm treatment with antimicrobial agents. At the cellular level, nanoparticles can penetrate into cells to deliver substances into targeted areas directly. In Section 14.5, we will discuss two models for intercellular and surface delivery, respectively.

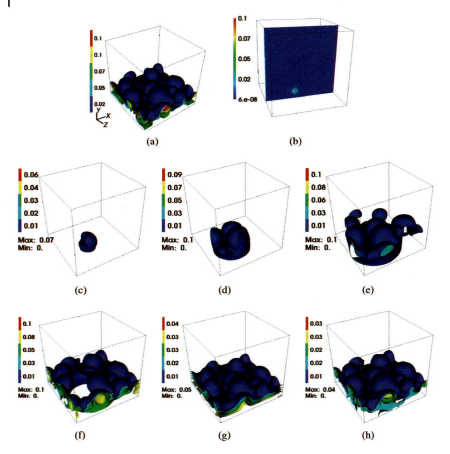

Figure 14.2 Antimicrobial treatment of biofilms by antimicrobial agents released by delivered nanoparticles. A highly heterogeneous distribution of dead bacteria is observed due to the pinpointed dosing delivery method. (a) The initial profile of the biofilm and the release position are chosen at (0.6, 0.1, 0.4). (c–e) The profiles of dead bacteria at selected time slot $t = 124$, 1463, respectively; (b) 2D slice at $z = 0.4$ of the concentration of antimicrobial agents at $t = 310$. The volume fractions of susceptible cells, persisters, and EPS at time $t = 1463$ are given in parts (f–h), respectively.

14.5
Models for Intercellular and Surface Delivery by Nanoparticles

The way how a nanoparticle penetrates into a cell membrane is another interesting topic related to biofilm treatment, but on a different scale (a single-cell scale). There are two distinctive mechanisms: one is known as endocytosis and the other is penetration. Endocytosis is a process in which the cell membrane wraps the nanoparticle gradually and engulfs it into the cell. When the nanoparticle is sufficiently small, it is believed that the particle can slip through the membrane known as penetration [87].

Here, we formulate a three-phase model to study the nanoparticle penetration process. We focus on the portion of the cell where the nanoparticle penetrates. We use ϕ_1, ϕ_2, ϕ_3 to represent the volume fraction of the nanoparticle, cell, and buffer, respectively. Then, the cell membrane is traced by the level set $\phi_2 = 1/2$. We further assume the fluid mixture is incompressible, that is,

$$\phi_1 + \phi_2 + \phi_3 = 1. \tag{14.11}$$

As a generalization from the biphasic model to the ternary phase model, the free energy is proposed as follows:

$$E(\phi_1, \phi_2, \phi_3) = \int_\Omega \left(\frac{3}{8}\Sigma_1 \epsilon |\nabla \phi_1|^2 + \frac{3}{8}\Sigma_2 \epsilon |\nabla \phi_2|^2 + \frac{3}{8}\Sigma_3 \epsilon |\nabla \phi_3|^2 + \frac{12}{\epsilon} F(\phi_1, \phi_2, \phi_3) \right) d\mathbf{x}, \tag{14.12}$$

where one possible choice for F is

$$F = \sum_{i=1}^{3} \frac{\Sigma_i}{2} \phi_i^2 (1 - \phi_i)^2 + 3\Lambda \phi_1^2 \phi_2^2 \phi_3^2. \tag{14.13}$$

To be algebraically consistent with the biphasic systems of surface tensions σ_{12}, σ_{13}, σ_{23}, the following conditions must hold [88]:

$$\Sigma_i = \sigma_{ij} + \sigma_{ik} - \sigma_{jk}, \quad i = 1, 2, 3. \tag{14.14}$$

The governing system of equations of the multiphasic model is summarized as follows:

$$\begin{aligned} \rho(\partial_t \mathbf{v} + \mathbf{v} \cdot \nabla \mathbf{v}) &= -\nabla p + \nabla \cdot (\beta \mathbf{D}) - \sum_{i=1}^{3} \phi_i \nabla \mu_i + \mathbf{f}_e(\phi_1), \\ \frac{\partial \phi_i}{\partial t} + \nabla \cdot (\mathbf{v}\phi) &= \Delta \cdot \left(\frac{M_0}{\Sigma_i} \Delta \mu_i \right), \\ \mu_i &= -\frac{3}{4}\epsilon \Sigma_i \Delta \phi_i + \frac{12}{\epsilon} \partial_i F + \beta, \quad i = 1, 2, 3, \end{aligned} \tag{14.15}$$

where ρ is the mass-averaged density, β is the mass-averaged viscosity, $\mathbf{f}_e(\phi_1)$ is the active force due to the propulsion of the nanoparticle, $\mathbf{D} = (1/2)(\nabla \mathbf{v} + \nabla \mathbf{v}^T)$ is the rate of strain tensor, and the initial

$$\mathbf{v} = 0, \quad p = 0, \quad \phi_i|_{(t=0)} = \phi_i^0, \quad \phi_1^0 + \phi_2^0 + \phi_3^0 = 1, \tag{14.16}$$

β is the Lagrange multiplier to ensure the incompressible condition (14.1):

$$\beta = -\frac{4\Sigma_T}{\epsilon} \left(\frac{1}{\Sigma_1} \partial_1 F + \frac{1}{\Sigma_2} \partial_2 F + \frac{1}{\Sigma_3} \partial_3 F \right), \tag{14.17}$$

where Σ_T is defined by $3/\Sigma_T = (1/\Sigma_1) + (2/\Sigma_2) + (3/\Sigma_3)$

Figure 14.3 is a simulation of the nanoparticle penetration into the cell using the three-phase hydrodynamic model. The model is applied locally to a cell membrane to highlight the process near the nanoparticle. This process is analogous to endocytosis but with a completely different mechanism.

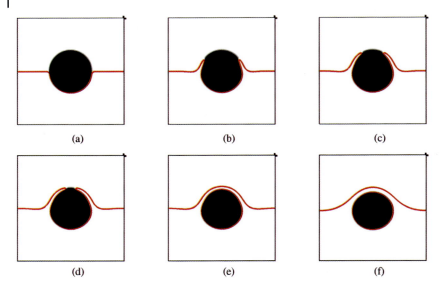

Figure 14.3 Endocytosis of a nanoparticle into a cell. This figure shows the endocytosis process of a nanoparticle via the cell membrane and penetrating into the cell. Here the nanoparticle is shown in black and the cell membrane is in red. The buffer is above the red line and the cytoplasm is below the red line.

Finally, we simulate nanoparticle delivery to the surface of the cell where the substance inside the nanoparticle is released as surfactant materials at the cell membrane. Figure 14.4 depicts a simulation of the type of surface delivery after which a layer of surfactant material is coated on the cell membrane surface. This is another means that antimicrobial agents can be delivered to combat bacterial cells.

14.6
Conclusion

We have reviewed a few families of biofilm models chronically. The models described in this chapter have already been used to explain many complicated phenomena in biofilm dynamics. However, there are still many unanswered questions in mathematical modeling of biofilms, especially in the effort of incorporating all the physical, chemical, biological, and ecological processes occurring at various timescales and length scales into a comprehensive analytical or computational model. With the improvement of biofilm dynamics, fast advancing computing technology, and experimental techniques, we anticipate a surge in research in this scientifically and technologically important and economically even more important topic. We then extend a biofilm model to study antimicrobial treatment of biofilms by antimicrobial substrate surfaces and the targeted nanoparticle delivery, pertinent to food storage and transportation. At a smaller

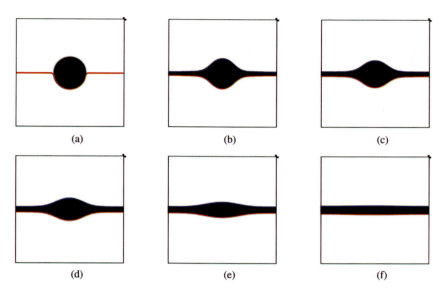

Figure 14.4 Nanosurfactant spreading on cell membrane. Here the red line represents the cell membrane (the interface between cell cytoplasm and surfactant), and blue line represents the interface between buffer and surfactant. Upper part above the blue line is the bulk of buffer; lower part below the red line is the cell cytoplasm.

scale, we devised a multiphasic model to study nanoparticle delivery into the cell and onto the cell surface as surfactants. All these modeling and computational studies can be used to study pertinent cases in food and medical sciences as well.

Acknowledgments

Jia Zhao is partially supported by an ASPIRE grant from the Office of the Vice President for Research at the University of South Carolina. Qi Wang is partially supported by NSF-DMS-1200487 and DMS-1517347, NIH-2R01GM078994-05A1, and SC EPSCOR/IDEA award. Qi Wang's research is also partially supported by NSFC award (#11571032). Tianyu Zhang is partially supported by NSF-DMS-1516951.

References

1 Berekaa, M. (2015) Nanotechnology in food industry; advances in food processing, packaging and food safety. *Int. J. Curr. Microbiol. Appl. Sci.*, **4** (5), 345–357.
2 Ikuma, K., Madden, A.S., Decho, A.W., and Lau, B.L.T. (2014) Deposition of nanoparticles onto polysaccharide-coated surfaces: implication for nanoparticle–biofilm interactions. *Environ. Sci. Nano*, **1**, 117–122.
3 Wang, L., Chen, Y., Milller, K., Cash, B., Jones, S., Glenn, S., Benicewicz, B., and

Decho, A. (2014) Functionalised nanoparticles complexed with antibiotic efficiently kill MRSA and other bacteria. *Chem. Commun.*, **50**, 12030–12033.

4 Decho, A.W. (2013) The EPS matrix as an adaptive bastion for biofilms: introduction to special issue. *Int. J. Mol. Sci.*, **14**, 23297–23300.

5 Costerton, J.W., Lewandowski, Z., Caldwell, D.E., Korber, D.R., and Lappin-Scott, H.M. (1995) Microbial biofilms. *Annu. Rev. Microbiol.*, **49**, 711–745.

6 Davey, M.E. and O'Toole, G.A. (2000) Microbial biofilms: from ecology to molecular genetics. *Microbiol. Mol. Biol. Rev.*, **64**, 847–867.

7 O'Toole, G.A., Kaplan, H.B., and Kolter, R. (2000) Biofilm formation as microbial development. *Annu. Rev. Microbiol.*, **54**, 49–79.

8 Picioreanu, C., Xavier, J.B., and van Loosdrecht, M.C.M. (2004) Advances in mathematical modeling of biofilm structure. *Biofilm*, **1**, 337–349.

9 Kissel, J.C., McCarty, P.L., and Street, R.L. (1984) Numerical simulation of mixed-culture biofilm. *J. Environ. Eng.*, **110**, 393–411.

10 Rittmann, B.E. (1982) The effect of shear stress on biofilm loss rate. *Biotechnol. Bioeng.*, **24**, 501–506.

11 Rittmann, B.E. and McCarty, P.L. (1980) Evaluation of steady-state- biofilm kinetics. *Biotechnol. Bioeng.*, **22**, 2359–2373.

12 Rittmann, B.E. and McCarty, P.L. (1980) Model of steady-state-biofilm kinetics. *Biotechnol. Bioeng.*, **22**, 2343–2357.

13 Wanner, O. and Gujer, W. (1984) Competition in biofilm. *Water Sci. Technol.*, **17**, 27–44.

14 Wanner, O. and Gujer, W. (1986) Multispecies biofilm model. *Water Sci. Technol.*, **28**, 314–328.

15 Wanner, O. and Riechert, P. (1996) Mathematical modeling of mixed-culture biofilms. *Biotechnol. Bioeng.*, **49**, 172–184.

16 Witten, T.A. and Sander, L.M. (1981) Diffusion-limited aggregation, a kinetic critical phenomenon. *Phys. Rev. Lett.*, **47**, 1400–1403.

17 Fujikawa, H. and Matsushita, M. (1989) Fractal growth of *Bacillus subtilis* on agar plates. *J. Phys. Soc. Jpn.*, **58**, 3875–3878.

18 Fujikawa, H. (1994) Diversity of the growth patterns of *Bacillus subtilis* colonies on agar plates. *FEMS Microbiol. Ecol.*, **13**, 159–158.

19 Matsushita, M. and Fujikawa, H. (1990) Diffusion-limited growth in bacterial colony formation. *Physica A*, **168**, 498–506.

20 Berlekamp, E.R., Conway, J.H., and Guy, R.K. (1982) *Winning Ways for Your Mathematical Plays*, Academic Press, New York, NY.

21 Ermentrout, G.B. and Edelstein-Keshet, L. (1993) Cellular automata approaches to biological modeling. *J. Theor. Biol.*, **160**, 97–133.

22 Green, D.G. (1990) Cellular automata models in biology. *Math. Comput. Model.*, **13**, 69–74.

23 Wolfram, S. (1984) Cellular automata as models of complexity. *Nature*, **311**, 419–424.

24 Barker, G.C. and Grimson, M.J. (1993) A cellular automaton model of microbial growth. *Binary Comput. Microbiol.*, **5**, 132–137.

25 Ben-Jacob, E., Schochet, O., Tenenbaum, A., Cohen, I., Czirok, A., and Tamas, V. (1994) Generic modelling of cooperative growth patterns in bacterial colonies. *Nature*, **368**, 46–49.

26 Colasanti, R.L. (1992) Cellular automata models of microbial colonies. *Binary Comput. Microbiol.*, **4**, 191–193.

27 Pizarro, G., Garcia, C., Moreno, R., and Sepulveda, M.E. (2004) Two-dimensional cellular automaton model for mixed-culture biofilm. *Water Sci. Technol.*, **49**, 193–198.

28 Pizarro, G., Griffeath, D., and Noguera, D.R. (2001) Quantitative cellular automaton model for biofilms. *J. Environ. Eng.*, **127**, 782–789.

29 Wimpenny, J.W.T. and Colasanti, R. (1997) A unifying hypothesis for the structure of microbial biofilms based on cellular automaton models. *FEMS Microbiol. Ecol.*, **22**, 1–16.

30 Picioreanu, C., Loosdrecht, M.C.M., and Heijnen, J.J. (1998) A new combined differential–discrete cellular automaton approach for biofilm modeling: application for growth in gel beads. *Biotechnol. Bioeng.*, **57**, 718–731.

31 Wijffels, R.H. (1994) Nitrification by immobilized cells, Ph.D. thesis, Wageningen Agricultural University, Wageningen and The Netherlands.

32 Wijffels, R.H., de Gooijer, C.D., and Kortekass, S. (1991) Growth and substrate consumption of *Nitrobacter agilis* cells immobilized in carrageenan: part 2. Model evaluation. *Biotechnol. Bioeng.*, **38**, 224–231.

33 Wijffels, R.H., de Gooijer, C.D., and Kortekass, S. (1991) Simulation of multispecies biofilm development in three dimensions. *Biotechnol. Bioeng.*, **38**, 232–240.

34 Noguera, D.R., Pizarro, G., Stahl, D.A., and Rittman, B.E. (1999) Simulation of multispecies biofilm development in three dimensions. *Water Sci. Technol.*, **39**, 123–130.

35 Hermanowicz, S.W. (2001) A simple 2D biofilm model yields a variety of morphological features. *Math. Biosci.*, **169**, 1–14.

36 DeBeer, D., Stoodely, P., and Lewandowski, Z. (1994) Liquid flow in heterogeneous biofilms. *Biotechnol. Bioeng.*, **44**, 636–641.

37 DeBeer, D., Stoodely, P., and Lewandowski, Z. (1996) Liquid flow and mass transport in heterogeneous biofilms. *Water Res.*, **30**, 2761–2765.

38 DeBeer, D., Stoodely, P., Roe, F., and Lewandowski, Z. (1994) Effects of biofilm structures on oxygen distribution and mass transport. *Biotechnol. Bioeng.*, **43**, 1131–1138.

39 Hunt, S.M., Hamilton, M.A., Sears, J.T., and Harkin, G. (2003) A computer investigation of chemically mediated detachment in bacterial biofilms. *Microbiology*, **149**, 1155–1163.

40 Hunt, S.M., Werner, E.M., Huang, B., Hamilton, M.A., and Stewart, P.S. (2004) Hypothesis for the role of nutrient starvation in biofilm detachment. *Appl. Environ. Microbiol.*, **72**, 7418–7425.

41 Chambless, J.D., Hunt, S.M., and Stewart, P.S. (2006) A three-dimensional computer model of four hypothetical mechanisms protecting biofilms from antimicrobials. *Appl. Environ. Microbiol.*, **72**, 2005–2013.

42 Chambless, J.D. and Stewart, P.S. (2007) A three-dimensional computer model analysis of three hypothetical biofilm detachment mechanisms. *Biotechnol. Bioeng.*, **97**, 1573–1584.

43 Gjaltema, A., Arts, P.A.M., van Loosdrecht, M.C.M., Kuenen, J.G., and Heijnen, J.J. (1994) Heterogeneity of biofilms in rotating annular reactors: occurrence, structure, and consequence. *Biotechnol. Bioeng.*, **44**, 194–204.

44 Zhang, T.C. and Bishop, P.L. (1994) Density, porosity, and pore structure of biofilms. *Water Res.*, **28**, 2267–2277.

45 Zhang, T.C. and Bishop, P.L. (1994) Evaluation of tortuosity factors and effective diffusivities in biofilms. *Water Res.*, **28**, 2279–2287.

46 Stoodley, P., Jorgensen, F., Williams, P., and Lappin-Scott, H.M. (1999) *Biofilms: The Good and the Bad and the Ugly*, Bioline Press, Cardiff, UK.

47 Stoodley, P., Lewandowski, Z., Boyle, J.D., and Lappin-Scott, H.M. (1999) The formation of migratory ripples in a mixed species bacterial biofilm growing in turbulent flow. *Environ. Microbiol.*, **1**, 447–457.

48 Stoodley, P., Lewandowski, Z., Boyle, J.D., and Lappin-Scott, H.M. (1999) Structural deformation of bacterial biofilms caused by short-term fluctuations in fluid shear: an *in situ* investigation of biofilm rheology. *Biotechnol. Bioeng.*, **65**, 83–92.

49 Picioreanu, C., Loosdrecht, M.C., and Heijnen, J.J. (1999) Discrete–differential modelling of biofilm structure. *Water Sci. Technol.*, **39**, 115–122.

50 Picioreanu, C., Loosdrecht, M.C., and Heijnen, J.J. (2000) Effect of diffusive and convective substrate transport on biofilm structure formation: a two-dimensional modeling study. *Biotechnol. Bioeng.*, **69**, 504–515.

51 Picioreanu, C., Loosdrecht, M.C., and Heijnen, J.J. (2001) Two-dimensional model of biofilm detachment caused by internal stress from liquid flow. *Biotechnol. Bioeng.*, **72**, 205–218.

52 Dillon, R. and Fauci, L. (2000) A microscale model of bacterial and biofilm dynamics in porous media. *Biotechnol. Bioeng.*, **68**, 536–547.

53 Dillon, R., Fauci, L., Fogelson, A., and Gaver, D. (1996) Modeling biofilm processes using the immersed boundary method. *J. Comput. Phys.*, **129**, 57–73.

54 Eberl, H.J., Parker, D.F., and van Loosdrecht, M.C.M. (2001) A new deterministic spatio-temporal continuum model for biofilm development. *J. Theor. Med.*, **3**, 161–175.

55 Dockery, J. and Klapper, I. (2001) Finger formation in biofilm layers. *SIAM J. Appl. Math.*, **62**, 853–869.

56 Alpkvist, E. and Klapper, I. (2007) A multidimensional multispecies continuum model for heterogeneous biofilm development. *Bull. Math. Biol.*, **69**, 765–789.

57 Cogan, N.G. and Keener, J.P. (2004) The role of the biofilm matrix in structural development. *Math. Med. Biol.*, **21**, 147–166.

58 Beris, A.N. and Edwards, B. (1994) *Thermodynamics of Flowing System*, Oxford University Press.

59 Fux, C.A., Wilson, S., and Stoodley, P. (2004) Detachment characteristics and oxacillin resistance of *Staphylococcus aureus* biofilm emboli in an *in vitro* catheter infection model. *J. Bacteriol.*, **186**, 4486–4491.

60 Klapper, I. and Dockery, J. (2006) Role of cohesion in the material description of biofilms. *Phys. Rev. E*, **74**, 031902.

61 Towler, B.W., Rupp, C.R., Cunningham, A., and Stoodley, P. (2003) A computer investigation of chemically mediated detachment in bacterial biofilms. *Biofouling.*, **19**, 279–285.

62 Klapper, I., Rupp, C.J., Cargo, R., Purevdorj, B., and Stoodly, P. (2002) Viscoelastic fluid description of bacterial biofilm material properties. *Biotechnol. Bioeng.*, **80**, 289–296.

63 Zhang, T., Cogan, N., and Wang, Q. (2008) Phase field models for biofilms: I. Theory and one-dimensional simulations. *SIAM J. Appl. Math.*, **69**, 641–669.

64 Zhang, T., Cogan, N., and Wang, Q. (2008) Phase-field models for biofilms: II. 2D numerical simulations of biofilm-flow interaction. *Commun. Comput. Phys.*, **4**, 72–101.

65 Wang, Q. and Zhang, T. (2009) Kinetic theories for biofilms. *Discrete Continuous Dyn. Syst. Ser. B*, **17** (3), 1027–1059.

66 Kreft, J.-U., Booth, G., and Wimpenny, J.W.T. (1998) BacSim, a simulator for individual-based modelling of bacterial colony growth. *Microbiology*, **144**, 3275–3287.

67 Kreft, J.-U., Picioreanu, C., Wimpenny, J.W.T., and van Loosdrecht, M.C.M. (2001) Individual-based modelling of biofilms. *Microbiology*, **147**, 2897–2912.

68 Picioreanu, C., Kreft, J.-U., and Loosdrecht, M.C. (2004) Particle-based multidimensional multispecies biofilm model. *Appl. Environ. Microbiol.*, **70**, 3024–3040.

69 DeBeer, D., Srinivasan, R., and Stewart, P.S. (1994) Direct measurement of chlorine penetration into biofilms during disinfection. *Appl. Environ. Microbiol.*, **60**, 4339–4344.

70 Gordon, C.A., Hodges, N.A., and Marriot, C. (1988) Antibiotic interaction and diffusion through alginate and exopolysaccharide of cystic fibrosis-derived *Pseudomonas aeruginosa*. *J. Antimicrob. Chemother.*, **22**, 667–674.

71 Nichols, W.W., Dorrington, S.M., Slack, M.P.E., and Walmsley, H.L. (1988) Inhibition of tobramycin diffusion by binding to alginate. *Antimicrob. Agents Chemother.*, **32**, 518–523.

72 Stewart, P.S. (1996) Theoretical aspects of antibiotic diffusion into microbial biofilms. *Antimicrob. Agents Chemother.*, **40**, 2517–2522.

73 Stewart, P.S. (2002) Mechanisms of antibiotic resistance in bacterial biofilms. *Int. J. Med. Microbiol.*, **292**, 107–113.

74 Stewart, P.S. and Raquepas, J. (1995) Implications of reaction–diffusion theory for the disinfection of microbial biofilms by reactive antimicrobial agents. *Chem. Eng. Sci.*, **50**, 3099–3104.

75 Keren, I., Kaldalu, N., Spoering, A., Wang, Y., and Lewis, K. (2004) Persister cells and tolerance to antimicrobials. *FEMS Microbiol. Lett.*, **230**, 13–18.

76 Lewis, K. (2001) Riddle of biofilm resistance. *Antimicrob. Agents Chemother.*, **45**, 999–1007.

77 Cogan, N.G. (2006) Effects of persister formation on bacterial response to dosing. *J. Theor. Biol.*, **238**, 694–703.

78 Klapper, I., Gilbert, P., Ayati, B.P., Dockery, J., and Stewart, P.S. (2007) Senescence can explain microbial persistence. *Microbiology*, **153**, 3623–3630.

79 Poplawski, N.J., Shirinifard, A., Swat, M., and Glazier, J.A. (2008) Simulation of single-species bacterial-biofilm growth using the Glazier–Graner–Hogeweg model and the CompuCell3D modeling environment. *Math. Biosci. Eng.*, **8**, 355–388.

80 Picioreanu, C., Loosdrecht, M.C., and Heijnen, J.J. (1998) Mathematical modeling of biofilm structure with a hybrid differential-discrete cellular automation approach. *Biotechnol. Bioeng.*, **58**, 101–116.

81 Rizzello, L., Cingolani, R., and Pompa, P. (2013) Nanotechnology tools for antibacterial materials. *Nanomedicine*, **8** (5), 807–821.

82 Feng, G., Cheng, Y., Wang, S., Hsu, L.C., Feliz, Y., Borca-Tasciuc, D., Worobo, R., and Moraru, C. (2014) Alumina surfaces with nanoscale topography reduce attachment and biofilm formation by *Escherichia coli* and *Listeria* spp. *Biofouling*, **30** (10), 1253–1268.

83 Epstien, A., Hochbaum, A., Kim, P., and Aizenberg, J. (2011) Control of bacterial biofilm growth on surfaces by nanostructural mechanics and geometry. *Nanotechnology*, **22** (49), 494007.

84 Nevius, B.A., Chen, Y.P., Ferry, J., and Decho, A.W. (2012) Surface-functionalization effects on uptake of fluorescent polystyrene nanoparticles by model biofilms. *Ecotoxicology*, **21** (8), 2205–2213.

85 Zhao, J, Seeluangsawat, P, and Wang, Q. (2016) Modeling antimicrobial tolerance and treatment of heterogeneous biofilms, **282**, 1–15.

86 Zhao, J. Shen, Y., Happasalo, M., Wang, Z., and Wang, Q. (2016) A 3D numerical study of antimicrobial persistence in heterogeneous multi-species biofilms, **392**, 83–98.

87 Chen, X., Tian, F., Zhang, X., and Wang, W. (2013) Internalization pathways of nanoparticles and their interaction with a vesicle. *Soft. Matter*, **9**, 7592–7600.

88 Boyer, F. and Lapuerta, C. (2006) Study of a three component Cahn–Hilliard flow model. *ESAIM: Math. Model. Numer. Anal.*, **40** (4), 653–687.

Part Three
Technical Challenges of Nanoscale Detection Systems

15
Smart Systems for Food Quality and Safety

Mark Bücking,[1] Andreas Hengse,[2] Heinrich Grüger,[3] and Henning Schulte[4]

[1]*Fraunhofer Institute for Molecular Biology and Applied Ecology IME, Department. Evaluation of Food Safety and Consumer Risk Assessment, Auf dem Aberg 1, Schmallenberg 57392, Germany*
[2]*Fraunhofer Food Chain Management Alliance, Fraunhofer Food Chain Management Alliance, Auf dem Aberg 1, Schmallenberg 57392, Germany*
[3]*Fraunhofer Institute for Photonic Microsystems IPMS, Department of Photonic Sensing, Maria-Reiche-Street 2, Dresden 01109, Germany*
[4]*Fraunhofer Institute of Optronics, System Technologies and Image Exploitation IOSB, Department of Visual Inspection, Fraunhoferstraße 1, Karlsruhe 76131, Germany*

15.1
Introduction

The European food sector is the second biggest manufacturing sector in Europe with more than 270 000 companies employing around 4.3 million persons and generating an annual turnover of around 1000 billion euros. The sector is also facing several challenges that require innovation and new technological solutions: the food industry needs to guarantee food safety, to improve the quality of the food products, to decrease its impact on the environment while continuing to provide affordable food supply to a growing population.

Microsystems have the potential to provide a wide range of technological solutions for the food industry. Micro- and Smart Systems are suitable for the following.

- *In situ*, noninvasive, fast, and automatic measurements.
- The building of multisensing platforms and use less sample and reagents.
- Low consumption of energy and enables them, unlike traditional laboratory equipment, to be distributed in space and time [1].

Project "FoodMicroSystems" [2]
Despite the potential of microsystems, only a few applications have been developed so far. The objective of the FP7 project "FoodMicroSystems" was to

improve this situation by promoting the implementation of smart systems in the food industry. The project provided review of the possibilities offered by microsystems to the food sector. The needs have been identified in several food chains (meat, dairy, beverage, fruits, and vegetable). Building on these results, three technological roadmaps have been prepared. This article looks at some selected question. For more details, please refer to Ref. [2].

Microsystems are also known as ICT/MST or MST or MEMS or integrated systems or lab-on-a-chip systems. The systems involve at least one component built by microtechnologies or nanotechnologies. Additives in food are not in the scope of microsystems in this chapter.

15.2
Overview [3]

The food sector represents a significant market for developers of microsystem solutions. It includes a limited number of large companies controlling many production sites and a high number of small producers supplied by a very large number of farmers (e.g., 1 million farms delivering milk to 5000 processors in the European Union). Key driver for implementing microsystems in the food sector are as follows:

- Food safety: the food industry needs solutions that guarantee that the food is safe for the consumers. There is a demand for in-line solutions (continuous measurement), portable, and easy-to-use devices.
- Food quality: a better monitoring of the quality parameters of the raw materials, of materials during processing and of the final products is needed to optimize the processes, mostly for solutions allowing continuous and simultaneous measurements of several parameters. Portable devices for in situ measurements are also in demand.
- The "horse meat" crisis in 2013 also reveals that there is a market for solutions allowing authentication (solutions to detect the origin of the food products), traceability, and detection of fraud and adulteration.
- Solutions to optimize the water and energy consumption are needed to decrease the impact of the industry on the environment and to reduce the production costs. Tools allowing the optimization of cleaning operations are particularly in demand. Intelligent packaging can monitor the deterioration of food products and increase their shelf life.
- In-line, on-line, and at-line measurement solutions will help the industry to address the four key challenges of the sector (safety, quality, sustainability, and efficiency).
- The main technological constraints are related to (i) the robustness of the devices (ii) the reliability of the measures (iii) the compatibility with food processes, and (iv) the time to process information. The sampling strategy (how many measures, when, where, etc.) can also be a challenge.

The food sector is an important pillar in the European economy, as one of the most successful and dynamic business sectors. Microsystems can provide a wide range of technological solutions to make food safer and of better quality, and can contribute to convenience, shelf life, and freshness.

The main demands for application concerns are as follows:

- The assessment of quality and safety of food products
- The monitoring processes in food industry and an improved control of the end quality of the product
- Food packaging applications
- Technologies to develop new food products

15.3
Roadmapping of Microsystem Technologies Toward Food Applications

The "FoodMicroSystems" project has developed roadmaps on three application sectors and on three key microsystem technology areas. The challenges to be addressed are more in innovation than research. For example, a pure technology push approach from microsystem providers will usually not be taken up by the food and beverage sector. With "price/cost" as the most important decision factor for the deployment of new technologies in the food sector, the current situation can be considered a chicken and egg situation. It is expected that, once an initial success story for microsystems deployment in the food supply chain is achieved, many others will follow [4–6].

15.3.1
Implementation of Microsystems in the Dairy Sector [4]

The economic importance of the dairy sector in Europe is underlined by more than 11 000 enterprises representing a total production value of more than 100 B€. Dairy is not only the largest segment of the food industry but also represents 17% of the total European food export.

There risks and issues associated with the consumption of dairy products (e.g., contamination, allergies) can be addressed by the introduction and appropriate use of microsystems throughout the supply chain, starting from milking, transport, and dairy processing, to logistics and sales, and further on to consumption at the customer's site.

Areas that stakeholders (dairy companies, consumers, and microsystems technology providers) consider being the top most challenges in the dairy industry are as follows:

- Guaranteeing product safety by preventing contaminated food reaching the market.

- Improving product quality by introducing more homogenous and optimized production.
- Increasing equipment utilization by using condition-based maintenance methods.
- Increasing process efficiency by using less energy and decreasing product loss.

15.3.1.1 Measurement of Contamination

Standardized measurement/or detection platform, where only small components (e.g., sensor elements) need to be adapted to enable different types of measurements. Sensors will be required that can detect a range of pesticides, herbicides, pathogens, metals, and milk ingredients like proteins. As cost is an important issue in the dairy industry, these sensors should be multifunctional and affordable, for example, integrated sensors based on low-cost disposables or, preferably, sensors that do not need disposables.

When it comes to sensors in packages, these sensors need to be of very low cost and will likely be in a "printed technology." There is also a need for sensor systems that can assess the status of the inner surfaces of process equipment in a dairy plant. Even more difficult to introduce then sensors will be advanced processes that modify the product itself, for example, lactose-free milk or double emulsions that will minimize the fat content.

15.3.1.2 List of Specific Items to be Tested in the Dairy Industry

The list of specific items to be tested for in the dairy industry is large and diverse, for example,

1) Pathogens and other organic ingredients and contaminants: herbicides, pesticides, antibiotics, dioxins/PCBs, allergens, mycotoxins, native protein versus denatured protein, lactose.
2) Inorganic ingredients and contaminants: pH, detergents/residual chemicals from cleaning, heavy metals, water activity: linked water versus free water.
3) Physical parameters: pressure and temperature during pasteurization, viscosity measurement before separation, humidity (in cheese), density.
4) Information about cells, somatic cell content, fat content, dry matter (quantity).

Food safety is high on the agenda in many countries. The most common contaminants to be tested for are pathogens (*Salmonella*, *Listeria*, *E. coli*, etc.), pesticides, allergens, and, at least in Europe, GMOs (genetically modified organisms).

The most critical "unwanted" substances, the contaminations, can be grouped in the following categories:

A) Microbiological contamination
B) Chemical contamination, residues, and other unwanted ingredients

Examples of microsystem roadmaps for the detection of pathogens have been shown in Figure 15.1 and for chemical contaminants, residues, and others in Figure 15.2 [4].

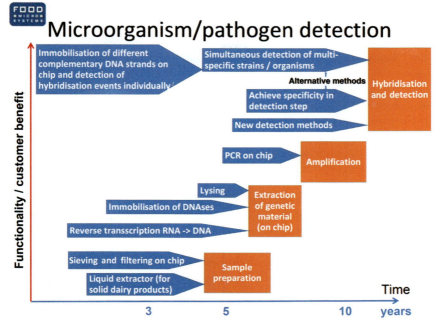

Figure 15.1 Roadmap for microdevices/components needed for a new concept of a μPCR with the whole process completely integrated in one microfluidic chip. (From Ref. [4], p. 33.)

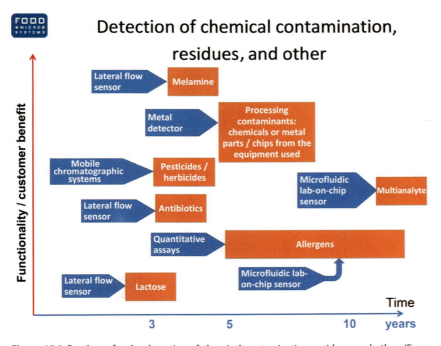

Figure 15.2 Roadmap for the detection of chemical contamination, residues, and other. (From Ref. [4], p. 36.)

15.3.2
Implementation of Microsystems in the Meat Sector [5]

The roadmap developed in the FoodMicroSystems project addresses the development of microsystems for (1) fresh meat sold by butchers or in packaged meat, (2) cold chain and (3) processed meat, analyzing example taken in the beef, pork, and poultry meats.

The most advanced deployment in food and food related industries is the use of RFID tags in processing lines and tracking of containers or pallets within a network of actors working for a given corporation. Smart phones and other connected cells are deeply modifying the consumer's acceptance.

The spreading of connected tools at the end user side can boost new emerging handheld equipment and bring a valuable benefit. The meat sector will turn their innovation effort toward the use of new Microsystems, for example, molecular or biodegradable markers, electronic on flexible foils, smart skins, and so on.

Beyond H2020, bioresorbable electronics will bring technologies that will allow a much more massive use of low-cost microsystems while preserving environment. Due to high economic and environmental pressure, smart devices integrated in packages should rely on cheap markers, for example, ink printed systems, battery-less and biodegradable electronics, and biomarkers. Figure 15.3 shows the microsystems roadmap for Beyond 2020.

Figure 15.3 Microsystems roadmap for the meat sector – beyond 2020. (From Ref. [5], p. 55.)

15.3.3
Implementation of Microsystems in the Food and Beverage Sector

European wine and beer producers are today facing new challenging issues due to the evolution of the consumption, changes in terms of the organization of the production structures, and increasing environmental concerns. The modern beverage industry needs tools for process control and quality assessment in order to better manage the main process steps, like the fermentation process. It is known that careful control of the fermentation process (considered one of the most important steps of the production process) can improve wine and beer quality and help producers to develop a product that satisfies evolving consumer tastes.

Sensors and microsystems may provide excellent solutions to complement laboratory analysis, by creating faster and portable systems for on-line and at-line use that will help on the full production process control of beverages. This better control may produce at the same time better products and at lower production costs.

Looking at all types of MST devices identified in the different roadmaps, they can be classified in six main groups:

- Gas and aroma sensing systems (volatiles, gases) may be of capital importance in the fast screening of raw materials and on the quality and authenticity of the final product and in the logistics sector.
- Chemical and electrochemical devices (pH, redox, ions, metals) are good candidates for on-line monitoring of the fermentation in the tank, of the maturation in barrel, of quality control in bottle and also for logistics, safety issues, and water control monitoring via the detection of heavy metals and other important ions.
- Biorecognition (immunosensors, genosensors, DNA) of compounds in wine and beer: that is, for safety, for water quality assurance but also during fermentation and aging monitoring.
- Physical sensors: temperature, pressure, ultrasounds, turbidity, color, and so on.
- RFID and other traditional electronic circuits.
- Artificial vision systems: CMOS cameras and image processing.

Developing new devices and smart systems for prompt monitoring following different strategies such as miniaturized multisensing platform based on gas microsensors, gas and liquid microchromatography, and mid-infrared microspectrometry should be of interest.

Microsystems may provide beverage producers with key information at critical check points in their process, supplementing their knowledge and experience, helping them to make the right decisions in pursuit of a product of quality.

15.4
Microsystem Technology Areas

15.4.1
Detection Methods [4]

The following technologies were taken into account, being regarded as potentially able to solve some of the more urgent problems in the dairy sector. Moreover, these technologies will also be used for the analysis of different foods, for example, meat, fruit and vegetables, beverages, and so on.

15.4.1.1 Near-Infrared Spectroscopy (NIRS)
Rational behind the measurement: determination of composition as well as the quality and authenticity, for example, of milk, butter, and cheese. The majority of the instruments are not particularly suited to in-line or on-line implementation yet, owing to issues such as sample size, scans time, sample preparation, and so on.

15.4.1.2 Mid-Infrared Spectroscopy
It is necessary to control the composition as well as processing conditions of the cheese in the process line itself. Miniature mid-infrared spectroscopy equipment is used in cheese production plants for determining composition and textural properties, mainly used for the rapid characterization of cheese and milk.

15.4.1.3 Imaging Techniques
In dairy processing, information about the surface and internal properties of the products are often judged by humans. Several efforts have been made to mimic human sight to judge the quality and composition of dairy foods in the process line with high speed. This has improved considerably with the introduction of high-resolution cameras, and with using monochrome or visible infrared light.

15.4.1.4 Hyperspectral Imaging
Hyperspectral imaging (HSI), like other spectral imaging, collects and processes information from across the electromagnetic spectrum. The optical properties of turbid liquids including milk can be well determined by HSI. The scattering effects or optical properties of milk are correlated with the fat content in milk. By analyzing the spatial images, the distribution of food constituents are detected. The amount of protein, fat, and carbohydrate can be determined by using various image data and analyzing these data in the perspective of partial least square regression errors. Products with inhomogeneous distribution of the constituents can be assessed by changed colouration of spatial images. Especially for cheese quality control, this method produces spatial images to determinate the fat distribution pattern in the cheese.

15.4.1.5 Ultrasound Imaging

Ultrasound imaging is also called ultrasound scanning or sonography. It produces real images by exposing the objects to ultrasound waves. Based on the acoustic properties of the objects, some of the waves are reflected while others are transmitted with different speed depending on the impedance throughout the object. These properties are useful to assess the different objects as well as difference within the same objects. Thus, the sonography or ultrasound scanning is useful in detecting extraneous materials in the objects as well as quality of the dairy products in a very short time. An important advantage is that it does need direct contact with the product. The sonography can also be utilized for mapping the internal structure of cheese. The formation of cracks, eyes, cheese matrix, and ripening age of hard cheese is better assessed by the three dimensional (3D) ultrasound images.

15.4.1.6 Magnetic Resonance Imaging (MRI) and X-ray scanning

Magnetic resonance imaging (MRI) is also known as "nuclear magnetic resonance imaging" (NMRI) or "magnetic resonance tomography" (MRT). This can be used for visualizing the internal structure, phase separation, component distribution, rheology and basic structure, and composition of dairy food. X-rays are suitable for the detection of foreign bodies, holes, cracks, and so on.

15.4.1.7 Dielectric sensor

Dielectric spectroscopy is a low cost sensing quality system for monitoring dairy products. Dielectric methods use the complex impedance (which is made up of the capacitance and the conductance) as a function of frequency to determine the dielectric spectrum. Dielectric spectroscopy is a real time, very rapid and noninvasive technology for measuring the moisture and salt content of food products, especially cheese and butter. The continuous on-line control of moisture and salt in butter and cheese is very important for controlling the quality of the product.

15.4.1.8 Process Viscometer

Viscosity is the resistance felt by a liquid during flowing and considered as important quality parameter of several dairy products. The process viscometer measures the viscosity of dairy products on-line or in-line, and thus, helps in the production of good quality milk products as well as initiation of corrective measures to fix a problem during the processing. The viscosity is correlated with the consistency of food and thus directly addresses the preferences of a consumer.

15.4.1.9 Direct Sensing with Electronic Nose Technology

In dairy processing, electronic nose technology can be used to assess aromatic quality or different classifications. Electronic noses, comprising sampling technology, sensor technology, and data evaluation, are rather a complementary technology to human sensor panels or sophisticated laboratory analyses. They offer a comparably easy and quick way to assess the volatile compound profile of a substance or product, and therefore, can be used in a large variety of applications. Electronic noses are found in many applications in food and beverage, yet only a limited number of studies report the use in dairy processing, probably due to the complexity of their matrices.

15.4.1.10 Chemical and Biochemical Electronic sensors and systems

A challenging problem in food processing industry is quality and safety of food products and thus the use of time consuming analytical tools, mostly based in complex and expensive laboratories has been a must for the food and beverage producers. Chromatography, HPLC, mass spectrometry, FTIR, enzymatic assays, and so on are traditional methods that can be complemented by new tools based on new technologies.

New solutions based on sensing devices and systems of different nature, for both monitoring quality and safety of the product during processing and for optimizing the production process. Such solutions could cover the following:

- Physical sensors: temperature, pressure, ultrasounds, turbidity, color, and so on
- Gas and aroma sensors: volatiles, gases, and so on
- Chemical sensors: pH, redox, specific ions, and so on
- Biosensing systems: antibody, enzymatic, genosensors, DNA, and so on
- Artificial vision: CMOS cameras and image processing
- RFID and other traditional electronics circuits

Finally, there is the group of chemical and biochemical sensors for gases and liquids that, despite having been the focus of research in the last decades, have not reached the degree of maturity necessary for being well accepted and marketable in the food sector. To be useful, these devices will have to answer the following expectations:

- Speed of the analysis, automation, and simplicity of use
- Autonomy and portability
- Sensitivity and reproducibility at least comparable with existing solutions
- With some added value
- Low cost

15.4.2
Gas Sensing Devices and Systems [7]

Many quality attributes and safety problems in food and beverage products and intermediate processing steps may be monitored by detecting their associated aromas, volatiles compounds, and gases.

Some gas sensing systems are available in the market for environmental, industrial safety, and other niche applications. E-nose systems, which are based on gas devices, have been developed for other applications and could be adapted to the food and beverage process monitoring. In food and beverage

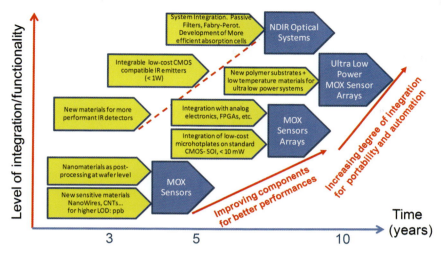

Figure 15.4 Technological trends for gas sensors FoodMicroSystems. (From Ref. [7], p. 17.)

applications we usually have a need to detect one or more components of a mixture of gases.

Generically, it can be stated that it is of interest for the following:

- To use cheap single sensors or simple systems with high sensitivity to one gas, for simple single point measurements of simple gases or vapours.
- To use optical sensors and systems when more selectivity and stability are required.
- To develop and integrate multi-sensing systems (e-noses . . .) that, with the help of data processing, may be useful in the case of detecting gases and volatile organic compounds (VOCs) in mixtures, or fingerprint such mixtures.
- To develop more advanced systems (e.g., micro gas chromatographers, ion mobility spectrometers) when high discrimination is required and not achieved with more simple systems.

Among the different technologies for gas sensors, metal oxide (MOX) gas sensors and arrays seem to be the more appropriate to meet the specifications required. For being useful in portable systems for food and beverage process control, such devices have to measure at ppb level and with ultralow power (<10 mW) consumption. Other interesting technologies, but less developed and more complex could be IMS (ion mobility systems) and SAW (surface acoustic wave) devices combined with polymers or infrared (NDIR) for more selective systems (Figure 15.4).

15.4.2.1 E-Nose Instruments

Information can also be obtained from arrays of nonspecific sensors with partially overlapping selectivity and treating the data obtained with pattern recognition software. These systems are often referred to as artificial senses.

An electronic nose is an alternative to single gas sensors for applications for which it is necessary to characterize complex mixtures that are more conveniently identified by an overall fingerprint rather than by the individual identification of their constituents. Electronic noses are found in many applications in food and beverage.

E-noses have been tested in a variety of applications in the food and beverage sector like classification, detection of aromatic profiles of products for monitoring their processing, discrimination, authentication, determination of attributes of different brands, comparison between beverages of different origins, optimization of the production process, and several other applications.

E-noses in meat have also been studied at R&D level for monitoring the evolution of meat freshness and bacterial spoilage during storage of beef, discrimination of batches of meat, authentication of meat, determination of meat species in processed products, classification of poultry depending on the storage conditions, detection of sex-linked differences, detection of boar taint in pork meat, fermentation of dried products, monitoring of microbiological changes in fresh meat, authentication of pig feeding and ripening time in Iberian hams, and so on.

E-noses still show good potential for the food and beverage sector if better and more stable sensor matrixes are used.

15.4.2.2 Microchromatographers

Many low-cost gas detection sensors have a broadband sensitivity to a range of volatile organic compounds and gases but suffer from poor selectivity. A system approach can improve it if separation elements are used prior to reaching the sensing devices. Gas chromatography (GC) is considered one of the most reliable separation methods. But GC systems are very expensive and laboratory based. Research on the application of microsystem technologies to down-size GC systems has been done in the last decade but because of the small dimensions they still show poor performance. However, micro-GC is being studied for other applications. The food and beverage industry may benefit of such developments.

15.4.3
NIR-Spectroscopy [7–9]

A very important indicator for the quality of food is the composition. A few major components, typically represented in the percent range, contribute to the quality of food. It is a well-established technology to analyze the composition of organic matter using near-infrared (NIR) spectroscopy. From the chemometric analysis, it is possible to evaluate the content of water, starch, sugar, protein, fat, and others. This can be correlated to parameters like ripeness, freshness as well as quality indicators like water content or other added ingredients.

Mobile measurements on-site require new approaches compared to state of the art systems. Higher demand for size, weight, and cost arises. A first device could be a mobile scanner unit for professional use. Later, similar systems might enter applications for customer information or mobile units will be realized to operate in combination with wireless interfaces. These devices can also be included into consumer products like mixers, scales, or a "check-in" device for the refrigerator. Finally, spectral analysis with the mobile phone could be realized in a not too far future.

Relevant progress for the spectrometer technology will be required. Serious composition correlated measurements need important band structures around 1450 nm for water and from 1650 to 1780 nm for carbon–carbon and carbon–hydrogen bonds. A promising approach is the use of MEMS (micro electromechanical systems) technologies in combination with sophisticated single detectors.

MEMS scanning grating spectrometers have entered the market more than 10 year ago. The latest developments aim at an optical bench with a size comparable to a sugar cube. Due to the high accuracy of the MEMS process, it turns out to be advantageous to include as many optical components as possible into the MEMS chip.

References [8,9] revealed a new modified Czerny Turner type spectrometer operating on the first diffraction order. The system can be build starting with a printed board, where the detector is mounted directly and then buried in a cavity of the MEMS chip. Readout and drive electronics have been adjusted for operation from a single 5 V DC source, enabling either USB or Bluetooth communication. The system (Figure 15.5) has been investigated in the lab. It has been proven to meet relevant specifications like 10 nm resolution in a 950–1900 nm range, high stability, and a good signal to noise ration.

Composition measurements on site require a database for the evaluation of the spectra. For several examples, investigations have been performed. As the chemometric models can be quite complex, online access to relevant database

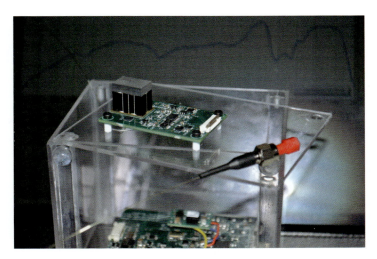

Figure 15.5 MEMS scanning grating spectrometers. (© Fraunhofer Institute for Photonic Microsystems IPMS.)

services will be necessary. First implementations for test use have been realized [10].

Future work is intended for the use with mobile phones. A combination with the camera for an estimation of the volume of food would enable applications like a nutrition advisor. From the already existing activity analysis, a helpful tool can be developed for a well-balanced diet against overweight.

15.4.4
Biochemical Sensors [7]

In food and beverage, a noticeable amount of quality attributes and safety problems cannot be related to an aroma but to the taste or to a marker that has to be measured in liquid media. In this case, chemical and biochemical devices and systems are the most appropriate.

Possible applications in the beverage sector are: identification of the aromatic richness of grapes and materials (e.g., barrels), authentication, classification, polyphenols contents, dissolved gases in beverages, pesticides, pH, redox potential, detergents and ions in water, and so on.

For dairy and meat, many analytical requirements for safety assessment may be based on biosensors. Tetracyclines, aflatoxins, and other mycotoxins pathogens are examples of concerns of the food producers (Figure 15.6).

15.4.4.1 E-Tongue Systems

Concepts similar to e-nose, but for use in liquid media, are related to the sense of taste in the same way that the electronic nose relates to olfaction and are usually known as taste sensors or electronic tongues. The study of taste sensors is still at an early stage.

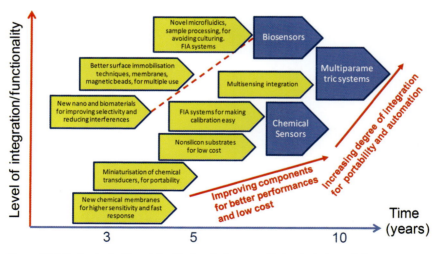

Figure 15.6 Technological trends for biochemical sensors. Technological trends for gas sensors. (From Ref. [7], p. 24.)

An electronic tongue is defined as a multisensory system, which consists of a number of low-selective chemical or biological sensors and a signal-processing module based on pattern recognition and/or multivariate analysis.

The main applications are as follows:

- Beverage sector: evaluation of raw materials and evolution of taste and flavor of beverages at different stages of their production, taste objectivation, classification and discrimination of beverages, authentication, detection of alcohol contents, quantify bitterness in beers, differentiate varieties and food and beverage (including mineral water) brands, determining flavors of commercially available fruit juices and soda drinks, and so on.
- Dairy sector: discriminate fresh from spoiled milk, analyze flavor aging in flavored milk, and so on.
- Meat sector: R&D studies have been done for the determination of salts, nitrates and nitrites in minced and processed meat and sausages.

15.4.5
Microorganism Detection [7]

There is a growing need for biosensors along the whole food chain in microbial food safety as well as by legal/governmental inspection agencies. The biosensor market is rapidly expanding with estimates of a growth from € 5 100 million in 2009 to € 11 000 million in 2016 [11]. The main areas for applications of biosensors include product safety (detection of pathogenic microorganisms in food and food contact material) and product quality (detection of commercial microorganisms, e.g., starter cultures in dairy products).

The main motivations for their utilization include the following:

- Increasing automation of food production including increasing time efficiency due to in-, on-, or at line control.
- Presence of standards/food laws and regulation that addresses increasingly more topics following continuous decrease of LOD and LOQ (limit of detection/quantification). In addition, food industry has much more regulations with respect to microorganisms than any other industry.
- Changes in the demands of consumers require new methods to check differentiated streams of raw materials, for example, organic food can contain much more pathogens (e.g., *Salmonella* in eggs or *S. aureus* in milk).
- In the spirit of sustainability and demand for less processed foods by, consumers it can be anticipated that new processing technologies – with less energy consumption – will be implemented in food industry. These technologies, for example, some new pasteurization strategies, might lead to an increase of pathogens.
- Increasing consumer concern with regard to food safety due to food scandals were microorganism-caused serious health problems.

So far researchers have not been successful in producing biosensors that meet the high expectations of the food industry [11]. Fast detection methods are not accepted by the regulatory authorities.

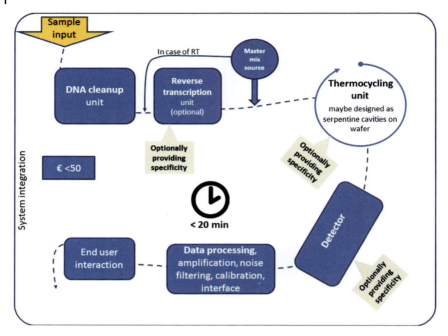

Figure 15.7 Functionality of a "µPCR" device. Technological trends for biochemical sensors. (From Ref. [7], p. 27.)

Regulations define specific reference methods that have to be applied for food law issues. This means that novel, fast detection methods cannot be used to replace current methods before they are defined as an accepted reference method.

In practice, this implies that new methods have to prove their quality for many years before they even have the chance to be considered as a reference method. Although the fast detection methods are not allowed to be regulatory authorities, they can be used for screening of products and materials to reduce the number of expensive tests.

The following idea – a µPCR – fulfils the needs of industry in terms of regulations, speed, price, and robustness, keeping the specific detection of pathogens in mind (Figure 15.7).

Figure 15.7 provides a schematic representation of the functionality of a "µPCR" device for miniaturized detection of food pathogens based on genetic material, either DNA or more sophisticated RNA. The latter approach would allow to distinguish whether a detected organism is still alive or not since RNA is readily degraded in an *ex vivo* environment or in a dead organism. However, this would effort an extra step in which RNA is reversely transcribed into DNA. In the future, modern polymerases will allow a decrease of required time dramatically, allowing for cycling intervals of a few seconds.

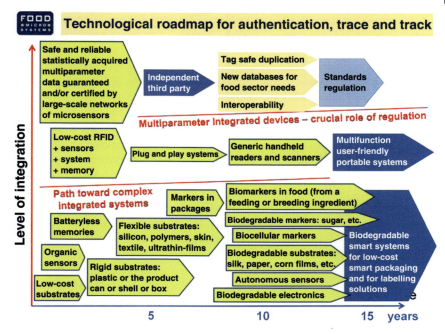

Figure 15.8 Technological roadmap for authentication, trace & track (From Ref. [7]) p. 38.

15.4.6
Tracking and Tracing

Food traceability is part of logistics management that captures, stores, and transmits adequate information about a food product at all stages in the food supply chain so that the product can be checked for safety and quality control, traced upward, and tracked downward at any time required. Tracking and tracing also helps to determine the correct origin of food products.

Microsystem technologies are often involved in the first steps of the authentication and tracking and tracing systems. Companies can also use tracking and tracing tools to analyze key data at each step of manufacturing process (Figure 15.8).

Thanks to low-cost integrated sensors embedded in (i) identification tags (on live cattle, food raw materials, packaging of the final products, etc.), (ii) in-line process equipment, (iii) handheld scanners, (iv) transport boxes, manufacturers, and retailers will continuously gather measured data about their production yield and reliability.

References

1 CORDIS (2013) FoodMicroSystems: Project Final Report, European Union. Available at http://cordis.europa.eu/project/rcn/99999_en.html.

2 CORDIS (2013) FoodMicroSystems, European Union. Available at http://cordis.europa.eu/project/rcn/99999_en.html.

3 CORDIS (2013) FoodMicroSystems: Final Report of Industrials Needs, Demands and Constraints, European Union. Available at http://cordis.europa.eu/project/rcn/99999_en.html.
4 CORDIS (2013) FoodMicroSystems: Roadmap Sector 1 – Implementation of Microsystems in the Dairy Sector, European Union. Available at http://cordis.europa.eu/project/rcn/99999_en.html.
5 CORDIS (2013) FoodMicroSystems: Roadmap Sector 2 – Implementation of Microsystems in the Meat Sector, European Union. Available at http://cordis.europa.eu/project/rcn/99999_en.html.
6 CORDIS (2013) FoodMicroSystems: Roadmap Sector 3 – Implementation of Microsystems in the Wine and Beer Sector, European Union. Available at http://cordis.europa.eu/project/rcn/99999_en.html.
7 CORDIS (2013) FoodMicroSystems: Final Report – Synthesis of the Roadmaps for the Implementation of Microsystems in the Food and Beverage Sectors, European Union. Available at http://cordis.europa.eu/project/rcn/99999_en.html.
8 Grüger, H., Pügner, T., Knobbe, J., and Schenk, H. (2013) *Next-Generation Spectroscopic Technologies VI*, vol. **8726**, Proceedings SPIE, DSS 2013, pp. 1–9.
9 Pügner, T., Knobbe, J., and Lakner, H. (2016) *Appl. Optics.*, **50** (24), 4894–4902.
10 Literature Fraunhofer Institute of Optronics, System Technologies and Image Exploitation IOSB (2015), System Technologies and Image Exploitation, IOSB.
11 Frost & Sullivan (2010) Analytical Review of World Biosensors Market, June.

16
Nanoelectronics: Technological Opportunities for the Management of the Food Chain

Kris Van De Voorde,[1] *Steven Van Campenhout,*[2] *Veerle De Graef,*[2] *Bart De Ketelaere,*[3] *and Steven Vermeir*[3]

[1]*IMEC, Smart Applications and Innovation Services Flanders, Kapeldreef 75, 3001 Leuven, Belgium*
[2]*Flanders' FOOD, Wetenschapsstraat 14 A, 1040 Brussel*
[3]*Katholieke Universiteit Leuven, Faculty of Bioscience Engineering, BIOSYST – MeBioS, Kasteelpark Arenberg 30 – Box 2456, 3001 Leuven, Belgium*

16.1
Technological Needs and Trends in the Food Industry

Food quality assurance is a key issue in the food chain. First of all, this encompasses assurance of food safety from "field to fork." The prevention of foodborne illnesses due to contamination of food by pathogenic microorganisms or chemical hazards is a task for every link in the chain. Microbial and chemical analyses focusing on these contaminants play an essential role in preventative food safety management systems like HACCP, ISO 22000, BRC, Global Gap, and so on. These kinds of analyses are planned according to sampling-based control programs and performed by time-consuming tests in specialized laboratory environments. In food industry, there is a demand for microbial and chemical tests that can be applied in or as close as possible to the process line, not only for product checking throughout the process (from raw material to end product) but also for validation of cleaning actions. In this way, contamination problems could be detected in an early stage and remediation measures taken directly. As a consequence, the tests should be fast, reliable, user-friendly, and robust.

A special type of food safety risk is posed by physical hazards (e.g., pieces of metal, plastic, wood, glass) that accidently end up in food products. In this case, prevention cannot be handled via statistics-based screenings. Rather, a continuous process monitoring is needed and in food companies metal detectors, X-ray inspection, or visual controls are typically applied for this purpose. Here, more technological improvements are required to enable better detection (small objects with comparable density or color than the food product) in a fast way.

Nanotechnology in Agriculture and Food Science, First Edition. Edited by Monique A.V. Axelos and Marcel Van de Voorde.
© 2017 Wiley-VCH Verlag GmbH & Co. KGaA. Published 2017 by Wiley-VCH Verlag GmbH & Co. KGaA.

Furthermore, in view of the high cost of the current systems, the technological updates should not result in significant cost increase.

From the viewpoint of producers and consumers, the most important food quality aspect relates to the perceived quality of intrinsic attributes (e.g., color, flavor, smell, appearance) that in turn is determined by the textural properties and the composition of the food. Standardization of food quality to comply with the specification of customers is a challenge as variability in natural products (raw material) is high while at the same time the product portfolio is often diverse.

Currently, sensory analysis (trained taste panels) and instrumental analyses (e.g., texture measurements, volatile analysis) are in place. These approaches are sampling based, as the analysis is destructive for the product under evaluation. The implementation of new type of food quality checks in food process lines necessitates contactless measurements in a continuous and fast way. Wave-based technologies (e.g., spectroscopy, vision systems) offer possibilities, but again attention should be paid to robustness and cost. The appearance of product quality-directed screening systems opens up a novel opportunity for food companies: to get their processes steered based on product status instead of the environmental conditions around the product (e.g., temperature, pressure, humidity). In other words, this would allow the coupling of product quality optimization with process efficiency. In this way, the food industry could follow the trend of "process analytical technology" in the pharmaceutical and chemical industries.

Although a wide variety of technologies and systems are commercially available, the industry has questions with respect to a correct choice of sensor technology to be implemented for the measurement of specific food quality parameters. Food products can be considered as multiscale entities with a complex chemical composition in combination with a complicated physical structure. For that, often specific sensor technologies and algorithms are needed since differences in chemical composition and physical structure have an influence on the sensor output (as is the case of, for example, vis/NIR spectroscopy). In this way, there are no generic solutions for food products and often specific methodologies or algorithms need to be designed per product type.

As a result of a finished project (Sensors For Food) in Flanders (Belgium), a general overview is obtained on the different needs from this industry. During the 4-year project, that was partially funded by industry, companies had the possibility to ask for questions related to the use of sensors for quality and safety assessment of foodstuff. A central contact point from the project consortium then formulated an advise on the choice and implementation of appropriate sensor technologies for the given case. As one of the outcomes of this service, we found that the most frequently asked questions were related to the detection of foreign objects in food, the quantification of the moisture content in food, methods for confirmation of food authenticity and integrity, and finally technologies related to the measurement of microstructure and particle sizes.

Nanoelectronics enable the development of analytical systems that are in line with the above-mentioned needs and opportunities of food companies regarding food quality and safety monitoring. At the heart of these systems are sensors that are produced using nanotechnology, offering the possibility to miniaturize the systems (so that they "can leave the laboratory") and also to lower the price of these systems. The latter is especially the case for sensor chips that have a market in the telecom or other sectors. For these markets, sensor chips can be fabricated via CMOS-based technologies, which scale with production numbers. Above all, sensor systems may offer new quality measurements (e.g., by combining visual appearance with composition screening, contactless determination of frozenness state) or combine sensing with on-chip computer power to determine the remaining shelf life of packed food products. New evolutions toward organic (plastic) and printed electronics will speedup price reduction and increase usability in food industry.

16.2 Cooperation Model to Stimulate "The Introduction of New Nanoelectronics-Based Technologies in Food Industry": An Engine for Innovation and Bridging the Gap

There is a need to join forces between food institutes, suppliers, and knowledge centers to evaluate, optimize, and validate innovative sensors for applications in the food industry. A common *focus is on sensors that allow for a better, faster, and more accurate evaluation of food quality, safety, and processing in a food production environment.*

To achieve this goal toward the use of innovative sensors during daily operations, we recommend a four-step model (Figure 16.1):

1) Awareness
2) Platform creation
3) Validation
4) Implementation

The aim of the four-step model is to improve and increase the awareness, and explore the application of existing, new and upcoming sensor systems for the food industry followed by a concrete implementation.

16.2.1
Awareness

Creating awareness is a continuous process during the whole technology integration process.

The key aspects in this process are as follows:

- Identification of intermediate players and sector organizations.
- Inform and convince intermediate players and sector organizations to join forces.

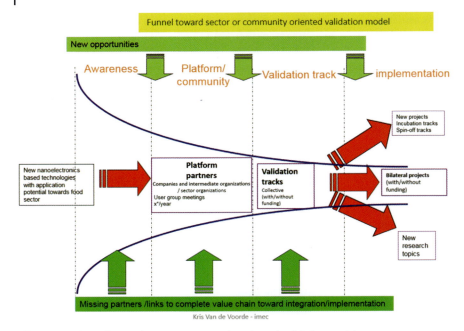

Figure 16.1 Funnel toward sector or community oriented validation model.

- Identification of top players (food and technology supplier companies) with support of intermediate players and sector organizations.
- Common visits from a *technology and sector representative defining relevant cooperation topics*.

16.2.2
Platform Creation

The goal of platform creation is to bring companies together in a forum with focus on sensor systems!

We can realize this by creating a nanoelectronics-based "Sensors For Food" platform run by a cross-sectorial team and on top a personal advisory service (http://www.sensorsforfood.com/en).

The platform should consist of different kinds of partners: knowledge centers (including intermediate organizations focusing on innovation and sector-related organizations), food companies, system integrators, and technology suppliers.

The target groups of the platform are the food producing companies as well as the technology providers for whom the food industry is an important partner.

The (Sensors For Food) platform brings food manufacturers and technology providers together in a forum for food industry sensor systems. The aim is to improve, increase the awareness, and explore the application of existing, new, and upcoming sensor systems for the food industry. Through a centralized contact point, services are provided concerning sensor systems that are already available for the food industry.

Activities should include the following:

- A screening of needs and opportunities for the food industry.
- A technology watch on emerging sensor innovations.
- Generation and support of innovative ideas, networking, and partner matching between food companies and technology suppliers.
- Through a centralized contact point, services are provided concerning sensor systems that are already available for the food industry.

For example, assistance can be offered for issues regarding sensor calibration and selection.

- Furthermore, a number of thematic seminars, workshops, and training courses will be organized to promote knowledge transfer.

For end users, the advantages of participating in such a platform are multiple:

- They will be informed on the needs and opportunities of the food industry.
- They will be kept up-to-date regarding the newest technological evolutions.
- They will be part of a unique network that will enhance your innovation process and facilitate partner matching.
- They will benefit from our service concerning sensor systems.
- They may attend the seminars, workshops, and trainings at a reduced price to stimulate participation.
- They get personal advice.

Whereas the platform favors a collective company approach, the advisory service starts from the need of individual companies. This can be done via individual advice on company-specific questions and challenges in the field of innovative nanotechnologies, partner matching, assistance and support in rolling out an implementation strategy, and initiating company-specific innovation projects. In addition, specific training sessions on the use of innovative technologies can be organized.

16.2.3
Validation of New Technologies

The target groups of validation tracks are the food producing companies as well as the technology providers for whom the food industry is an important partner. Awareness to validation is a collective process, reaching common goals between all partners: demonstrate/test/feasibility checks of new technologies toward all kinds of possible applications. The process of validation gives insight in the technical specifications needed for industrial applications.

During validation, knowledge centers can bring technology to a level when technology suppliers can take over for implementation toward end users, which is very important in convincing these end users (food-producing companies) to move toward the implementation phase.

16.2.4
Implementation of New Technologies

The role of system integrators is key during the implementation phase. They make the link from prototype to industrial process. Knowledge partners will give advice and further the support during this phase. Technologies can be licensed by knowledge partners to make integration possible by system integrators toward end users such as food producing companies.

Realizations in the industry create visibility for all stakeholders and will help to attract new partners for newly defined validation tracks, which is a supporting evolution to reinforce the funnel.

16.3
Existing Technologies That Can Be Used in a Wide Range of Applications: The Present

16.3.1
Characteristics

16.3.1.1 Compact, State-of-the-Art Technology
We all want the best, most precise information at our fingertips. Process automation in the food industry is a growing area with a substantial business potential.

16.3.1.2 User-Friendly Technology
Plug-and-play sensors are the key to a successful application. Thanks to easy-to-understand procedures and the Ethernet IP connection, installing and operating the new sensors is easier. The user can set up the commissioning remotely from the control room and in clear text. The data stored in the control system enable quick sensor changing for calibration or repair.

16.3.1.3 Standardization
Designing a sensor that is able to withstand the conditions of a food-processing factory, that operates intelligently, is outstandingly robust, high quality, and available for a reasonable price is a challenge.

The customer must choose the best options for the system application at hand, for example, with stainless steel or aluminum housing, with or without a display, or equipped with the latest communications technology.

16.3.1.4 Integration

A sensor must always be fully integrated in a system for maximum performance: in-line, online or at-line if possible. Only if these are not possible, laboratory off-line analysis will be performed.

16.3.2
Some Examples of Existing Technologies and Suppliers

As the introduction of at-line, in-line, or online sensor technologies becomes more and more routine in the European food industry, innovations in the field of nanotechnologies and nanofabrication have paved the way for the development of miniaturized sensor technologies, providing the extra asset of mobility/portability, compared to the aforementioned systems. In this way, analysis can take place at several locations of the production floor, starting with quick screenings of incoming products (either raw materials from the field or intermediate products from other production facilities), monitoring (and possibly adapting) the different production processes, and performing final quality control checks on the released products.

Different technologies have evolved in mobile/portable applications and a classification can be made based on the transducer mechanism, integrated in the sensor device to translate the interaction of the sensor with the product into a readable electronic signal, indicating the physical or the chemical status of this product. With respect to applications on (agro)food products, the following subclasses can be differentiated: (i) spectral systems, based on the interaction of light with matter; (ii) aroma analyzers, either in the form of miniaturized GC/MS systems or in the form of electronic noses; (iii) biosensor technologies; and (iv) lab-on-a-chip systems. Whereas for the latter two classes, applications related to food analysis are merely situated in a research context, several sensor technologies based on the first two classes have for some years been on the commercial market. Another specific class of nanotechnology sensors is the temperature–time loggers but this class will not be discussed further.

16.3.2.1 Spectral Systems

In the class of the spectral systems, several subclasses can be differentiated such as mobile point spectroscopy systems (in the vis/NIR region or the MIR region), portable hyperspectral camera technologies, miniaturized Raman spectroscopy systems, and portable X-ray fluorescence (XRF) systems [1]. Although most of the mobile systems were developed for applications in another field (mostly related to pharmaceutical or biomedical applications), they can also be used for the execution of *in situ* quality control checks on food during various stages of the production.

In scientific research for food applications, especially the potential of portable spectroscopy devices in the NIR wavelength region has been investigated. Such systems are used for the characterization of internal/external quality parameters of fruit (such as titratable acidity, soluble solid content, firmness), the storage condition and fat content in meat and fish, the quality evaluation of drinks and dairy products (in casu fat, protein, and moisture content), and the composition of grain products (protein and moisture content). This is not a limiting list and due to developments in the field of hardware (source, detector, optical elements, etc.) and software (chemometrics, multivariate data analysis, and mathematical algorithms), other applications are becoming possible. Most of these applications are used for the execution of quantitative chemical analyses while the systems can also be used for product identification (based on spectral fingerprinting) or for physical structure analyses.

To indicate the potential of the technology, the portable systems were benchmarked in different studies with state-of-the-art benchtop NIR devices on the identical sample set. From these studies, it became clear that the benchtop devices were superior to the portable systems; however, this difference was minimal compared to the other advantages of the portable systems such as better flexibility, *in situ* use, and speed. In some conditions, these advantages overrule the small difference in performance. Companies developing those systems include ASD, Avantes, Brimrose, Thermo Scientific, Polychromix, Ocean Optics, and JDSU. Next to the mobile point spectroscopy systems, a first example of a mobile hyperspectral camera is also commercially available. This will be explained further in Section 16.4.

16.3.2.2 Portable Aroma Systems

Traditional aroma analysis is done with large benchtop devices, based on separation of the individual components through gas chromatography (GC) and identification of these components with mass spectrometry (MS) methodologies. In addition, specific sample preparation methods (such as solid-phase microextraction) need to be optimized toward a correct analysis of the aroma components. This approach has a drawback that only a limited number of samples can be analyzed in a quality laboratory. But thanks to a further miniaturization of the components, the benchtop devices have evolved into mobile systems, to be used on the production floor for the online measurement of the aroma headspace of the food products, for example, for a quick detection of off-flavor components. Devices that are commercially available include the Torion T-9 GC/MS and the Syft system. With the latter, a specific profile will be obtained without a quantification of the concentration of the individual components in the aroma space. Hence, specific (multivariate) statistical techniques are necessary to analyze the profiles and to extract essential information from these profiles. Another class of sensors forms the group of e-nose technologies, consisting of an array of nonspecific individual sensors that result in specific digital patterns. Although promising results are obtained with this technology in scientific literature, some important challenges remain when they are implemented in an industrial production context, such as limited stability and repeatability (drift), the strong effect of interferential parameters (humidity), and the long analysis time.

16.3.2.3 Biosensor Technologies

Biosensor technologies are often considered as emerging examples of the potential of nanotechnology in the context of food applications [2–6]. Although most of the elaborated systems (for food applications) have not entered the commercial market as yet, several reviews indicate its flexibility for the detection of a broad range of components such as (pathogenic) microorganisms and chemical components (such as pesticide residues, mycotoxins, and other chemicals). One of the clear advantages of this type of technology is that a single transducer platform can serve as a general base for the quantification of the components by the combination of this platform with (disposable) component-specific sensing elements. Recent reviews, discussing the applications on food products with biosensor applications, can be found in Refs [7,8]. A specific biosensor, based on optical fiber technology, is discussed in more detail in Section 16.4.1.2.

16.3.2.4 Lab-on-a-Chip Systems

The lab-on-a-chip concept involves the integration of all the manual operations, executed in a laboratory, such as mixing, diluting, and analysis, in one single chip of one or a few square centimeter. This approach offers the advantage of a significant reduction in analysis volume and time. Based on the actuation methods of the fluid propagation within the chip, several subclasses of lab-on-a-chip systems can be differentiated. In 2011, a review [9] discussed the possible applications of lab-on-a-chip technology related to food analysis applications. This review indicates the wide potential of this technology, especially in the field of electrophoresis applications. An important point of attention is the integration of the appropriate detector systems on the chip to avoid that a lab-on-a-chip methodology finally results in a chip-in-a-lab.

16.4
New Technology Developments: The Future

We can split "new technology developments" in two tracks: short-term available technology (2016–2018) and long-term available technology (2018 and further).

16.4.1
Short-Term New Technologies: Recently Validated

Principles and examples (see also http://www.sensorsforfood.com/en).

16.4.1.1 New Reflection-Based Camera Technologies: Hyperspectral Imaging

A new technology with a lot of potential is called "hyperspectral imaging."

Miniaturized hyperspectral camera systems were identified as promising innovative sensor systems for the detection of the following:

- Foreign objects.
- Latent defects.

- For thorough quality inspection of product surfaces.
- For visualization of product composition (water, sugar, fat, protein).
- This technique also offers possibilities for hygiene screening as some invisible aspects can be visualized.

Hyperspectral cameras are the vision systems of the future as they may extract much more detailed information from a scene and even from one pixel, in real-time and at high speeds. In fact, *they combine traditional image processing with spectroscopy. In contrast to classical color camera systems* that use only three broad wavelength ranges in the visible light (RGB – red, green, blue), hyperspectral cameras generate a detailed spectral signature for every pixel in the image. Such a sensor thus splits light into many narrow wavelength bands, resulting in a particular spectral fingerprint of an object.

Taking into consideration the costs and benefits, it will be determined which system can be implemented in the company; a flexible but high-cost spectral camera or a low-cost vision system based on standard cameras and LED illumination on specific wavelengths. Using line scanners, this technology is suited to be implemented above transportation belts or sorting lines. A complete system architecture will be developed that can be elaborated by an integrator (third party) into a specific product.

Two types of miniaturized hyperspectral cameras are developed in the visual domain by IMEC (VNIR domain (400–1000 nm)) (Figure 16.2):

> Line scan cameras suited to be implemented above transportation belts or sorting lines and snapshot cameras that allow taking a whole image at a specific wavelength.
>
> http://www2.imec.be/be_en/research/image-sensors-and-vision-systems/hyperspectral-imaging.html

Miniaturization is a must to make the cameras useful for integration in food industry. This is realized by processing a spectral filter on the image sensor.

Note: In the future, miniaturized hyperspectral imagers in the NIR domain (1000–2500 nm) will open extra opportunities. However, this future research is costly and will proceed only when there is enough industrial interest. Tests with commercially available systems (not miniaturized) have already been performed.

Companies should evaluate hyperspectral imaging starting with feasibility tests on their own products, feasibility tests for thorough quality inspection, visualization of product composition, as well as detection of foreign objects in raw materials, intermediate, and final products and/or processes. These tests should be followed by further optimizations and a tailor-made design considering the balance between cost and performance. After validation of this concept, several system integrators are ready to offer extended advice for the implementation into an actual monitoring device.

There is also a possibility to do feasibility tests to determine the moisture content of your raw materials, intermediate products, and end products and to

(a)

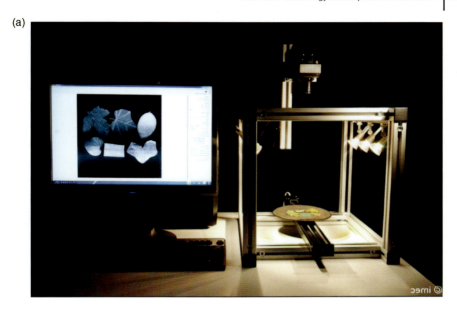

(b)

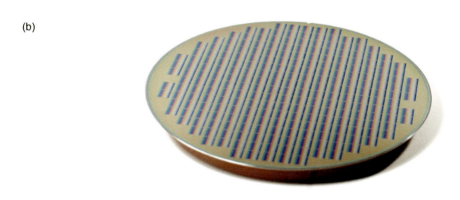

© imec

Figure 16.2 (a) Miniaturized hyperspectral camera IMEC setup. (b) Wafer with hyperspectral filter.[1] (Reproduced with permission.)

evaluate freezing and drying processes, followed by further optimization and advice on an integrated sensor solution based on your needs, again taking into account cost versus measuring precision. After validation, in-depth advice on an implementation is the next step and challenge.

1) https://www.flickr.com/photos/imec_int/25398461204/in/album-72157661604794973/ https://www.flickr.com/photos/imec_int/25760289574/in/album-72157661604794973/.

Figure 16.3 FOx Diagnostics platform. (Reproduced with permission.)

16.4.1.2 Optical Fiber Biosensor Technology

As indicated in Section 16.3.2, biosensor technologies are considered as emerging sensor technologies with a lot of potential applications for the food industry. In the Sensors For Food project, a prototype biosensor technology based on optical fiber technology was used for the detection of allergens and pathogenic microorganisms. The biosensor device consists of a fixed hardware platform (including light source, optical fibers, detector) and a modular tip specifically designed for the detection of the individual biomolecules of interest. Depending on the application, specific labeling steps (with nanoparticles) or heating steps need to be included in the analysis protocol to achieve the analytic goals. Due to its flexibility, the system can be used for multiplex screening of raw materials, intermediate and end products, and production processes and cleaning procedures for various specific biomolecules and microorganisms at an affordable cost. Today prototypes are developed that can perform fully automated immuno-based assays and that will be benchmarked to industrial technologies. Platforms and consumables will be available for sales by early 2017 for R&D laboratories. First applications can be generated by the client and support is available from the FOx Diagnostics spin-off company (Figure 16.3) (www.foxdiagnostics.eu coming soon).

16.4.1.3 New Transmission-Based Technology: Millimeter Wave Sensors (GHz–THz Sensor)

The millimeter wave sensors developed *by the research group ETRO of the VUB (Vrije Universiteit Brussel)* are identified as promising innovative sensor systems for the optimization of *industrial drying and freezing processes* in the food industry.

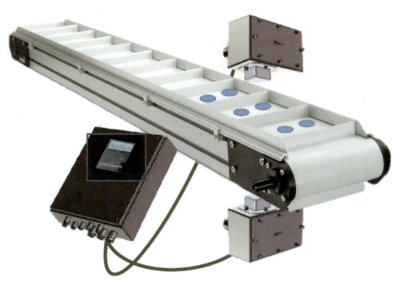

Figure 16.4 Setup sensor configuration Aquantis. (Reproduced with permission.)

These sensors allow for contactless measurements into the core of the food product. They combine the penetration depth and the high resolution of millimeter waves. These waves have a very specific interaction with water: they are absorbed by free water molecules and the degree of absorption depends on the residual moisture in the food product.

As such, these sensors can be applied to measure the moisture content of, for example, fresh vegetables and fruits or dried nuts. With specific sensor configurations the drying temperature and drying speed can be monitored and the final moisture content determined. In contrast, upon freezing the millimeter waves are no longer absorbed that makes these sensors also suitable for monitoring freezing processes. Furthermore, millimeter waves are not influenced by dust, dirt, fog, or temperature effects, which makes this system suitable for use on the production floor. Today, prototypes are developed by a spin-off company Aquantis (www.aquantis.org) that can reach the optimal sensitivity and accuracy and will be tested on practical industrial drying and freezing applications (Figure 16.4).

These compact sensor configurations can be installed above a production line and used for online monitoring of freezing and drying processes.

The validation process combines technical evaluations (sensitivity, detection limit, robustness, measuring rate, etc.) with a comparative study of other measuring techniques and known destructive sampling experiments that will be used as references. After this validation process, advice can be offered on the implementation of affordable integrated sensor solutions. Engineering companies that can build these more integrated versions are available.

OCI™-2000 snapshot
handheld
hyperspectral imager

Figure 16.5 Bayspec portable hyperspectral camera with IMEC sensor chip.[2] (Reproduced with permission.)

16.4.2
Long-Term New/Future Technologies: To Be Validated

Principles and examples (See also www.ifastproject.com).

The acronym for the i-FAST project with different validation tracks stands for "In-factory food analytical systems and technologies".

16.4.2.1 Portable Hyperspectral Camera Technology

Point spectroscopy is already well known in the food industry for applications such as determining the composition of bulk product, however, the full potential of the technology is yet to be discovered. Especially, the available information on the product microstructure is nowadays not used. Therefore, technology that aims at separating the information on microstructure and composition present in the spectra is the key and can be used for maximizing product information.

The first portable (miniature) point spectrophotometers are already on the market, but real applications in the food industry are almost nonexisting.

Portable hyperspectral cameras are just emerging and the potential for the food industry still needs to be explored. Added value can be sought in flexible quality control on the factory floor and process optimization such as

- fast screening of incoming products to bypass time delays caused by sampling and laboratory analysis (authenticity control, freshness, composition, etc.) and
- monitoring the evolution in composition and microstructure of products during the production process (emulsions, suspensions, etc.).

In Belgium, IMEC has a validation track on portable hyperspectral camera technology as part of the i-FAST project. Companies like Bayspec are entering the market with first miniaturized hyperspectral tools (Figure 16.5). A comparison between different technologies (Figure 16.6) is always relevant to make the right choice.

2) http://www.bayspec.com/category/spectroscopy/hyperspectral-imaging/.

GHz–THz sensor

- High penetration depth
- → Volume information
- Low spatial resolution
- Dedicated conversion algorithms
  (Based on S-parameters)

Hyperspectral camera

- Low penetration depth
- → Surface information
- High spatial resolution (Imaging)
- Spectrum per pixel

Figure 16.6 Comparison between a GHz–THz sensor and a hyperspectral sensor. (Reproduced with permission.)

16.4.2.2 NMR

The physical structure of concentrated emulsions (minarines, dressings, recombined cream, etc.) is an important parameter that influences stability, sensory properties, and functionality. Not only the droplet size distribution but also the homogeneity is very important for this type of emulsions. Technically all tools are available to determine both parameters with low-resolution NMR (Figure 16.7).

Benchtop NMR Analyser-MQC

A range of cost-effective systems based on nuclear magnetic resonance (NMR) for fast, easy, solvent-free measurement of oil, fat moisture, fluorine, and emulsions.

Figure 16.7 Benchtop NMR analyzer.[3] (Reproduced with permission.)

3) http://www.oxford-instruments.com/products/analysers/stationary-benchtop-analyser/maran-ultra.

Low-resolution NMR has the advantage that is does not require any sample preparation and the measurement is nondestructive, meaning that one sample can be followed over time to assess the stability. However, to implement this technology, a thorough understanding of the physical parameters is necessary to adequately set the instrumental parameters (e.g., for W/O or O/W characterization). NMR technology aims at removing these obstacles in the future for implementation with the goal to make low-resolution NMR a user-friendly and easy-to-operate technique. Added value can be found in a faster monitoring system of products and processes that will allow noticing possible deviations in an early stage. In Belgium, UGent (http://www.ugent.be/bw/tafc/en/research-groups/paint) has a validation track on the use of low-resolution NMR for controlling food related emulsions as part of the i-FAST project.

16.4.2.3 3D X-Ray

Both structure and microstructure have a significant influence on general food quality parameters such as texture and also on the presence of internal defects, stability, and the shelf life potential of food products. Consumer demands for an improved functional and nutritional quality (composition), sensory properties (texture, internal defects), and safety (foreign objects) are strong drivers for food companies to optimize their products and processes or to apply innovations in food structure composition. X-ray tomography allows the visualization and the quantification of the three-dimensional structure of food products in a nondestructive way. The technology is based on the detection of local density differences, inducing a difference in the absorption of X-rays and allowing the generation of submicron resolution 3D images. X-ray tomography has a lot of potential to be applied on products with high-density differences. As a consequence, the technology is perfectly suited to visualize porous products or layered products, composed of different materials. The technology was already applied successfully for structure analysis of cheese, vegetables and fruits, bakery, and ice cream products. Intrinsically, the technology has the possibility for the high-resolution visualization of the internal microstructure of food products, and in addition the innovative application to scan online products with internal defects or foreign objects inside. In Belgium, KULeuven (MeBioS; http://www.biw.kuleuven.be/biosyst/mebios) has a validation track on the use of 3D X-ray as part of the i-FAST project.

The analysis of the X-ray images provides quantitative information on the three-dimensional internal composition and structure of the food products to be used for both product and process optimization or for online nondestructive quality control. In the first case, a test facility for microstructure will be developed, based on X-ray tomography, offering the possibility to study the relation between (micro)structural properties (dimensions, form, internal distribution of the different components, and layers) and the sensory or functional microscale properties of products as a function of different ingredients or process methods. In the latter case, the potential of X-ray tomography is explored toward the execution of online nondestructive quality control by visualizing the internal composition and dimensions and by detecting defects and foreign objects.

16.4.3
The IoT (R)evolution and Big Data

Producing food products is a challenging business, where highly variable raw material is processed into foodstuff for which the consumer requires a high quality and no (or a minimal) variation. Quantifying the quality and the variability of the incoming material requires adequate sampling and quality assessment procedures to be in place, and also for the final products a quality check is indispensable. During many years, food companies measured a limited number quality attributes with a low sampling rate, and used (extrapolated) this information to judge products and processes. As extensively described in the above paragraphs, new developments in sensor technologies give companies tools to measure more quality attributes of their products with a higher accuracy and sampling rate than ever before.

This evolution on how we measure is highly encouraged, but brings an important challenge: how to turn the massive amount of data into valuable information that improves business? This question is essential, and if not properly addressed, it will result in companies that invested heavily in high-tech sensor technologies having no return on investment. Key to translating the vast amount of data into valuable information is *data aggregation and analysis*, two aspects that will be briefly discussed below.

Data need to be *aggregated* to have a global overview of the whole production process, which is usually a collection of unit operations. Too often, process and quality characteristics of the different processing steps are stored in separate data sets that are used for optimization on the level of the unit operation. Unfortunately, those local optima almost never result in finding the global optimum, which is the ultimate goal. In order to be aggregated (connected), recent developments in the concept of Internet of Things (IoT) are very promising. In IoT, all sorts of "smart devices" can be interconnected allowing to gain valuable information. Some key technologies that found their way into the food industry include radio frequency identification (RFID), infrared sensors, wireless sensor networks, and machine-to-machine communications. An important attention point in this aggregation step is matching data from the different sources, since often they are sampled at different times at different places. In a recent paper, Ref. [10] conclude that the penetration of IoT in the food supply chain is still in its infancy and deserves further attention, seen its enormous benefit in other sectors where IoT has been deployed more broadly.

IoT and all data aggregation solutions are catalyzing the *analysis* of those data sets to yield valuable information. This "mining of large data sets"–with terms such as big data and predictive analytics being often touted–has been paying off for many sectors, and also for the food industry, where they are not yet deeply penetrated and hold a huge potential. Big data are large volumes of data that offer the opportunity for better insight and improved decision making that could not be accomplished analyzing smaller data sets separately. In order to exploit the hidden patterns, trends, and correlations, the management and analysis of such large data require careful consideration of several important aspects that are often described by the four V's: volume, velocity, variety, and veracity. Variety stands for the aggregation of data from different sources, where often different sampling intervals are used, so that alignment of the data are required. Veracity embraces aspects such as the validity of the data – are thedata that we measured or stored in the database valid and secure? In order to cope with such data, new developments are required in database architectures that are capable of handling data that are unstructured, as well as in data analysis tools that can handle a vast amount of data [11].

For the food industry, the creation of an IoT and big data environment is a huge challenge, and we think that a down to the earth, stepwise approach will be most rewarding in a first phase. Indeed, many companies still produce handwritten data, and the interpretation of them is performed in the minds of experienced personnel, rather than through objective functions and models. Removing those two bottlenecks is probably the most important step to take, and involves the investment in IT solutions (e.g., data acquisition platforms and data analysis software) as well as in data analysts that are capable of handling food-related issues. This last aspect requires dedicated education and training programs to be set up that cover food science as well as data science. We feel that in the future, educational programs, existing or to be set up, should better reflect this industrial demand. From the data analysis side, a tremendous amount of work has been spent by researchers as well as by software companies to bring data analysis and representation closer to practice. For example, where the design of experiments was considered to be a purely academic topic by the food industry during many years because they did not have the flexibility to cope with the specific settings encountered (limited resources, parameter restrictions, hard-to-change factors, among others), it is now seen that modern software (JMP, Minitab and SPSS, to name a few) offer excellent and easy to use options for tailor-made designs. Also from the analysis side, user-friendliness of commercial software packages has increased dramatically – both from the user interface side as well as from the output and visualization side, so that data scientists have valuable tools at hand to convince management.

When the food industry succeeds in applying those concepts in daily practice, they make a leap forward and get closer to the long-term goal of completely integrated, understood, and controlled processes, an aspect that forms the basis of the Process Analytical Technologies (PAT) initiative that was set up by the Food and Drug Administration (FDA) back in the early 2000s. Finally, as mentioned in the column of Ref. [12], partnerships between companies and well-positioned public (research) centers are considered very useful to guide companies through this data-driven process for which they have limited experience. In such public–private partnerships, research centers can educate companies in the landscape of potential methods, and guide them towards the right methods for their own business. Inversely, practical challenges that are encountered during such partnerships should fuel future (basic) research at the institutes to find adequate solutions for the future.

References

1 Li, Y.-S. and Church, J.S. (2014) Raman spectroscopy in the analysis of food and pharmaceutical nanomaterials. *J. Food Drug Anal.*, **22**, 29–48.
2 Duran, N. and Marcato, P.D. (2013) Nanobiotechnology perspectives. Role of nanotechnology in the food industry: a review. *Int. J. Food Sci. Technol.*, **48**, 1127–1134.
3 Ravichandran, R. (2010) Nanotechnology applications in food and food processing: Innovative green approaches, opportunities and uncertainties for global market. *Int. J. Green Nanotechnol. Phys. Chem.*, **1**, P72–P96.
4 Neethirajan, S. and Jayas, D.S. (2011) Nanotechnology for the food and bioprocessing industries. *Food Bioprocess. Technol.*, **4**, 39–47.
5 Cushen, M., Kerry, J., Morris, M., Cruz-Romero, M., and Cummins, E. (2012) Nanotechnologies in the food industry – recent developments, risks and regulation. *Trends Food Sci. Technol.*, **24**, 30–46.
6 Ranjan, S., Dasgupta, N., Chakraborty, A.R., Samuel, S.M., Ramalingam, C., Shanker, R., and Kumar, A. (2014) Nanoscience and nanotechnologies in food industries: opportunities and research trends. *J. Nanopart. Res.*, **16**, 2464.
7 Narsaiah, K., Jha, S.N., Bhardwaj, R., Sharma, R., and Kumar, R. (2012) Optical biosensors for food quality and safety assurance-a review. *J. Food Sci. Technol.*, **49** (4), 383–406.
8 Thakur, M.S. and Ragavan, K.V. (2013) Biosensors in food processing. *J. Food Sci. Technol.*, **50** (4), 625–641.
9 Atalay, Y., Vermeir, S., Witters, D., Vergauwe, N., Verbruggen, B., Verboven, P., Nicolai, B., and Lammertyn, J. (2011) Microfluidic analytical systems for food analysis. *Trends Food Sci. Technol.*, **22** (7), 386–404.
10 Pang, Z., Chen, Q., Han, W., and Zheng, L. (2015) Value-centric design of the Internet-of-things solution for food supply chain: value creation, sensor portfolio and information fusion. *Inf. Syst. Front.*, **17**, 289–319.
11 Strawn, L.K., Brown, E.W., David, J.R.D., den Bakker, H.C., Vangay, P., Yiannas, F., and Wiedmann, M. (2015) Big data in food safety and quality. *Food Technol.*, **69** (2), 42–49.
12 Wiedmann, M. (2015) Can big data revolutionize food safety? Food Quality & Safety. Available at: http://www.foodqualityandsafety.com/article/can-big-datarevolutionize-food-safety/.

Part Four
Nanotechnology: Toxicology Aspects and Regulatory Issues

17
Quality and Safety of Nanofood

Oluwatosin Ademola Ijabadeniyi

Durban University of Technology, Faculty of Applied sciences, Department of Biotechnology and Food Technology, PO Box 1334, Durban 4001, South Africa

17.1 Introduction

17.1.1 Nanotechnology and Nanofood: Background and Definition

Nanotechnology was first used in 1974 to describe production technology at ultrafine dimensions by Norio Taniguchi [1]. The prefix "nano" is a Greek word meaning dwarf [2]. Today, nanotechnology has developed into a multidisciplinary research sector with a lot of potential for industrial applications. However, the commercial application of nanotechnology is at the moment more highly developed in areas such as material science, microelectronics, aerospace, and pharmaceutical industries unlike in the food industry [3]. Dingman [4] also reported that nanotechnology is used in self cleaning glass and in army uniforms that monitor the health of the wearer to camouflage those changes to match its surroundings.

According to FSAI [2], the major areas where nanotechnology has potential for use in the food sector are encapsulation and emulsion formation, in food contact materials and sensor development but it is expected that this novel technology will be employed in much more areas as the years go by.

Reference [5] also defined nanotechnology as "the understanding and control of matter at the dimensions of roughly 1–100 nm, where unique phenomena enable novel applications. Encompassing nanoscale science, engineering and technology; nanotechnology involves imaging, measuring, modeling, and manipulating matter at this length scale." The International Standards Organization has also recently defined nanotechnology as "understanding and control of matter and processes at the nanoscale, typically, but not exclusively, below 100 nm in one or more dimensions where the onset of size-dependent phenomena usually enables novel applications, where 1 nm is one thousand millionth of a meter." [6].

Nanotechnology in Agriculture and Food Science, First Edition. Edited by Monique A.V. Axelos and Marcel Van de Voorde.
© 2017 Wiley-VCH Verlag GmbH & Co. KGaA. Published 2017 by Wiley-VCH Verlag GmbH & Co. KGaA.

In layman terms, it can be said that nanotechnology is the use of engineered nanomaterial that has been intentionally synthesized or incidentally produced to exploit functional properties exhibited on the nanoscale [2].

So what is nanofood? According to Garber [7], a food is termed "nanofood" when nanoparticles, nanotechnology techniques, or tools are used during cultivation, production, processing, or packaging of the food. Also, nanofood can be derived after foods, seeds, chemical pesticides, and food packaging have been broken down and manipulated at the microscale level through nanotechnology [8].

17.2
Current and Future Application of Nanotechnology in the Food Industry

Nanotechnology has diverse application in the sector at the moment and it may likely change the whole agrifood sector in the nearest future. Furthermore, food nanotechnology has the potential to open up whole new possibilities for the food industry [9]. FSAI [2] grouped the global applications of nanotechnology into six categories which included the following:

1) Sensory improvements (flavor/color enhancement and texture modification)
2) Increased absorption and targeted delivery of nutrients and bioactive compounds
3) Stabilization of active ingredients such as nutraceuticals in food matrices
4) Packaging and product innovation to extend shelf-life
5) Sensors to improve food safety
6) Antimicrobials to kill pathogenic bacteria in food

From the categories listed above, it can be inferred that nanotechnology will improve food processing, packaging, and safety, it will enhance flavor and nutrition, it will lead to production of more functional foods from everyday foods with added medicines and supplements, and it will result in increased food production and cost effectiveness.

In packaging, for example, bionanocomposites which are hybrid nanostructured materials with improved mechanical, thermal, and gas properties may be used to package the food, increase its shelf and also provide a more environment-friendly solution because of reduction of reliance on plastics as packaging materials [10,11]. An example of such bionanocomposites is zein, a prolamin, and the major component of corn protein. When dissolved in ethanol or acetone, a biodegradable zein films with good tensile and water-barrier properties can be derived [11]. Also, Emamifar *et al.* [12] showed that Packaging materials made from nanocomposite film containing nanosilver and nano zinc oxide were significantly able to reduce microorganisms that could cause spoilage in orange juice.

Novel food packaging technology may in fact be the most promising benefit of nanotechnology in the food industry in the near future and food companies are said to have started producing packaging materials based on nanotechnology that are delaying spoilage and improving microbial food safety [7].

Another area of interest to me and equally of great importance for the food industry is food safety and preservation. Food pathogens will be detected with the aid of nanosensors that had been placed directly into the packaging material serving as "electronic tongue" or noses by detecting chemicals release during food contamination and spoilage [13]. According to Lilie and Cantini [14], nanosenser will help to recover even just one *E. coli* bacterium located in ground beef. Another advantage of nanosensor is that it can measure safety at real time and the procedures are quick, sensitive, and less labor-intensive [15].

Figure 17.1 displays examples of other nanofood applications.

Food companies especially in the United States are already getting ready for the future application of nanotechnology. First among them is Kraft Foods that established Nanotech Consortium (consisting of 15 universities and national research labs) in 2000 [16]. Their goal was to produce food products that are customized to fit the individual tastes and needs of consumers. Future products, that is, drinks will change colors and flavors to foods that can recognize and

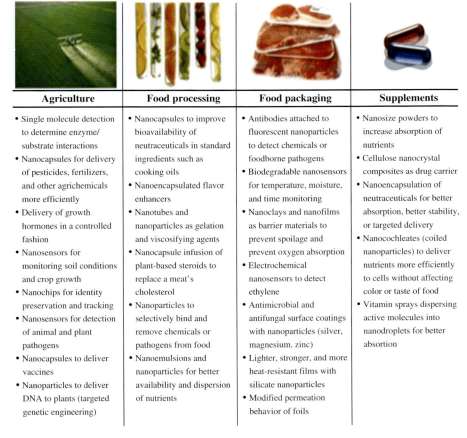

Agriculture	Food processing	Food packaging	Supplements
• Single molecule detection to determine enzyme/substrate interactions • Nanocapsules for delivery of pesticides, fertilizers, and other agrichemicals more efficiently • Delivery of growth hormones in a controlled fashion • Nanosensors for monitoring soil conditions and crop growth • Nanochips for identity preservation and tracking • Nanosensors for detection of animal and plant pathogens • Nanocapsules to deliver vaccines • Nanoparticles to deliver DNA to plants (targeted genetic engineering)	• Nanocapsules to improve bioavailability of neutraceuticals in standard ingredients such as cooking oils • Nanoencapsulated flavor enhancers • Nanotubes and nanoparticles as gelation and viscosifying agents • Nanocapsule infusion of plant-based steroids to replace a meat's cholesterol • Nanoparticles to selectively bind and remove chemicals or pathogens from food • Nanoemulsions and nanoparticles for better availability and dispersion of nutrients	• Antibodies attached to fluorescent nanoparticles to detect chemicals or foodborne pathogens • Biodegradable nanosensors for temperature, moisture, and time monitoring • Nanoclays and nanofilms as barrier materials to prevent spoilage and prevent oxygen absorption • Electrochemical nanosensors to detect ethylene • Antimicrobial and antifungal surface coatings with nanoparticles (silver, magnesium, zinc) • Lighter, stronger, and more heat-resistant films with silicate nanoparticles • Modified permeation behavior of foils	• Nanosize powders to increase absorption of nutrients • Cellulose nanocrystal composites as drug carrier • Nanoencapsulation of neutraceuticals for better absorption, better stability, or targeted delivery • Nanocochleates (coiled nanoparticles) to deliver nutrients more efficiently to cells without affecting color or taste of food • Vitamin sprays dispersing active molecules into nanodroplets for better absortion

Figure 17.1 Examples of nanofood applications [7].

adjust to a consumer's allergies or nutritional needs. It seems unbelievable and unachievable but that is the target of this food giant. Trust them, they may achieve it.

Nestle and Unilever are also reported to be researching on improved emulsifiers that will make food texture more homogeneous.

In other continents such as Australia, nanocapsules are used to add omega-3-fatty acids to one of the country's most popular brands of white bread. In Asia (joint effort between Mexico and India), researchers are developing nontoxic nanoscale herbicides to attack weed's seed coating and preventing from germinating.

Weiss *et al.* [3] have also reported that the use nanotechnology in agriculture and food systems will lead to great advancements in the food industry. Specific examples of advancements that will likely take place according to the researchers are as follows:

- Increased security of manufacturing, processing, and shipping of food products through sensors for pathogen and contaminant detection.
- Devices to maintain historical environmental records of a particular product and tracking of individual shipments.
- Systems that provide integration of sensing, localization, reporting, and remote control of food products (smart/intelligent systems) and that can increase efficacy and security of food processing and transportation.
- Encapsulation and delivery systems that carry, protect, and deliver functional food ingredients to their specific site of action.

The outcome of the above will lead to a range of benefits to the food industry, that is, there will be range of products with new tastes, textures, and sensation and also less use of fat and sugar. Furthermore, there will be enhanced absorption of nutrients, improved packaging, traceability, and security of food products [17]

Friends of the Earth; a popular organization with offices in Australia, the United States, Europe, and Germany have also reported that foods that contain manufactured nanomaterial ingredients are being sold to consumers in supermarkets [18]. A total of 150–600 nanofoods and 400–500 nanofood applications may in fact be on the market according to FOE [18]. However, in spite of the many benefits and potential of nanotechnology, many consumers are pessimistic about nanotechnology applications in the food system [19]. This also signifies that there is a lot of work to be done to change public perceptions of food nanotechnology. Furthermore, there could be a change of perception if the industry proactively communicate transparently on the use of nanotechnology in food [20].

Examples of foods, food packaging, and agriculture products that now contain nanomaterials are given on Table 17.1.

Having discussed the numerous benefits of nanotechnology in the food sector, it is important to consider the possible dangers and safety concerns that may be associated with consumption of nanofood, that is, foods that have been produced directly or indirectly with nanomaterials.

Table 17.1 Examples of foods, food packaging, and agriculture products that now contain nanomaterials [18].

Type of product	Product name and manufacturer	Nano content	Purpose
Beverage	Oat chocolate and oat vanilla nutritional drink mixes; TODDLER Health	300 nm particles of iron (SunActive Fe)	Nanosized iron particles have increased reactivity and bioavailability
Food additive	Aquasol preservative; AquaNova	Nanoscale micelle (capsule) of lipophilic or water insoluble substances	Nanoencapsulation increases absorption of nutritional additives, increases effectiveness of preservatives and food processing aids Used in wide range of foods and beverages
Food additive	Bioral™ Omega-3 nanocochleates; Bio-Delivery Sciences International	Nanocochleates as small as 50 nm	Effective means for the addition of highly bioavailable omega-3 fatty acids to cakes, muffins, pasta, soups, cookies, cereals, chips, and confectionery
Food additive	Synthetic lycopene; BASF	LycoVit 10% (<200 nm synthetic lycopene)	Bright red color and potent antioxidant. Sold for use in health supplements, soft drinks, juices, margarine, breakfast cereals, instant soups, salad dressings, yoghurt, crackers, and so on
Food contact material	Nanosilver cutting board; A-Do Global	Nanoparticles of silver	"99.9% antibacterial"
Food contact material	Antibacterial kitchenware; Nano Care Technology/NCT	Nanoparticles of silver	Ladles, egg flips, serving spoons and so on have increased antibacterial properties
Food packaging	Food packaging Durethan® KU 2-2601 plastic wrapping; Bayer	Nanoparticles of silica in a polymer-based nanocomposite	Nanoparticles of silica in the plastic prevent the penetration of oxygen and gas of the wrapping, extending the product's shelf life. To wrap meat, cheese, long-life juice, and so on
Food packaging	Nano-ZnO plastic wrap; SongSing Nanotechnology	Nanoparticles of zinc oxide	Antibacterial, UV-protected food wrap
Plant growth treatment	PrimoMaxx, Syngenta	100 nm particle size emulsion	Very small particle size means mixes completely with water and does not settle out in a spray tank

17.3
Food Quality and Food Safety

Food safety, which is an integral aspect of food quality deals with those requirements affecting the characteristics or properties that have the potential to be harmful to health or cause illness or injury. Food quality on the other hand is the ability of the food to meet the established requirements affecting the characteristics of a food. A food that does not meet the food safety requirements automatically fails the food quality requirements.

Food quality and safety are paramount from farm to fork and in particular in food manufacturing. According to Alli [21], 12 factors have influenced food industry operations including food quality and safety in the last 40 years and the factors include: consumer expectations (e.g., nutrition, convenience, and additives), incidents relating to food safety, environmental concerns, changes in governmental regulatory processes, traceability in food production and processing, technological changes, foods derived from biotechnology, irradiated foods, organic foods, economic factors, issues relating to international trade, and food security concerns related to bioterrorism. Although food nanotechnology was not listed, it is however an emerging factor that will significantly impact the food industry. In fact, a recent report (Collab4Safety) sponsored by European Commission [22], regarded unintended effects of nanotechnology as an emerging risk. The report also showed that the effect of drivers of food safety risks, such as new technologies (e.g., biotechnology, nanotechnology) are experienced around the world, but not at different levels and with different health, environmental, and economic impacts. Though they may not be felt at the same level, globalization remains a challenge to safety and quality in the food supply chain [23].

17.4
How Safe is Nanofood?

Nanofoods will most likely have longer shelf life, better taste, and not cause food borne illnesses because of the advantage of early detection of pathogens as result of the embedded nanosensers. According to Ref. [24] nanotechnology will make more hygienic food/feed processing (better food and feed safety and quality, reducing food-borne illnesses in developing countries) possible; however, no one is sure of the problem that may arise if there is a reaction between the cells of the body and the nanomaterials. Experiment has also shown that nanoparticles can enter in to plants and gastrointestinal region and also take part in transmembrane [18,25].

Nanoparticles equipped with new chemical and physical properties that vary from normal macroparticles of the same composition may interact with the living systems thereby causing unexpected toxicity [15]. Also, nanomaterials present different hazards from those of the same material in a micro- or macro form [2]. The use of nanoparticles in foods or food contact materials appears at the moment to consist of uncertainties and safety concerns [26].

Furthermore, indepth potential risks of nanomaterials to human health and even to the environment are still unknown [17,27]. This is because there is little or no scientific information on the effects of nanotechnology applications on human, animal health, and also the environment [28]. On the problems of uncertainty about nanotechnology and food safety, Bosso [29], wrote that nanoparticles may cause disease and death and that regulators are doing little to respond coupled with this is the obsession with technology innovation and economic growth over safety by the big companies.

Concerns that may arise as a result of consumption of nanofoods have been grouped into three major areas: least concern, some concern, and major concern. According to Ref. [24], area of least concern is where processed food with nanomaterials that are not biopersistent are digested or solubilized in the gastrointestinal tract. However, where food products also contain nonbiopersistent nanomaterials but carry across the gastrointestinal tract are areas of some concern. Other areas of some concern are increased bioavailability of vitamins and minerals may not always be beneficial for consumer health. Also a greater uptake of food colors or preservatives could take the application outside of the conditions under which the ADI (acceptable daily intake value) was set for the additive. The areas of major concern however are where foods include insoluble, indigestible, and potentially biopersistent nanoadditives (e.g., metals or metal oxides), or functionalized nanomaterials. Such applications may pose a risk of consumer exposure to "hard" nanomaterials – the ADME profile (adsorption, distribution, metabolism, and elimination) and toxicological properties of which are not fully known at present [24]. Another concern according to Mukul *et al.* [30] is that most nanomaterials used in foods are organic moieties and may contain and carry other foreign substances into the blood through the nutrient delivery system.

Powell *et al.* [31] have also reported that it is possible under certain conditions, for very small nanoparticles to gain access to the gastrointestinal tissue via paracellular transcytosis across tight junctions of the epithelial cell layer. However, whether there are realistic situations of nanoparticle exposure that lead to significantly abnormal reactive oxygen species (ROS) and inflammasome activation responses *in vivo* in the gut remains have not been established.

It is important at this junction to state that all the negative assumptions about the risk and dangers of nanomaterials in food may not be true but it is ethically necessary to fill the present knowledge gaps through more research and thorough risk assessment of nanofoods and nanopackaging materials. FSAI [2] reported on the lack of knowledge regarding the effect on pharmoacokinetics and bioavailability of changes in the physicochemical properties of normally inert and nonbiodegradable materials such as inorganic particles, for example, titanium dioxide and biological polymers in moving to the nanoscale. It is of great concern because changes may occur with potential cascade effects on cellular homeostatis when they get into the body system [2].

17.5
The Need for Risk Assessment

Nanotechnology has become important development and innovation in the food industry. Promising results and applications will be achieved especially in the areas of food packaging and microbial food safety. Although there are concerns, it will be out of place to be totally against the use of nanomaterials in foods as being suggested by some pressure groups. What should be done is for a proper risk assessment framework to be provided. However, it is sad that limited toxicological/safety assessments have been carried out for a few nanoparticles; hence, studies relevant to oral exposure risk assessment are required for particles to be used in food [15]. Sozer and Kokini [11] also suggested that governments should set down regulations and appropriate labeling that will help to increase consumer acceptability. Labeling and legislation will be a good development because skeptical consumers will have a choice not to buy nanofood.

The 2006 Institute of Food Science and Technologists report, which recommended that nanomaterials should only be used in the food industry after they have been proven following vigorous testing, should be taken serious [27].

The food safety authority of Ireland has written on the importance of risk assessment of food containing nanomaterials [2]. According to the report, a proper risk assessment will consist of four components, that is, hazard identification, hazard characterization, exposure assessment, and risk characterization.

The risk assessment of silver nanoparticles used as antimicrobial agents and the effect on the environment have been carried out [32], however what is necessary is the risk assessment of nanotechnology in food and the effect on humans and animals.

17.6
Regulations for Food Nanotechnology

Apart from the European Union who in 2011 adopted a mandatory labeling regulation requiring food ingredients to be listed as "nano" if they fit with their definition of engineered nanomaterials few countries only have some regulations for nanotechnology however there is no regulation focusing on foods [33,34]. According to Sekhon [9], there is little or no EU/WE/global legislation for regulation of nanotechnology in food. This however must change. Regulatory development for nanofoods should be carried out through a three phase process described by Jones [35] and they include: (1) the use of research and development database to assess applications of nanotechnology to food and agriculture, (2) selecting particular products to assess and identifying the risks and benefits, and (3) extrapolating to analyze appropriate regulatory or nonregulatory governance systems for the applications of nanotechnology in foods. Furthermore, criteria such as particular size range and measure, physical and chemical properties, processing, and safety concerns are needed to be considered for the development of the standard, definition, control measure, and regulation of nanofood [36].

17.7
Conclusion

The benefits of nanotechnology and nanofoods can never be overemphasized. Nanotechnology has come to stay. It has been estimated that nanotechnology market will grow to USD 1 trillion by 2015 [37]. However, it is important that safety becomes the key word when trying to implement it in the food industry. It must be recognized that the presence of nanoparticles in food is not a new trend. According to FSAI [2], "People have always been exposed to very fine particles (nanoparticles) in their diet, without harmful effects, since many food and feed ingredients are comprised of proteins, carbohydrates, and fats with sizes extending from large biopolymers (macromolecules) down to the nanoscale. Even when food is consumed predominantly as macromolecules, the natural digestive processes of the body reduce these to the nanoscale in order to utilize the energy contained in the molecules for the maintenance of physiological processes." Furthermore, it is not actually nanoparticles in food that should be a source of concern but rather disputed nanoparticles. Materials such as silver, titanium, or zinc that may be hazardous to human and animals at macroscale should be subjected to thorough risk assessment if it must be used as nanoparticles in foods or packaging materials.

References

1 Taniguchi, N. (1974) On the basic concept of 'nanotechnology'. Proceedings of the International Conference on Production Engineering Tokyo, Part II. Japan Society of Precision Engineering.
2 FSAI (2008) The Relevance for Food Safety of Applications of Nanotechnology in the Food and Feed Industries, Dublin. p. 82.
3 Weiss, J., Takhistoy, P., and McClements, D.J. (2006) Functional materials in food nanotechnology. *J. Food Sci.*, **71**, R107–R116.
4 Dingman, J. (2008) Nanotechnology: it's impact on food safety. *J. Environ. Health*, **70**, 47–50.
5 NRC (National Research Council) (2006) *A Matter of Size: Triennial Review of the National Nanotechnology Initiative*, National Academies Press, Washington, DC.
6 ISO TC 229 (2008) Draft standard on nanotechnologies: terminology and definitions for nanoparticles. Available at http://Http://www.iso.org/iso/iso_catalogue/catalogue_tc/catalogue_detail.htm?csnumber=44278 (accessed June 17, 2011).
7 Garber, C. (2007) Nanotechnology food coming to a fridge near you. Available at http://www.nanowerk.com/spotlight/spotid=1360.php (accessed June 11, 2011).
8 Scrinis, G. (2010) Nanotechnology: transforming food and the environment. Availability at http://www.foodfirst.org/en/node/2862 (accessed June 15, 2011).
9 Sekhon, B.S. (2010) Food nanotechnology – an overview. *Nanotechnol. Sci. Appl.*, **3**, 1–15.

10 Perch, H. (2007) How is nanotechnology being used in food science? Available at www.understandingnano.com/food.html (accessed June 11, 2011).

11 Sozer, N. and Kokini, J.L. (2009) Nanotechnology and its applications in the food sector. *Trends Biotechnol.*, **27**, 82–89.

12 Emamifar, A., Kadivar, M., Shahedi, M., and Zad-Soleimanian, S. (2011) Effect of nanocomposite packaging containing Ag and ZnO on inactivation of *Lactobacillus plantarum* in orange juice. *Food Control*, **22**, 408–413.

13 Bhattacharya, S., Jang, J., Yang, L., Akin, D., and Bashir, R. (2007) Biomems and nanotechnology-based approaches for rapid detection of biological entities. *J. Rapid Meth. Aut. Mic.*, **15**, 1–32.

14 Lilie, M. and Anna Cantini, A. (2001) Nanotechnology in agriculture and food processing. Conference proceedings, University of Pittsburgh, Eleventh Annual Freshman Conference, April 9. pp. 1–9.

15 Das, M., Saxena, N., and Dwivedi, P.D. (2009) Emerging trends of nanoparticles application in food technology: safety paradigms. *Nanotoxicology*, **3**, 10–18.

16 Sanguansri, P. and Augustin, M.A. (2006) Nanoscale materials development – a food industry perspective. *Trends Food Sci. Tech.*, **17**, 547–556.

17 Chaudhry, Q., Scotter, M., Blackburn, J., Ross, B., Boxall, A., Castle, L., Aitken, R., and Watkins, R. (2008) Applications and implications of nanotechnologies for the food sector. *Food Addit. Contam.*, **25** (3), 241–258.

18 FOE (Friends of the Earth) (2008) Out of the laboratory and on to our plates: Nanotechnology in food and agriculture. Available at http://midgetechnology.com/Documents/Nano%20Out%20of%20the%20Lab%20On%20To%20Our%20Plstes.pdf (accessed June 16, 2011).

19 Vandermoere, F., Blanchemanche, S., Bieberstein, A., Marette, S., and Roosen, J. (2011) The public understanding of nanotechnology in the food domain. *Public Underst. Sci.*, **20**, 195–206.

20 Duncan, T.V. (2011) The communication challenges presented by nanofoods. *Nat. Nanotechnol.*, **6**, 683–688.

21 Alli, I. (2004) *Food quality Assurance; principles and practices*, CRC Press LLC, Boca Raton, Florida, p. 141.

22 European Commission (2014) Periodic Report Summary 1 – Collab4Safety (Towards sustainable global food safety collaboration). Available at http:cordis.europa.eu/result/rcn/141488_en.html (accessed May 31, 2016).

23 WHO (1998) Food safety and globalization of trade in food. A challenge to the public health sector. Available at http://www.who.int/foodsafety/publications/globalization-trade-food/en (accessed May 31, 2016).

24 Chaudhry, Q. and Castle, L. (2011) Food applications of nanotechnologies: an overview of opportunities and challenges for developing countries. *Trends Food Sci. Technol.*, **22**, 595–603.

25 Bhattacharyya, A., Datta, P.S., Chandhmi, P., and Barik, B.R. (2011) Nanotechnology – a new frontier for food security in socio economic development. Available at http://Http://disasterresearch.net/drvc2011/paper/fullpaper_22.pdf (accessed June 15, 2011).

26 Cheftel, C.J. (2011) Emerging risks related to food technology. Advances in Food Protection. NATO Science for peace. Available at http://www.springerlink.com/content/p585630701412061/ (accessed June 17, 2011).

27 Dowling, A.P. (2004) Development of nanotechnologies. *Mater. Today*, **7**, 30–35.

28 Casabona, R., Escajedo, C.M., Epifanjo, S., Emaldi, L., and Cirion, A. (2010) Safe and socially robust development of nanofood through ISO standards? Conference paper (EurSafe 2010, Bilbao, Spain, 16 – 81 September 2010) – Global food security: ethical and legal challenges 2010, pp. 521–526.

29 Bosso, C. (2010) Nanotechnology and environmental governance: the problem(s) of uncertainty. Available at http://eprints.internano.org/505/1/Bosso_NNN_2010.pdf (accessed June 15, 2011).

30 Mukul, D., Ansari, K.M., Anurag, T., and Dwivedi, P.D. (2001) Need for safety of nanoparticles used in food industry. *J. Biomed. Nanotech.*, **7**, 13–14.

31 Powell, J.J., Faria, N., Thomas-McKay, E., and Pele, C.L. (2010) Origin and fate of dietary nanoparticles and microparticles in the gastrointestinal tract. *J. Autoimmun.*, **34**, J226–J233.

32 Blaser, S.A., Scheringer, M., Macleod, M., and Hungerbuhler, K. (2008) Estimation of cumulative aquatic exposure and risk due to silver: contribution of nanofunctionalized plastics and textiles. *Sci. Total Environ.*, **390**, 396–409.

33 European Parliament and the Council (2011) Regulation (EU) No 1169/2011 of 25 October 2011. Official Journal of the European Union. L304/18. http://faolex.fao.org/docs/pdf/den108120.pdf. Accessed May 27, 2016.

34 Gruere, G.P. (2012) Implications of nanotechnology growth in food and agriculture in OECD countries. *Food Policy*, **37**, 191–198.

35 Jones, P.B.C. (2006) ISB News Report. A Nanotech Revolution in Agriculture and the Food Industry. Available at http://www.isb.vt.edu/news/2006/artspdf/jun0605.pdf (accessed April 13, 2016).

36 Chau, C., Wu, S., and Yen, G. (2007) The development of regulations for food nanotechnology. *Trends Food Sci. Technol.*, **18**, 269–280.

37 Bhat, J.S. (2003) Heralding a new future. *Nanotechnology. Curr. Sci.*, **85**, 147–154.

18
Interaction between Ingested-Engineered Nanomaterials and the Gastrointestinal Tract: *In Vitro* Toxicology Aspects

Laurie Laloux, Madeleine Polet, and Yves-Jacques Schneider

UC Louvain, Institut des Sciences de la Vie (ISV), Croix du Sud, 5, Louvain-la-Neuve 1348, Belgium

18.1
Introduction

We are constantly exposed to nanometer-sized particles. Most of them are found naturally in the environment, generated by processes such as volcanic eruptions, forest fires, photochemical reactions, or simple erosions. Nevertheless, DNA, proteins, complex carbohydrates, or viruses are other kinds of natural nanoscale structures. Moreover, humans have unintentionally produced nanomaterials (NMs) for centuries by cooking or simple combustion [1]. Since the industrial revolution, their quantity has dramatically risen because of some processes such as welding, ore refining and smelting, or combustion of coal and fuel in power plants and motor vehicles [1,2]. More recently, with the development of nanotechnologies, intentional manufactured and engineered nanomaterials (ENMs) became available on the market [2,3]. They are usually employed in medicine, cosmetics, electronics, textiles, food, and agriculture [4]. Among ENMs used in consumer products, silver, titanium oxide, silica, and carbon are the most represented [5].

The development of nanotechnology could be beneficial to society and economy [6]. Indeed, engineered nanomaterials have interesting physicochemical properties that could help to tackle some challenges of the new century. Their ultrasmall size, at the transition between individual atoms or molecules and the corresponding bulk material, confers a high surface-to-volume ratio giving them novel properties compared to larger particles [7–9]. In addition, their reactivity is enhanced as the proportion of energetically unstable atoms increases at their surface [1,8]. However, the unusual properties of ENMs, which make them attractive for many applications, could also lead to significant adverse health effects. ENMs may interact in new unpredicted ways with biological systems inducing cell damage and dysfunction that are not caused by the same material in larger form [2,6–9]. Moreover, due to their manufacture and use, ENMs are

Nanotechnology in Agriculture and Food Science, First Edition. Edited by Monique A.V. Axelos and Marcel Van de Voorde.
© 2017 Wiley-VCH Verlag GmbH & Co. KGaA. Published 2017 by Wiley-VCH Verlag GmbH & Co. KGaA.

released in the environment and could have significant impacts on ecosystems and society [6]. Therefore, it is crucial to set up analytical tools and elaborate regulations to address potential risks of ENMs on health and environment [4,10].

The agrofood sector is investigating the use of nanotechnology in many applications, from farm to fork [11]. Nanostructured food ingredients are being developed in order to improve taste, texture, or consistency of the product [12]. For instance, low-fat nanostructured mayonnaise was set up with no effect on taste, simply by including nanosized water droplets in fat droplets [13]. Nutraceutical compounds such as vitamins, probiotics, bioactive peptides, or antioxidants can be incorporated in nanodelivery systems in order to control their stability, solubility, bioavailability, and release [4,14]. ENMs can also be found unintentionally in food following migration from packaging, use of nanosized pesticides and veterinary medicines, or contact with nanocoatings. Food packaging is the most important application of the food sector that uses ENMs [12]. For example, ENMs are incorporated in packaging materials to improve their barrier properties. Nanosensors are also developed to respond to environmental modifications, food spoilage, or contamination by microorganisms [4]. Although the use of ENMs in the food sector is constantly growing, no specific regulation currently exists for it [15].

The gastrointestinal tract (GIT) is the most important route for the entry of macromolecules as well as particles, including microorganisms, in the body [1]. Once inside, macromolecules and particles are exposed to the digestion process consisting in physical, chemical, enzymatic, and microbiota exposure [16]. ENMs introduced in food, either intentionally or by contamination/migration from food contact materials, are expected to be ingested and pass into the gastrointestinal tract. In addition, ENMs contained in medicines, water, dental prosthesis debris, or cosmetics, such as toothpaste and lipstick, can be ingested. Finally, inhaled nanomaterials cleared by the mucociliary escalator are also released into the GIT [1,17]. It is worth noting that the gut itself also produces nanomaterials *de novo*, derived from intestinal phosphate and calcium secretions [1,11].

During the digestion process, the pH and ionic strength changes, the presence of digestive enzymes, food matrices, and microbiota alter the ENMs properties, that is, their size, shape, stability, or aggregation state [18]. Due to their high reactivity and surface-to-volume ratio, ENMs can also interact with proteins, carbohydrates, and fats in order to develop a dynamic corona coating [19]. All these changes may influence their effects on cells or tissues, and their biodistribution [18]. As digested ENMs could be very different than those that were ingested, it is important to take these modifications into account for a relevant risk assessment [16].

Therefore, the first part of this chapter will focus on the fate of ENMs once in the gastrointestinal tract. ENMs properties modifications and interactions with digestive enzymes and food matrices will be discussed as well as their impact on biological responses.

The Nanotechnology Consumer Products Inventory was created in 2005 by the Woodrow Wilson International Center for Scholars and the Project on

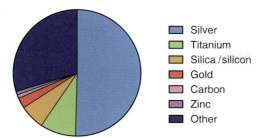

Figure 18.1 Types of ENMs found in consumer products and potentially ingested, based on the Nanotechnology Consumer Products Inventory [5].

Emerging Nanotechnology [20]. Consumer products containing nanoscale components are compiled in this inventory and classified under different categories [15]. According to this list, 111 consumer products are potentially ingested, releasing ENMs in the GIT. Among these products, a half contains silver nanoparticles (AgNPs) and 15% are composed of either titanium dioxide (TiO_2) or silica (SiO_2) (Figure 18.1) [5]. AgNPs are used in, for example, food packaging and food supplements, SiO_2 in food additives as anticaking agent while TiO_2 is mostly found in cosmetics and hygiene products, especially sunscreens and toothpastes [17]. Although it is important to approximate the intake of ENMs, this estimation is hard to make because quantitative consumer exposure data and contextual exposure information are very limited [17,21]. However, some studies tried to estimate ENMs dietary exposure. For example, Dekkers *et al.* [22] calculated a worst case intake of 1.8 mg nanosilica/kg BW/day for an average adult [15,22]. Concerning the food additive E171 consisting in TiO_2, Weir *et al.* [23] estimated an exposure to TiO_2 varying between 0.2 and 3 mg/kg BW/day. Based on 89 products, they have found that 36% of the E171 particles are in the nanosize, which would lead to an oral exposition between 0.07 and 1.08 mg/kg BW/day of nanoTiO_2. Children are even more exposed because of the higher occurrence of TiO_2 in candies, white icing, or powdered sugar toppings. Some dietary habits can dramatically increase the exposure up to several hundreds of mg/day [15,23]. Concerning AgNPs, the Danish Environmental Protection Agency has estimated an exposure to silver of 0.02 mg/kg BW/day for people taking food supplements containing colloidal silver [24].

Based on the Nanotechnology Consumer Products Inventory, applications of AgNPs are found mainly in the "Health and Fitness" and in the "Food and beverage" sectors (Figure 18.2), two categories that potentially lead to ingestion [5,25]. Some medical products such as toothpastes or dental amalgams [26] also contain AgNPs [5] and can lead to an oral exposure. Several unauthorized food supplements containing AgNPs claiming to have "immunobooster properties," help against "allergy prophylaxis," or protection "against severe illness," can be purchased online [27,28]. However, since they were not evaluated by control agencies, they are not considered as pharmaceutical products [28]. Furthermore, inhaled AgNPs after, for example, occupational exposure can be ingested after

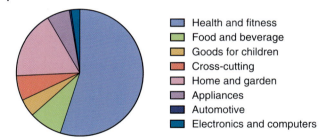

Figure 18.2 Categories of consumer products in which AgNPs can be found, based on the Nanotechnology Consumer Products Inventory [5].

their elimination by the mucociliary escalator [1,29]. Finally, the presence of AgNPs has been recently underlined in the E174 food additive, mainly present in pastry decorations such as silver pearls, chocolates, and liqueurs [30].

Aside from this last application, AgNPs are not directly incorporated in food products. Due to their antimicrobial properties, they can be found in food packaging and products such as cleaning sprays, coatings of cooking instruments, or storage boxes [28]. Exposure can occur after migration into the food, guarantying the claimed antimicrobial effect. The presence of silver in food after storage in an AgNPs-embedded packaging material was evidenced in orange juice [31] or in boneless chicken breasts [32]. The migration of silver in the food seems to be more efficient at acidic pH, depending thus on the food stored in the container [33–36]. As the oxidation of silver is more efficient at low pH the major part of silver seems to migrate as oxidized silver [33,37]. However, a nonnegligible proportion of AgNPs migrates in food [36,38], even being the major form of silver in some cases [33], suggesting that the oxidation is not the only process explaining the migration of silver from packages. Although estimated levels of exposure to Ag through this way are low (estimated at 4.2 µg silver for a new box in the worst case scenario [36], the number of applications for AgNPs and their increasing use in consumer products dramatically raise our exposure to AgNPs and render it difficult to estimate.

All these applications of AgNPs in food related products make them more susceptible to be ingested. After digestion, they come in contact with the intestinal barrier and potentially lead to adverse health effects. As a result, their toxicity has to be deeply evaluated. After a description of the *in vitro* models of the intestinal barrier and the possible drawbacks of *in vitro* nanotoxicology assays, the second part of this chapter will focus on the toxicity of AgNPs on these models.

18.2
Influence of the Gastrointestinal Tract on the Ingested Nanomaterials Characteristics

The human gastrointestinal tract (GIT) extends from the mouth to the anus. It is involved in food ingestion, digestion, and fermentation, allowing the formation of nutrients that are absorbed to ensure the body's metabolic requirements, as well as the handling of bioactive substances. In addition, the GIT and its microbiota maintain the water homeostasis, assure the removal of undigested and unabsorbed fractions and play an important role in endocrine and immune systems [19,39].

Once ingested, food is chewed and mixed with saliva in the buccal cavity. Saliva lubricates food, protects the mucous membranes, and contains immonocompetent substances and enzymes (α-amylase and lipase) starting the food biochemical digestion. Then, the food bolus is swallowed and transported through the esophagus to the stomach. In addition to food temporary storage, the stomach kneads the bolus and secretes enzymes that partially break down proteins to form the chyme. Afterward, the small intestine receives pancreatic juices that contain electrolytes neutralizing the acidic chyme coming from the stomach. Bile, secreted by liver and stored by gallbladder, is also released in the duodenum and is essential for fat digestion. Under the action of bile and enzymes coming from the pancreas and intestinal mucosa, the chyme is broken down into nutrients that are absorbed and delivered to the rest of the body. Undigested residues arrive then in the large intestine. A portion is used by the microbiota for its metabolism, which, in turn, provides short-chain fatty acids and vitamin K to the host. Finally, the waste are excreted after water and electrolytes absorption [39–41].

Besides the digestion and the absorption of nutrients, the intestine plays a role of protection against xenobiotics, thanks to several biotransformation activities as well as ATP-dependent extrusion systems of its mucosa. The microorganisms living in the gastrointestinal tract possess also various enzymes capable of biotransformation. These modifications may have a protective effect by enhancing polarity and water solubility of xenobiotics, which accelerate their excretion. However, in some cases, they may be ineffective or even harmful by activating nontoxic molecules into toxic ones [42–44].

Digestive enzymes and bile, ionic strength and pH variations, the composition of the food matrix, and microbiota form together a complex environment that affects ingested nanomaterials. All these parameters could alter their physicochemical properties and modify their interaction with the body's structures [1,16].

18.2.1
In vitro **Models of the Gastrointestinal Tract**

In vitro models may be useful to understand the behavior of ingested nanomaterials in the GIT. Among them, *in vitro* digestion processes are employed to mimic the gastrointestinal environment as closely as possible [45].

In vivo studies of the human digestion process are ethically and technically difficult to implement. They are expensive, time consuming, and it is difficult to sample digesta from the different sections of the GIT. Animal studies may not be a suitable alternative because their digestive system can strongly differ from human's one [46,47]. In contrast, *in vitro* models are less expensive, more rapid, and do not need ethical approval. Moreover, they allow to process a large number of trials with an easy sampling and a high reproducibility [45].

Two kinds of *in vitro* digestion models have been developed: static and dynamic. In the first one, materials are incubated sequentially with digestive fluids mimicking the different compartment of the GIT (buccal cavity, stomach, small intestine, and, sometimes, colon and its microbiota). Parameters such as the pH, digestive enzymes, salt concentrations, temperature, and duration of residence are taken into account, but physical processes like chewing, mixing, or peristalsis are not emulated [45,47,48]. In addition, the pH variations or augmentation of enzyme secretions following the introduction of food bolus are not simulated in static models [49]. On the other hand, dynamic models precisely mimic the gradual transit of ingested materials through the GIT [48]. They include physical processes and reproduce continuous changes of the luminal conditions to get closer to *in vivo* conditions [47,49]. However, these models are time-consuming and more expensive than the static models [49].

In vitro models of the digestion process have already been employed to study the influence of the GIT on various nanomaterials such as AgNPs, SiO_2, TiO_2, and ZnO nanoparticles (NPs), or nanosized engineered mineral clay particles [15,18,50–53].

18.2.2
Influence of pH and Ionic Strength on Ingested Nanomaterials

The ionic strength of the gastrointestinal fluids fluctuates from 0.0 to 0.4 M while the pH is slightly acidic to neutral in the buccal cavity and the small intestine, and is strongly acidic in the stomach [17,54]. These variations may have a significant impact on the surface charge and, as a result, the agglomeration state of ingested nanomaterials. Indeed, zeta-potential modifications by the ionic salts may influence electrostatic repulsion [17,19]. For example, the surface electrostatic repulsion of TiO_2 nanoparticles disappears as electrolyte concentration increases, leading to the aggregation of NPs [19,55]. The zeta-potential of AgNPs also decreases in acidic conditions, reducing the stability of the suspension. Metal-based nanomaterials are particularly responsive to pH variations. These NMs may be dissolved in acidic environment and release ions. For instance, in addition to aggregation, AgNPs can liberate silver ions when exposed to a simulated stomach fluid (HCl 0.42 M and glycine 0.40 M, pH 1.5). Chloride ions may then react with silver ions to form insoluble AgCl [19,56,57].

18.2.3
Influence of Digestive Enzymes and Food Matrices on Ingested Nanomaterials

In biological fluids, the surface of nanomaterials is altered by the adsorption of several kinds of biomolecules such as sugars, lipids, and, mainly, proteins, forming a dynamic biomolecular corona. The corona may be divided into two parts: the "hard" and "soft" coronas. Biomolecules with high affinities for the NM surface and with long residence time create the "hard" corona while those with fast, reversible binding, and quicker exchange rates form the "soft" corona [58–60]. Both properties of NMs (nature, size, shape, solubility, surface properties such as charge and surface curvature, dispersant media, etc.) and parameters of the biological fluid (pH, ionic strength, nature and salt concentration, biomolecular composition, etc.) are involved in the composition of the corona [58,59,61–63].

Food components as well as biomolecules released in the GIT may compete to create this dynamic corona. They will define the biological identity of the NM and influence their passage through the GIT as well as their ADME properties and the further biological responses of the body. Indeed, cellular barriers and membrane receptors can directly interact with the corona while the pristine NM surface is hidden and remains inaccessible [11,59,63]. Therefore, the biomolecules–nanomaterial complex must be treated as a novel entity with different properties than the pristine object and the "hard" corona must be considered in the definition of NMs in biological fluids [58,64].

Some studies suggest that the presence of digestive enzymes and food components modifies the fate of NMs during an *in vitro* digestion process. Contrary to the results obtained by Mwilu *et al.* [56] and Rogers *et al.* [57], Walczak *et al.* [53] observed totally dissolved AgNPs in the acidic condition of the stomach in the absence of proteins, while the presence of enzymes protected the NPs from the degradation and led to the formation of clusters composed of AgNPs and chlorine. The agglomeration of AgNPs resulted both from the acidic environment and the adsorption of proteins acting as junctions between the NPs [16,18,53,65]. Böhmert *et al.* [18] also noticed AgNPs agglomeration during the gastric digestion with an increase of the hydrodynamic radius that could be due to the adsorption of digestive enzymes [18]. In addition, based on infrared spectroscopy, bile salts and proteins were revealed in the corona of silica and TiO_2 nanoparticles treated with simulated intestinal fluid [15]. More recently, the influence of food components on the fate of AgNPs in the gastrointestinal tract was also emphasized: less aggregates were formed in the presence of food components that could act as colloidal stabilizers [66].

In the past few months, Walczak *et al.* [65] also demonstrated that the GIT environment modifies the composition of the corona: polystyrene nanoparticles (PS-NPs) exposed or not to a simulated digestion did not have the same corona when they were incubated in cell culture medium. More low molecular weight proteins were found in the digested PS–NPs corona compared to the pristine equivalents. In addition, bands of digested samples were smeary, suggesting enzymatic breakdown of proteins [65]. Amylase, lipase, nuclease, pepsin, and other proteases can denude ingested nanomaterials of their surface-adsorbed molecules and have an incidence on the kinetic of the dynamic corona. The activity of enzymes released in the gastrointestinal tract could also alter the integrity of some nanomaterials, such as those composed of starch, lipids, proteins, or nucleic acids that may be subject to digestion [19,67]. Inversely, the GIT environment could also be modified by the presence of ingested NMs. For instance, the activity and structure of proteins may be altered following adsorption to NMs. A protein could lose or increase its activity depending on the adsorption site [62].

18.2.4
Characterization Techniques of Ingested Nanomaterials

After their passage through the gastrointestinal tract, most of nanomaterials are mainly excreted in the feces while a low percentage is absorbed by mucosal cells, enters the lymphatic or blood systems, and gains access to peripheral tissues and organs. The absorption of ingested NMs depends on their properties (size, aggregation, shape, surface properties, and corona). We have previously shown that all of those physicochemical properties can be altered by interaction with the gastrointestinal environment. Therefore, it is crucial to assess the fate of NMs in the GIT to better understand their behavior and effect on the body [1,11,19].

Size, shape, agglomeration state, surface charge, coating, and corona are important parameters influencing the behavior of NMs. Several techniques may be used to investigate these physicochemical properties [11]. For instance, the number-based size distribution of particles may be assessed by single particle – inductively coupled plasma mass spectrometry (SP-ICP-MS), dynamic light scattering (DLS), or electron microscopy [16,53]. This last device may also be useful to observe the shape and the agglomeration state of NMs, even in complex environments such as food matrices [68,69]. Regarding the surface charge, zeta-potential measurement could provide some information. Finally, identification and even quantification of the proteins incorporated into the corona could be performed by mass spectrometry after the isolation of nanomaterial–proteins complex. Various methods exist to isolate this complex, centrifugation being the most conventional [64,70].

18.3
In Vitro Models of the Intestinal Barrier

Ex vivo models of the human gut are not widely used for absorption or cytotoxicity studies because of their difficulty to form a monolayer, their hyperpermeability, and their overproduction of mucus [16,67]. To overcome this situation, different *in vitro* models have been developed coming from cancerous cells. In this chapter, we will focus on Caco-2 cells, as they are the most studied for pharmacological and toxicological assessments. Moreover, in nanotoxicology, Caco-2 cells are the most widely used for nanomaterial translocation assessment [50]. We will also present some improvements leading to a more physiologically relevant situation and mimicking different cell types found *in vivo*.

Caco-2 cells were originally isolated from a colonic adenocarcinoma [71]. They differentiate spontaneously at confluence into enterocyte-like cells presenting the same characteristics such as functional tight junctions, biotransformation enzymes, or efflux pumps [72–75]. Although their colonic origin, they lead to cells with small intestinal enterocyte phenotype [76]. The colonic enterocyte phenotype can be achieved by the use of T84 cells, originally isolated from a pulmonary metastasis of a colon carcinoma [16].

Despite their extensive use since 20 years, the major drawback of Caco-2 cells is the absence of mucus at their surface, which is, however a barrier of special interest for cytotoxicity assessment. In particular, for the study of NMs, mucus could form a barrier entrapping them or could modify them. To overcome this limitation, mucus-secreting cells coming from another cancerous cell line, HT-29 were developed. This cell line, originally isolated from a human colorectal cancer [71] leads to differentiation in an heterogeneous cell layer containing, among others, some mucus-secreting cells [77]. These cells were isolated by a treatment with the anticancer drug methotrexate to form the new cell line HT29-MTX producing mucus at differentiation [78]. This capacity has been exploited to develop an *in vitro* coculture of Caco-2 cells and HT29-MTX cells, mimicking the two most representative cell types in the intestinal epithelium: enterocytes and goblet cells [79–81]. This model provides a continuous mucus layer of 10 µm, which is a two-third thickness of the physiological situation and is more resistant to washings than an artificially added mucus layer [82]. Moreover, the permeability of this model is higher than Caco-2 cells and more physiological, counteracting another disadvantage of Caco-2 cells that present a very limited paracellular transport [80,82,83]. This point is however not so important in nanotoxicological studies as this pathway is largely negligible for the uptake of nanomaterials [17]. Although of higher physiological relevance, this coculture model is not extensively used in nanotoxicology, only uptake studies were performed in a pharmaceutical context such as in Behrens *et al.* [84].

Investigating the impact of the presence of M cells in the intestinal mucosa and in particular in the follicle-associated epithelium covering, for example, Peyer's patches, requires more complex models as no cell line giving M cells is known and as they can only survive a few hours in *ex vivo* cultures. Because of their physiological role, delivering luminal antigens to the underlying immune system [85], they are of particular interest in nanotoxicology, being a potential entrance for NMs [17]. The current existing models are based on the conversion of some Caco-2 cells in the presence of lymphocytes to a M cell phenotype. The first *in vitro* model was set up by Kerneis *et al.* [86] using immune cells isolated from murine Peyer's patches and Caco-2 cells seeded on inverted culture inserts [17,86]. Gullberg *et al.* [87] have on their side developed a model of normally seeded Caco-2 cells cultivated in the presence of Raji B cells, a continuous line of human lymphoblastoid cells isolated from a Burkitt lymphoma [87]. des Rieux *et al.* [88] combined the two previous models, using inverted inserts covered by Caco-2 and Raji B cells, the inversion leading to a better contact between the two cell types. This model leads to the appearance of 15–30% of M cell-like cells in the monolayer, overestimating the physiological situation where around 10% of the cells have a M-cell phenotype. This model allows a better study of transport mechanism, the latter being 50 times higher than in a monoculture model [17,88,89].

A combination of the two previous models, leading to a monolayer containing M-cells and goblet cells embedded in a Caco-2 layer has also been developed [90,91]. In Mahler *et al.* [91], this model was used to study the effect of polystyrene nanoparticles on iron absorption [91]. The use of this model is however very limited in literature.

Finally, even if macrophages are not directly incorporated in the intestinal epithelium, different coculture models have been developed to mimic their presence in the subjacent mucosa. These models are based either on the murine cell line RAW264.7 [92,93], or on the human THP-1 cell line [94–96]. A more complex model, comprising macrophages and dendritic cells has even been established [97].

18.4
Cytotoxicity Assessment and Application to Silver Nanoparticles

Because of their high reactivity, NMs are more susceptible to interact with cytotoxicity assays and lead to different kinds of bias in the toxicity assessment. The techniques have thus to be adapted or modified, considering the nanomaterial specific properties such as their intrinsic absorbance or fluorescence [98], their high capacity of adsorption, which could lead to an adsorption of assay components or analytes [99], or their high catalytic activity that could interfere if the assay is based on the oxidation/reduction of a substrate [100]. Moreover, ROS induced by NMs could also possibly inactivate proteins [101]. Concerning AgNPs, studies have underlined that AgNPs could affect the result of lactate dehydrogenase leakage assay. A proposed mechanism is the combination of the adsorption of lactate dehydrogenase by AgNPs and inhibition of its activity by ROS generated after AgNPs exposure [101].

Different controls must be performed to avoid bias in the nanotoxicity measurement. First, the absorbance/fluorescence of NMs solutions has to be measured. Then, the interference between NMs and assay components should be assessed. Finally, when possible, the analyte should be added to NMs and assay components [99]. Moreover, the toxicity or the innocuity has to be confirmed with at least two concording and different cytotoxicity assays [102]. Ultimately, NMs should be eliminated if possible either by multiple washings if cell layers are used for the assay, or by centrifugation if supernatants are utilized [99]. Although these recommendations are well known, Ong *et al.* [99] have shown through a literature study that about 95% of papers published in 2010 assessing NMs cytotoxicity thanks to biochemical techniques did not report any controls for interference. This was only slightly improved in 2012 papers [99]. This could explain a part of the contradictory results concerning NMs toxicity. Another drawback for nanotoxicology is the ambiguous definition of the dose. Contrary to soluble compounds, NMs are subjected to diffusion, settling, and aggregation/

agglomeration, all these processes affecting the quantity of NMs reaching cells. As they vary with different parameters of NMs such as their size, density, or surface, as well as parameters of the solution in which NMs are diluted, the definition of the dose delivered to cells is really difficult to establish [103]. According to these authors, the effective dose reaching cells would be really low compared to the nominal doses. However, according to Lison et al. [104], the convection process allows the access of the majority of particles to cells [104]. Moreover, as the surface area influences the interaction of NMs with cells, the nominal concentration (in µg/ml) could be converted to surface area concentration (cm^2/ml), a better metric taking both mass and size distribution into account [105]. The possible dissolution of NMs can affect not only the dosis but also the uptake pathway or the toxicity mechanisms. In nanotoxicology, it is thus important to know if the toxicity comes from ions, nanomaterials, or a combination of them [106]. This phenomenon is however difficult to control, as the dissolution may vary with the nanomaterial physicochemical parameters, the environment, and the presence of organic compounds [106]. Concerning AgNPs, this phenomenon requires oxygen and is favored with the acidity and the temperature. As a result, AgNPs colloidal solutions contain a nonnegligible proportion of silver in the ionic form either free or on the surface of AgNPs. Van der Zande et al. [107] have shown by ultrafiltration that AgNPs solutions (NM-300 K from the JRC) contained in average 7% of silver in ionic form [107].

Finally, the different published studies are very difficult to compare, because different parameters can influence the toxicity of NMs other than their chemical composition, such as intrinsic parameters of the NMs (size, shape, surface charge, etc.), the target cells, the exposure duration, or the measured endpoints [2,15,108]. It is, for instance, the case for AgNPs cytotoxicity studies already published. Although being numerous, their lack of unicity concerning AgNPs type, cell models, or measured cytotoxicity endpoints renders their comparison very difficult.

Once all these considerations have been taken into account, *in vitro* cytotoxicity assays are really attractive being more rapid, less expensive, and without ethical restrictions compared to *in vivo* models. As a result, they are really suitable for both compounds screening and mechanistic studies, allowing a large number of replicates, an assessment of defined endpoints, and controlled conditions [100].

In vitro models of the gut have already been used for the cytotoxicity assessment of AgNPs. AgNPs affected cell morphology [109,110], decreasing cell adherence capacity [110]. Significant necrosis was observed after AgNPs incubation [111]. Apoptosis involvement in AgNPs toxicity is more controversial. According to Böhmert et al. [112], AgNPs toxicity did not involve an apoptotic process, which was not the case in Gopinath et al. [113] underlining an activation of p53-mediated apoptotic pathway by AgNPs in HT-29 cells [112,113]. Bouwmeester et al. [114] in their whole-genome gene expression study also pointed an activation of some apoptosis-related genes. The metabolic activity of cells treated with AgNPs was almost in all cases altered with a toxicity appearing around 10–30 µg/ml, for 3–48 h incubation, no matter the assay [109,111,114,115]. This is consistent with studies on other cellular models such as hepatic cells [116], neuroblastoma [117], or skin cells [118]. As expected, smaller AgNPs led to a higher cell toxicity [110]. AgNPs also seem to inhibit cell proliferation [119], and be less toxic than silver ions, at the same concentrations, affecting the morphology and apoptosis [111] as well as metabolic activity [110–112,115]. Martirosyan *et al.* [115] used a concentration of silver ions that could be expected in AgNPs colloidal solutions (10% w/w). Compared with AgNPs 10 times more concentrated, silver ions were less toxic, suggesting that AgNPs toxicity did not only come from silver ions free in colloidal solution [115]. However, silver ions could be involved in a Trojan-horse mechanism of toxicity [120].

The ability of the intestinal mucosa to maintain a functional integrity in order to modulate the permeability for water, electrolytes, nutrients, and bioactive substances, is a major gatekeeper of the intestinal function, as it gives its mechanical strength to the epithelial layer and restrict the passage of substances between cells [17]. This capacity is mainly maintained by extracellular structures surrounding each cell like tight junctions. Due to their presence in the extracellular space, these structures can be affected by substances present in the intestinal lumen, but also by compounds penetrated inside cells as they are tightly connected to the cytoskeleton [121]. However, little is known about the effect of AgNPs on these structures. Böhmert *et al.* [112], using microarrays analysis, predicted an influence on tight junction function and cytoskeleton signaling on Caco-2 cells [112]. Although Bouwmeester *et al.* [114] did not observe any functional alteration of tight junctions through electric resistance measurements, Martirosyan *et al.* [122] provided evidences that AgNPs can alter the intestinal permeability through structural alterations of tight junctions. The paracellular permeability was increased, as observed by an increased passage of the paracellular dye lucifer yellow. Moreover, in contrast to Bouwmeester *et al.* [114], the electric resistance of the layer was reduced after incubation with AgNPs. Finally, immunostainings of occludin and zonula-occludens 1, two proteins of tight junctions, suggested an aggregation and a disruption of continuity for both proteins [114,122]. Furthermore, in another study of Martirosyan *et al.* [115], the coincubation of AgNPs and quercetin or kaempferol, two antioxidant polyphenols, reduced these deleterious effects, suggesting that the alteration of tight junctions could at least partially be mediated by oxidative stress [115]. Tight junctions function impairment was also shown for alumina nanoparticles that were also inducing oxidative stress [123]. Oxidative stress is in fact known to alter the redox-responsive signaling pathways impairing tight junction proteins [124,125], damaging them, favoring their proteosomal degradation [126], and stimulating matrix metalloproteinases responsible for their hydrolysis [127]. Indeed, the ROS production by AgNPs was observed, probably mediated by a disturbance of the respiratory chain, disrupting the electron passage [128,129]. The resulting oxidative stress could damage lipids, proteins, and DNA, which could explain the observed toxicity [28,110,128].

In vitro gut models were also used to assess the absorption of AgNPs by the intestinal barrier. Different studies highlighted the presence of AgNPs in cells through different techniques: electron microscopy [111,112], confocal microscopy combined with reflectance analysis [25], or ICP-MS [119]. However, AgNPs were poorly taken up, in lower proportion than in hepatocytes, but higher than silver microparticles [25]. In cells, AgNPs were mainly located in the cytoplasm and near nuclear membrane as agglomerated forms, but none was found in nuclei, following an electron microscopy observation of Caco-2 cells [109]. Bouwmeester *et al.* [114] used the M-cell model cultivated in the bicameral culture system set up by des Rieux *et al.* [88], to estimate the translocation of AgNPs through intestinal cells. The amount of silver found in basolateral compartment, estimated by ICP-MS and by atomic absorption was the same upon exposure to AgNPs or to silver ions [88]. It suggests that AgNPs are transported in the ionic form, although the techniques used did not allow concluding about the silver form in the basolateral compartment [114].

18.5 Conclusion

In daily life, consumers are increasingly exposed to ENMs due to many interesting and useful applications. Ingestion is one of the major routes of exposure to ENMs due to their incorporation in some products such as food or health supplements, paramedical devices, and packaging materials. Among all ENMs, AgNPs are the most commonly found in consumer products. However, more and more evidences suggest a toxic effect of ENMs for human health. A regulation on their presence in food is necessary and is currently in revision. In its guidance on risk assessment of nanomaterials, the EFSA highlights that the fate of ENMs in the gastrointestinal tract need to be better understood [8]. ENMs can indeed interact with the food matrices and the complex environment of the GIT, which influence their structure, properties, and further biological response. *In vitro* models as well as various characterization methods may be useful to investigate the behavior of ENMs once ingested. Although EFSA recommendations and its relevance, this topic remains poorly studied and needs to be more explored.

Regarding ENMs toxicity, *in vitro* models, more cost-effective, reproducible, and without ethical restriction compared to *in vivo* methods, may also be helpful. However, results are sometimes difficult to explain because ENMs may interact with the classical cytotoxicity assays. As a result, adequate controls should be included in assays, avoiding drawing conclusions from a result distorted with interference. Different models of the intestinal gut have been developed, with different levels of complexity and physiological relevance. However, the simplest of them, based on Caco-2 cells, remains the golden standard and the majority of studies about the toxic effect of ingested AgNPs have been performed with this cell line. Although it is difficult to compare results coming from different researches due to the lack of homogeneity, AgNPs have shown an alteration of the cell viability. The effect of AgNPs on the tight junctions, gatekeepers of the intestinal homeostasis, as well as the absorption mechanisms are key factors in the understanding of the toxicity and have to be more investigated.

References

1 Buzea, C., Pacheco Blandino, I.I., and Robbie, K. (2007) Nanomaterials and nanoparticles: sources and toxicity. *Biointerphases*, **2**, MR17–MR172.

2 Oberdörster, G., Oberdörster, E., and Oberdörster, J. (2005) Nanotoxicology: an emerging discipline evolving from studies of ultrafine particles. *Environ. Health Perspect.*, **113**, 823–839.

3 Simon-Deckers, A. (2008) Effets biologiques de nanoparticules manufacturées: influence de leurs caractéristiques. Thesis, l'Institut des Sciences et Industries du Vivant et de l'Environnement, pp. 280.

4 Bouwmeester, H., Dekkers, S., Noordam, M.Y., Hagens, W.I., Bulder, A.S., de Heer, C., ten Voorde, S.E.C.G., Wijnhoven, S.W.P., Marvin, H.J.P., and Sips, A.J.A.M. (2009) Review of health safety aspects of nanotechnologies in food production. *Regul. Toxicol. Pharm.*, **53**, 52–62.

5 Project on Emerging Nanotechnologies (2013) Consumer Products Inventory. Retrieved from www.nanotechproject.org/cpi (August 27, 2015).

6 Oberdörster, G., Stone, V., and Donaldson, K. (2007) Toxicology of nanoparticles: a historical perspective. *Nanotoxicology*, **1**, 2–25.

7 Beer, C., Foldbjerg, R., Hayashi, Y., Sutherland, D.S., and Autrup, H. (2012) Toxicity of silver nanoparticles – nanoparticle or silver ion? *Toxicol. Lett.*, **208**, 286–292.

8 EFSA (2009) *Scientific Opinion of the Scientific Committee: The Potential Risks Arising from Nanoscience and Nanotechnologies on Food and Feed Safety* (eds S. Barlow, A. Chesson, J.D. Collins, A. Flynn, A. Hardy, K.-D. Jany, A. Knaap, H. Kuiper, J.C. Larsen, and P. Le Neindre et al.), European Food Safety Authority, pp. 1–39.

9 Nel, A., Xia, T., Madler, L., and Li, N. (2006) Toxic potential of materials at the nanolevel. *Supramol. Sci.*, **311**, 622–627.

10 Chaudhry, Q. and Castle, L. (2011) Food applications of nanotechnologies: an overview of opportunities and challenges for developing countries. *Trends Food Sci. Technol.*, **22**, 595–603.

11 Martirosyan, A. and Schneider, Y.J. (2014) Engineered nanomaterials in food: implications for food safety and consumer health. *Int. J. Environ. Res. Public. Health*, **11**, 5720–5750.

12 Chaudhry, Q., Scotter, M., Blackburn, J., Ross, B., Boxall, A., Castle, L., Aitken, R., and Watkins, R. (2008) Applications and implications of nanotechnologies for the food sector. *Food Addit. Contam.*, **25**, 241–258.

13 Smolkova, B., El Yamani, N., Collins, A.R., Gutleb, A.C., and Dusinska, M. (2015) Nanoparticles in food. Epigenetic changes induced by nanomaterials and possible impact on health. *Food Chem. Toxicol.*, **77**, 64–73.

14 Borel, T. and Sabliov, C.M. (2014) Nanodelivery of bioactive components for food applications: types of delivery systems, properties, and their effect on ADME profiles and toxicity of nanoparticles. *Annu. Rev. Food Sci. Technol.*, **5**, 197–213.

15 McCracken, C., Zane, A., Knight, D.A., Dutta, P.K., and Waldman, W.J. (2013) Minimal intestinal epithelial cell toxicity in response to short- and long-term food-relevant inorganic nanoparticle exposure. *Chem. Res. Toxicol.*, **26**, 1514–1525.

16 Lefebvre, D.E., Venema, K., Gombau, L., Valerio Jr, L.G., Raju, J., Bondy, G.S., Bouwmeester, H., Singh, R.P., Clippinger, A.J., Collnot, E.-M. et al. (2015) Utility of models of the gastrointestinal tract for assessment of the digestion and

absorption of engineered nanomaterials released from food matrices. *Nanotoxicology*, **9**, 523–542.

17 Fröhlich, E. and Roblegg, E. (2012) Models for oral uptake of nanoparticles in consumer products. *Annu. Rev. Pharmacol. Toxicol.*, **291**, 10–17.

18 Böhmert, L., Girod, M., Hansen, U., Maul, R., Knappe, P., Niemann, B., Weidner, S.M., Thunemann, A.F., and Lampen, A. (2014) Analytically monitored digestion of silver nanoparticles and their toxicity on human intestinal cells. *Nanotoxicology*, **8**, 631–642.

19 Bellmann, S., Carlander, D., Fasano, A., Momcilovic, D., Scimeca, J.A., Waldman, W.J., Gombau, L., Tsytsikova, L., Canady, R., Pereira, D.I.A. *et al.* (2015) Mammalian gastrointestinal tract parameters modulating the integrity, surface properties, and absorption of food-relevant nanomaterials. *Wiley Interdiscip. Rev. Nanomed. Nanobiotechnol.*, **7**, 609–622.

20 Vance, M.E., Kuiken, T., Vejerano, E.P., McGinnis, S.P., Hochella, M.F. Jr., Rejeski, D., and Hull, M.S. (2015) Nanotechnology in the real world: redeveloping the nanomaterial consumer products inventory. *Beilstein J. Nanotechnol.*, **6**, 1769–1780.

21 Clark, K., van Tongeren, M., Christensen, F., Brouwer, D., Nowack, B., Gottschalk, F., Micheletti, C., Schmid, K., Gerritsen, R., Aitken, R. *et al.* (2012) Limitations and information needs for engineered nanomaterial-specific exposure estimation and scenarios: recommendations for improved reporting practices. *J. Nanopart. Res.*, **14**, 1–14.

22 Dekkers, S., Krystek, P., Peters, R.J.B., Lankveld, D.P.K., Bokkers, B.G.H., van Hoeven-Arentzen, P.H., Bouwmeester, H., and Oomen, A.G. (2011) Presence and risks of nanosilica in food products. *Nanotoxicology*, **5**, 393–405.

23 Weir, A., Westerhoff, P., Fabricius, L., Hristovski, K., and von Goetz, N. (2012) Titanium dioxide nanoparticles in food and personal care products. *Environ. Sci. Technol.*, **46**, 2242–2250.

24 Larsen, P.B., Christensen, F., Jensen, K.A., Brinch, A., Mikkelsen, S.H., Clausen, A.J., Leck, F., Koivisto, A.J., and Nørgaard, A.W. (2015) *Exposure Assessment of Nanomaterials in Consumer Products*, The Danish Environmental Protection Agency.

25 Gaiser, B.K., Fernandes, T.F., Jepson, M.A., Lead, J.R., Tyler, C.R., Baalousha, M., Biswas, A., Britton, G.J., Cole, P.A., Johnston, B.D. *et al.* (2012) Interspecies comparisons on the uptake and toxicity of silver and cerium dioxide nanoparticles. *Environ. Toxicol. Chem.*, **31**, 144–154.

26 Correa, J.M., Mori, M., Sanches, H.L., da Cruz, A.D., Poiate, E., and Poiate, I.A.V.P. (2015) Silver nanoparticles in dental biomaterials. *Int. J. Biomater.*, **2015**, 1–9.

27 Silver, S. (2003) Bacterial silver resistance: molecular biology and uses and misuses of silver compounds. *FEMS Microbiol. Rev.*, **27**, 341–353.

28 Wijnhoven, S.W.P., Peijnenburg, W.J.G.M., Herberts, C.A., Hagens, W.I., Oomen, A.G., Heugens, E.H.W., Roszek, B., Bisschops, J., Gosens, I., Van De Meent, D. *et al.* (2009) Nano-silver – a review of available data and knowledge gaps in human and environmental risk assessment. *Nanotoxicology*, **3**, 109–138.

29 Arora, S., Rajwade, J.M., and Paknikar, K.M. (2012) Nanotoxicology and *in vitro* studies: the need of the hour. *Toxicol. Appl. Pharmacol.*, **258**, 151–165.

30 Verleysen, E., Van Doren, E., Waegeneers, N., De Temmerman, P.J., Abi Daoud Francisco, M., and Mast, J. (2015) TEM and SP-ICP-MS analysis of the release of silver nanoparticles from decoration of pastry. *J. Agric. Food Chem.*, **63**, 3570–3578.

31 Emamifar, A., Kadivar, M., Shahedi, M., and Solimanian-Zad, S. (2012) Effect of nanocomposite packaging containing Ag and ZnO on reducing pasteurization temperature of orange juice. *J. Food Process. Preserv.*, **36**, 104–112.

32 Cushen, M., Kerry, J., Morris, M., Cruz-Romero, M., and Cummins, E. (2014) Evaluation and simulation of silver and copper nanoparticle migration from polyethylene nanocomposites to food and

33 Echegoyen, Y. and Nerin, C. (2013) Nanoparticle release from nano-silver antimicrobial food containers. *Food Chem. Toxicol.*, **62**, 16–22.

34 Hauri, J.F. and Niece, B.K. (2011) Leaching of silver from silver-impregnated food storage containers. *J. Chem. Educ.*, **88**, 1407–1409.

35 Song, H., Li, B., Lin, Q.B., Wu, H.J., and Chen, Y. (2011) Migration of silver from nanosilver–polyethylene composite packaging into food simulants. *Food Addit. Contam. Part A*, **28**, 1758–1762.

36 von Goetz, N., Fabricius, L., Glaus, R., Weitbrecht, V., Gunther, D., and Hungerbuhler, K. (2013) Migration of silver from commercial plastic food containers and implications for consumer exposure assessment. *Food Addit. Contam. Part A*, **30**, 612–620.

37 Liu, J., Sonshine, D.A., Shervani, S., and Hurt, R.H. (2010) Controlled release of biologically active silver from nanosilver surfaces. *ACS Nano*, **4**, 6903–6913.

38 Huang, Y., Chen, S., Bing, X., Gao, C., Wang, T., and Yuan, B. (2011) Nanosilver migrated into food-simulating solutions from commercially available food fresh containers. *Packaging Technol. Sci.*, **24**, 291–297.

39 Marieb, E.N., Hoehn, K., Moussakova, L., and Lachaine, R. (2010) *Anatomie et Physiologie Humaines*, Pearson, pp. 1293.

40 Campbell, N. and Reece, J. (2007) *Biologie*, Pearson, pp. 1334.

41 Silbernagl, S. and Despopoulos, A. (2003) *Color Atlas of Physiology*, Thieme.

42 Deshpande, S.S. (2002) *Handbook of Food Toxicology*, Taylor & Francis.

43 Hoensch, H.P., Hutt, R., and Hartmann, F. (1979) Biotransformation of xenobiotics in human intestinal mucosa. *Environ. Health Perspect.*, **33**, 71–78.

44 Mózsik, G., Hänninen, O., and Jàvor, T. (2013) Gastrointestinal Defence Mechanisms: Satellite Symposium of the 28th International Congress of Physiological Sciences, Pécs, Hungary, 1980, Elsevier Science.

45 Minekus, M., Alminger, M., Alvito, P., Ballance, S., Bohn, T., Bourlieu, C., Carriere, F., Boutrou, R., Corredig, M., Dupont, D. *et al.* (2014) A standardised static *in vitro* digestion method suitable for food – an international consensus. *Food Funct.*, **5**, 1113–1124.

46 Hur, S.J., Lim, B.O., Decker, E.A., and McClements, D.J. (2011) *In vitro* human digestion models for food applications. *Food Chem.*, **125**, 1–12.

47 Wickham, M., Faulks, R., and Mills, C. (2009) *In vitro* digestion methods for assessing the effect of food structure on allergen breakdown. *Mol. Nutr. Food Res.*, **53**, 952–958.

48 Torres-Escribano, S., Denis, S., Blanquet-Diot, S., Calatayud, M., Barrios, L., Vélez, D., Alric, M., and Montoro, R. (2011) Comparison of a static and a dynamic *in vitro* model to estimate the bioaccessibility of As, Cd, Pb and Hg from food reference materials Fucus sp. (IAEA-140/TM) and *Lobster hepatopancreas* (TORT-2). *Sci. Total Environ.*, **409**, 604–611.

49 Alminger, M., Aura, A.M., Bohn, T., Dufour, C., El, S.N., Gomes, A., Karakaya, S., Martínez-Cuesta, M.C., McDougall, G.J., Requena, T. *et al.* (2014) *In vitro* models for studying secondary plant metabolite digestion and bioaccessibility. *Compr. Rev. Food Sci. Food Saf.*, **13**, 413–436.

50 Braakhuis, H., Kloet, S., Kezic, S., Kuper, F., Park, M.D.Z., Bellmann, S., van der Zande, M., Le Gac, S., Krystek, P., Peters, R.B. *et al.* (2015) Progress and future of *in vitro* models to study translocation of nanoparticles. *Arch. Toxicol.*, **89**, 1469–1495.

51 Newsome, R. (2014) 2013 IFT International Food Nanoscience Conference: Proceedings. *Compr. Rev. Food Sci. Food Saf.*, **13**, 190–228.

52 Peters, R., Kramer, E., Oomen, A.G., Rivera, Z.E.H., Oegema, G., Tromp, P.C., Fokksik, R., Rietveld, A., Marvin, H.J.P., Weigel, S. *et al.* (2012) Presence of nano-sized silica during *in vitro* digestion of foods containing silica as a food additive. *ACS Nano*, **6**, 2441–2451.

53 Walczak, A.P., Fokkink, R., Peters, R., Tromp, P., Herrera Rivera, Z.E., Rietjens, I.M.C.M., Hendriksen, P.J.M., and

Bouwmeester, H. (2013) Behaviour of silver nanoparticles and silver ions in an *in vitro* human gastrointestinal digestion model. *Nanotoxicology*, **7**, 1198–1210.

54 Asare-Addo, K., Conway, B.R., Larhrib, H., Levina, M., Rajabi-Siahboomi, A.R., Tetteh, J., Boateng, J., and Nokhodchi, A. (2013) The effect of pH and ionic strength of dissolution media on *in vitro* release of two model drugs of different solubilities from HPMC matrices. *Colloids Surf. B Biointerfaces*, **111**, 384–391.

55 Zhou, D., Ji, Z., Jiang, X., Dunphy, D.R., Brinker, J., and Keller, A.A. (2013) Influence of material properties on TiO(2) nanoparticle agglomeration. *PLoS One*, **8**, e81239.

56 Mwilu, S.K., El Badawy, A.M., Bradham, K., Nelson, C., Thomas, D., Scheckel, K.G., Tolaymat, T., Ma, L., and Rogers, K.R. (2013) Changes in silver nanoparticles exposed to human synthetic stomach fluid: effects of particle size and surface chemistry. *Sci. Total Environ.*, **447**, 90–98.

57 Rogers, K.R., Bradham, K., Tolaymat, T., Thomas, D.J., Hartmann, T., Ma, L., and Williams, A. (2012) Alterations in physical state of silver nanoparticles exposed to synthetic human stomach fluid. *Sci. Total Environ.*, **420**, 334–339.

58 Monopoli, M.P., Walczyk, D., Campbell, A., Elia, G., Lynch, I., Baldelli Bombelli, F., and Dawson, K.A. (2011) Physical–chemical aspects of protein corona: Relevance to *in vitro* and *in vivo* biological impacts of nanoparticles. *J. Am. Chem. Soc.*, **133**, 2525–2534.

59 Monopoli, M.P., Wan, S.H.A., Bombelli, F.B., Mahon, E., and Dawson, K.A. (2013) Comparisons of nanoparticle protein corona complexes isolated with different methods. *Nano Life*, **3**, 1343004.

60 Saptarshi, S., Duschl, A., and Lopata, A. (2013) Interaction of nanoparticles with proteins: relation to bio-reactivity of the nanoparticle. *J. Nanobiotechnology*, **11**, 1–12.

61 Akesson, A., Cardenas, M., Elia, G., Monopoli, M.P., and Dawson, K.A. (2012) The protein corona of dendrimers: PAMAM binds and activates complement proteins in human plasma in a generation dependent manner. *RSC Adv.*, **2**, 11245–11248.

62 Devineau, S., Boulard, Y., and Labarre, J. (2013) Protéines et nanoparticules, ça colle . . . ou pas. *Biofutur*, **347**, 34–38.

63 Tenzer, S., Docter, D., Rosfa, S., Wlodarski, A., Kuharev, J., Rekik, A., Knauer, S.K., Bantz, C., Nawroth, T., Bier, C. *et al.* (2011) Nanoparticle size is a critical physicochemical determinant of the human blood plasma corona: a comprehensive quantitative proteomic analysis. *ACS Nano*, **5**, 7155–7167.

64 Docter, D., Distler, U., Storck, W., Kuharev, J., Wünsch, D., Hahlbrock, A., Knauer, S.K., Tenzer, S., and Stauber, R.H. (2014) Quantitative profiling of the protein coronas that form around nanoparticles. *Nat. Protoc.*, **9**, 2030–2044.

65 Walczak, A.P., Kramer, E., Hendriksen, P.J.M., Helsdingen, R., van der Zande, M., Rietjens, I.M.C.M., and Bouwmeester, H. (2015) *In vitro* gastrointestinal digestion increases the translocation of polystyrene nanoparticles in an *in vitro* intestinal co-culture model. *Nanotoxicology*, **9**, 886–894.

66 Lichtenstein, D., Ebmeyer, J., Knappe, P., Juling, S., Böhmert, L., Selve, S., Niemann, B., Braeuning, A., Thunemann, A.F., and Lampen, A. (2015) Impact of food components during *in vitro* digestion of silver nanoparticles on cellular uptake and cytotoxicity in intestinal cells. *Biol. Chem.*, **396**, 1255–1264.

67 Powell, J.J., Faria, N., Thomas-McKay, E., and Pele, L.C. (2010) Origin and fate of dietary nanoparticles and microparticles in the gastrointestinal tract. *J. Autoimmun.*, **34**, 226–233.

68 Dudkiewicz, A., Tiede, K., Loeschner, K., Jensen, L.H.S., Jensen, E., Wierzbicki, R., Boxall, A.B.A., and Molhave, K. (2011) Characterization of nanomaterials in food by electron microscopy. *Trends Analyt. Chem.*, **30**, 28–43.

69 Linsinger, T., Roebben, G., Gilliland, D., Calzolai, L., Rossi, F., Gibson, P., and Klein, C. (2012) *Requirements on Measurements for the Implementation of the European Commission Definition of*

the Term 'Nanomaterial', JRC, Publications Office of the European Union, pp. 1–56.

70 Rhaman, M., Laurent, S., Tawil, N., Yahia, L.H., and Mahmoudi, M. (2013) *Protein-Nanoparticle Interactions*, Springer-Verlag Berlin Heidelberg, Heidelberg.

71 Fogh, J. and Trempe, G. (1975) New human tumor cell lines, in *Human Tumor Cells in Vitro*, (ed. J. Fogh), Springer, US, pp. 115–159.

72 Artursson, P. (1990) Epithelial transport of drugs in cell culture. I: a model for studying the passive diffusion of drugs over intestinal absorptive (Caco-2) cells. *J. Pharm. Sci.*, **79**, 476–482.

73 Lampen, A., Bader, A., Bestmann, T., Winkler, M., Witte, L., and Borlak, J.T. (1998) Catalytic activities, protein- and mRNA-expression of cytochrome P450 isoenzymes in intestinal cell lines. *Xenobiotica*, **28**, 429–441.

74 Sambuy, Y., De Angelis, I., Ranaldi, G., Scarino, M.L., Stammati, A., and Zucco, F. (2005) The Caco-2 cell line as a model of the intestinal barrier: influence of cell and culture-related factors on Caco-2 cell functional characteristics. *Cell Biol. Toxicol.*, **21**, 1–26.

75 Smetanova, L., Stetinova, V., Svoboda, Z., and Kvetina, J. (2011) Caco-2 cells, biopharmaceutics classification system (BCS) and biowaiver. *Acta Medica (Hradec Kralove)*, **54**, 3–8.

76 Delie, F. and Rubas, W. (1997) A human colonic cell line sharing similarities with enterocytes as a model to examine oral absorption: advantages and limitations of the Caco-2 model. *Crit. Rev. Ther. Drug*, **14**, 221–286.

77 Gagnon, M., Zihler Berner, A., Chervet, N., Chassard, C., and Lacroix, C. (2013) Comparison of the Caco-2, HT-29 and the mucus-secreting HT29-MTX intestinal cell models to investigate *Salmonella* adhesion and invasion. *J. Microbiol. Methods*, **94**, 274–279.

78 Lesuffleur, T., Barbat, A., Dussaulx, E., and Zweibaum, A. (1990) Growth adaptation to methotrexate of HT-29 human colon carcinoma cells is associated with their ability to differentiate into columnar absorptive and mucus-secreting cells. *Cancer Res.*, **50**, 6334–6343.

79 Hilgendorf, C., Spahn-Langguth, H., Regårdh, C.G., Lipka, E., Amidon, G.L., and Langguth, P. (2000) Caco-2 versus caco-2/HT29-MTX co-cultured cell lines: permeabilities via diffusion, inside- and outside-directed carrier-mediated transport. *J. Pharm. Sci.*, **89**, 63–75.

80 Nollevaux, G., Devillé, C., El Moualij, B., Zorzi, W., Deloyer, P., Schneider, Y.-J., Peulen, O., and Dandrifosse, G. (2006) Development of a serum-free co-culture of human intestinal epithelium cell-lines (Caco-2/HT29-5M21). *BMC Cell Biol.*, **7**, 1–11.

81 Walter, E., Janich, S., Roessler, B.J., Hilfinger, J.M., and Amidon, G.L. (1996) HT29-MTX/Caco-2 cocultures as an *in vitro* model for the intestinal epithelium: *in vitro–in vivo* correlation with permeability data from rats and humans. *J. Pharm. Sci.*, **85**, 1070–1076.

82 Mahler, G.J., Shuler, M.L., and Glahn, R.P. (2009) Characterization of Caco-2 and HT29-MTX cocultures in an *in vitro* digestion/cell culture model used to predict iron bioavailability. *J. Nutr. Biochem.*, **20**, 494–502.

83 Beduneau, A., Tempesta, C., Fimbel, S., Pellequer, Y., Jannin, V., Demarne, F., and Lamprecht, A. (2014) A tunable Caco-2/HT29-MTX co-culture model mimicking variable permeabilities of the human intestine obtained by an original seeding procedure. *Eur. J. Pharm. Biopharm.*, **87**, 290–298.

84 Behrens, I., Pena, A.I., Alonso, M.J., and Kissel, T. (2002) Comparative uptake studies of bioadhesive and non-bioadhesive nanoparticles in human intestinal cell lines and rats: the effect of mucus on particle adsorption and transport. *Pharm. Res.*, **19**, 1185–1193.

85 Gebert, A., Rothkotter, H.J., and Pabst, R. (1996) M cells in Peyer's patches of the intestine. *Int. Rev. Cytol.*, **167**, 91–159.

86 Kerneis, S., Bogdanova, A., Kraehenbuhl, J.P., and Pringault, E. (1997) Conversion by Peyer's patch lymphocytes of human enterocytes into M cells that transport bacteria. *Supramol. Sci.*, **277**, 949–952.

87 Gullberg, E., Leonard, M., Karlsson, J., Hopkins, A.M., Brayden, D., Baird, A.W., and Artursson, P. (2000) Expression of specific markers and particle transport in a new human intestinal M-cell model. *Biochem. Biophys. Res. Commun.*, **279**, 808–813.

88 des Rieux, A., Fievez, V., Theate, I., Mast, J., Preat, V., and Schneider, Y.J. (2007) An improved *in vitro* model of human intestinal follicle-associated epithelium to study nanoparticle transport by M cells. *Eur. J. Pharm. Sci.*, **30**, 380–391.

89 des Rieux, A., Ragnarsson, E.G., Gullberg, E., Preat, V., Schneider, Y.J., and Artursson, P. (2005) Transport of nanoparticles across an *in vitro* model of the human intestinal follicle associated epithelium. *Eur. J. Pharm. Sci.*, **25**, 455–465.

90 Bazes, A., Nollevaux, G., Coco, R., Joly, A., Sergent, T., and Schneider, Y.-J. (2011) Development of a triculture based system for improved benefit/risk assessment in pharmacology and human food. *BMC Proc.*, **5**, 1–3.

91 Mahler, G.J., Esch, M.B., Tako, E., Southard, T.L., Archer, S.D., Glahn, R.P., and Shuler, M.L. (2012) Oral exposure to polystyrene nanoparticles affects iron absorption. *Nat. Nanotechnol.*, **7**, 264–271.

92 Kure, I., Nishiumi, S., Nishitani, Y., Tanoue, T., Ishida, T., Mizuno, M., Fujita, T., Kutsumi, H., Arita, M., Azuma, T. *et al.* (2010) Lipoxin A(4) reduces lipopolysaccharide-induced inflammation in macrophages and intestinal epithelial cells through inhibition of nuclear factor-kappaB activation. *J. Pharmacol. Exp. Ther.*, **332**, 541–548.

93 Tanoue, T., Nishitani, Y., Kanazawa, K., Hashimoto, T., and Mizuno, M. (2008) *In vitro* model to estimate gut inflammation using co-cultured Caco-2 and RAW264.7 cells. *Biochem. Biophys. Res. Commun.*, **374**, 565–569.

94 Ishimoto, Y., Satsu, H., Totsuka, M., and Shimizu, M. (2011) IEX-1 suppresses apoptotic damage in human intestinal epithelial Caco-2 cells induced by co-culturing with macrophage-like THP-1 cells. *Biosci. Rep.*, **31**, 345–351.

95 Moyes, S.M., Morris, J.F., and Carr, K.E. (2010) Macrophages increase microparticle uptake by enterocyte-like Caco-2 cell monolayers. *J. Anat.*, **217**, 740–754.

96 Satsu, H., Ishimoto, Y., Nakano, T., Mochizuki, T., Iwanaga, T., and Shimizu, M. (2006) Induction by activated macrophage-like THP-1 cells of apoptotic and necrotic cell death in intestinal epithelial Caco-2 monolayers via tumor necrosis factor-alpha. *Exp. Cell Res.*, **312**, 3909–3919.

97 Leonard, F., Collnot, E.M., and Lehr, C.M. (2010) A three-dimensional coculture of enterocytes, monocytes and dendritic cells to model inflamed intestinal mucosa *in vitro*. *Mol. Pharm.*, **7**, 2103–2119.

98 Davis, M.E., Chen, Z., and Shin, D.M. (2008) Nanoparticle therapeutics: an emerging treatment modality for cancer. *Nat. Rev. Drug Discov.*, **7**, 771–782.

99 Ong, K.J., MacCormack, T.J., Clark, R.J., Ede, J.D., Ortega, V.A., Felix, L.C., Dang, M.K.M., Ma, G., Fenniri, H., Veinot, J.G.C. *et al.* (2014) Widespread nanoparticle-assay interference: implications for nanotoxicity testing. *PLoS One*, **9**, e90650.

100 Kroll, A., Pillukat, M.H., Hahn, D., and Schnekenburger, J. (2009) Current *in vitro* methods in nanoparticle risk assessment: limitations and challenges. *Eur. J. Pharm. Biopharm.*, **72**, 370–377.

101 Oh, S.J., Kim, H., Liu, Y., Han, H.K., Kwon, K., Chang, K.H., Park, K., Kim, Y., Shim, K., An, S.S. *et al.* (2014) Incompatibility of silver nanoparticles with lactate dehydrogenase leakage assay for cellular viability test is attributed to protein binding and reactive oxygen species generation. *Toxicol. Lett.*, **225**, 422–432.

102 Worle-Knirsch, J.M., Pulskamp, K., and Krug, H.F. (2006) Oops they did it again! Carbon nanotubes hoax scientists in viability assays. *Nano Lett.*, **6**, 1261–1268.

103 Teeguarden, J.G., Hinderliter, P.M., Orr, G., Thrall, B.D., and Pounds, J.G. (2007) Particokinetics *in vitro*: dosimetry considerations for *in vitro* nanoparticle

toxicity assessments. *Toxicol. Sci.*, **95**, 300–312.

104 Lison, D., Thomassen, L.C.J., Rabolli, V., Gonzalez, L., Napierska, D., Seo, J.W., Kirsch-Volders, M., Hoet, P., Kirschhock, C.E.A., and Martens, J.A. (2008) Nominal and effective dosimetry of silica nanoparticles in cytotoxicity assays. *Toxicol. Sci.*, **104**, 155–162.

105 Elsaesser, A. and Howard, C.V. (2012) Toxicology of nanoparticles. *Adv. Drug Deliv. Rev.*, **64**, 129–137.

106 Misra, S.K., Dybowska, A., Berhanu, D., Luoma, S.N., and Valsami-Jones, E. (2012) The complexity of nanoparticle dissolution and its importance in nanotoxicological studies. *Sci. Total Environ.*, **438**, 225–232.

107 van der Zande, M., Vandebriel, R.J., Van Doren, E., Kramer, E., Herrera Rivera, Z., Serrano-Rojero, C.S., Gremmer, E.R., Mast, J., Peters, R.J., Hollman, P.C. *et al.* (2012) Distribution, elimination, and toxicity of silver nanoparticles and silver ions in rats after 28-day oral exposure. *ACS Nano*, **6**, 7427–7442.

108 Kong, B., Seog, J.H., Graham, L.M., and Lee, S.B. (2011) Experimental considerations on the cytotoxicity of nanoparticles. *Nanomedicine (Lond.)*, **6**, 929–941.

109 Aueviriyavit, S., Phummiratch, D., and Maniratanachote, R. (2014) Mechanistic study on the biological effects of silver and gold nanoparticles in Caco-2 cells – induction of the Nrf2/HO-1 pathway by high concentrations of silver nanoparticles. *Toxicol. Lett.*, **224**, 73–83.

110 Böhmert, L., Niemann, B., Thünemann, A., and Lampen, A. (2012) Cytotoxicity of peptide-coated silver nanoparticles on the human intestinal cell line Caco-2. *Arch. Toxicol.*, **86**, 1107–1115.

111 Kaiser, J.P., Roesslein, M., Diener, L., and Wick, P. (2013) Human health risk of ingested nanoparticles that are added as multifunctional agents to paints: an *in vitro* study. *PLoS One*, **8**, e83215.

112 Böhmert, L., Niemann, B., Lichtenstein, D., Juling, S., and Lampen, A. (2015) Molecular mechanism of silver nanoparticles in human intestinal cells. *Nanotoxicology*, **9**, 852–860.

113 Gopinath, P., Gogoi, S.K., Sanpui, P., Paul, A., Chattopadhyay, A., and Ghosh, S.S. (2010) Signaling gene cascade in silver nanoparticle induced apoptosis. *Colloids Surf B Biointerfaces*, **77**, 240–245.

114 Bouwmeester, H., Poortman, J., Peters, R.J., Wijma, E., Kramer, E., Makama, S., Puspitaninganindita, K., Marvin, H.J., Peijnenburg, A.A., and Hendriksen, P.J. (2011) Characterization of translocation of silver nanoparticles and effects on whole-genome gene expression using an *in vitro* intestinal epithelium coculture model. *ACS Nano*, **5**, 4091–4103.

115 Martirosyan, A., Bazes, A., and Schneider, Y.J. (2014) *In vitro* toxicity assessment of silver nanoparticles in the presence of phenolic compounds–preventive agents against the harmful effect? *Nanotoxicology*, **8**, 573–582.

116 Hussain, S.M., Hess, K.L., Gearhart, J.M., Geiss, K.T., and Schlager, J.J. (2005) *In vitro* toxicity of nanoparticles in BRL 3A rat liver cells. *Toxicol. In Vitro*, **19**, 975–983.

117 Schrand, A.M., Braydich-Stolle, L.K., Schlager, J.J., Dai, L., and Hussain, S.M. (2008) Can silver nanoparticles be useful as potential biological labels? *Nanotechnology*, **19**, 235104.

118 Arora, S., Jain, J., Rajwade, J.M., and Paknikar, K.M. (2008) Cellular responses induced by silver nanoparticles: *in vitro* studies. *Toxicol. Lett.*, **179**, 93–100.

119 Sahu, S.C., Roy, S., Zheng, J., Yourick, J.J., and Sprando, R.L. (2014) Comparative genotoxicity of nanosilver in human liver HepG2 and colon Caco2 cells evaluated by fluorescent microscopy of cytochalasin B-blocked micronucleus formation. *J. Appl. Toxicol.*, **34**, 1200–1208.

120 Park, E.J., Yi, J., Kim, Y., Choi, K., and Park, K. (2010) Silver nanoparticles induce cytotoxicity by a Trojan-horse type mechanism. *Toxicol. In Vitro*, **24**, 872–878.

121 Ward, P.D., Tippin, T.K., and Thakker, D.R. (2000) Enhancing paracellular permeability by modulating epithelial tight junctions. *Pharm. Sci. Technol. Today*, **3**, 346–358.

122 Martirosyan, A., Polet, M., Bazes, A., Sergent, T., and Schneider, Y.-J. (2012) Food nanoparticles and intestinal inflammation: a real risk?.

123 Chen, L., Yokel, R.A., Hennig, B., and Toborek, M. (2008) Manufactured aluminum oxide nanoparticles decrease expression of tight junction proteins in brain vasculature. *J. Neuroimmune Pharm.*, **3**, 286–295.

124 Andras, I.E., Pu, H., Tian, J., Deli, M.A., Nath, A., Hennig, B., and Toborek, M. (2005) Signaling mechanisms of HIV-1 Tat-induced alterations of claudin-5 expression in brain endothelial cells. *J. Cereb. Blood Flow Metab.*, **25**, 1159–1170.

125 Persidsky, Y., Heilman, D., Haorah, J., Zelivyanskaya, M., Persidsky, R., Weber, G.A., Shimokawa, H., Kaibuchi, K., and Ikezu, T. (2006) Rho-mediated regulation of tight junctions during monocyte migration across the blood–brain barrier in HIV-1 encephalitis (HIVE). *Blood*, **107**, 4770–4780.

126 Lui, W.Y. and Lee, W.M. (2005) cAMP perturbs inter-Sertoli tight junction permeability barrier *in vitro* via its effect on proteasome-sensitive ubiquitination of occludin. *J. Cell. Physiol.*, **203**, 564–572.

127 Yang, Y., Estrada, E.Y., Thompson, J.F., Liu, W., and Rosenberg, G.A. (2007) Matrix metalloproteinase-mediated disruption of tight junction proteins in cerebral vessels is reversed by synthetic matrix metalloproteinase inhibitor in focal ischemia in rat. *J. Cereb. Blood Flow Metab.*, **27**, 697–709.

128 AshaRani, P.V., Low Kah Mun, G., Hande, M.P., and Valiyaveettil, S. (2009) Cytotoxicity and genotoxicity of silver nanoparticles in human cells. *ACS Nano*, **3**, 279–290.

129 Park, M.V., Neigh, A.M., Vermeulen, J.P., de la Fonteyne, L.J., Verharen, H.W., Briede, J.J., van Loveren, H., and de Jong, W.H. (2011) The effect of particle size on the cytotoxicity, inflammation, developmental toxicity and genotoxicity of silver nanoparticles. *Biomaterials*, **32**, 9810–9817.

19
Life Cycle of Nanoparticles in the Environment

Nanomaterials and Agriculture Application – Life Cycle through Their Direct Application and Indirect Application in Biosludge-Amended Soils

Jean-Yves Bottero,[1,2] Mark R. Wiesner,[3,4] Jérôme Labille,[1,2] Melanie Auffan,[1,2] Vladimir Vidal,[1,2] and Catherine Santaella[5]

[1]*CNRS-AMU-IRD-CdF, CEREGE UMR 7330, Avenue Louis Philibert, 13545 Aix en Provence cedex 04, France*
[2]*CNRS-Duke University, International Consortium for the Environmental Implications of Nanotechnology, Avenue Louis Philibert, 13545 Aix en Provence cedex 04, France*
[3]*Duke University, Civil and Environmental Engineering, Hudson Hall, West Campus, Durham, NC 90287, USA*
[4]*Duke University, Center for the Environmental Implications of NanoTechnology (CEINT), Hudson Hall, West Campus, Durham, NC, 90287, USA*
[5]*Aix-Marseille Université, CNRS, CEA Cadarache DSV/IBEB/SBVME, Laboratoire d'écologie microbienne de la rhizosphère et d'environnements extrêmes, UMR 7265, 13115 Saint-Paul-lez-Durance, France*

19.1
Introduction

The applications of nanotechnology in agriculture include fertilizers to increase plant growth and yield, pesticides for pest and disease management, and sensors for monitoring soil quality and plant health. These approaches aim to enhance the efficiency and sustainability of agricultural practices by requiring less input and generating less waste than conventional products and approaches. A recent publication [1] shows the nature and number of nanomaterials (NMs) used in agriculture (plant protection and fertilizer) in terms of publications and patents. Carbon nanomaterials (CN) are the most cited (emulsions and carbon nanotubes, CNTs) followed by TiO_2, Ag^0, ZnO, SiO_2, CuO, CeO_2, and sulfur-containing NMs. The majority of the patents in this area are described as plant protection products rather than fertilizers. Nanomaterials were declared as active constituents in 41% of the patents (mainly Ag^0 and sulfur as constituents) while 57% of the patents declared nanomaterials as additives to control the release of some other active compound. TiO_2 is most often conceived in these applications for its photocatalyst properties to destroy pollutants, pesticides, and fungi [2,3]. However, TiO_2 (rutile) coated with Al_2O_2 is used to limit the photodegradation of the active substances. The quantity of TiO_2 as described in the patents is quite high and 4000-fold higher than the probabilist estimations [4]. The second most frequently cited NM is Ag, used as fungicide. The flux of engineered nanomaterials (ENMs) as described in the patents is 254-fold higher than calculated in Ref 2. Oxides such as SiO_2 and Al_2O_3 or clays are more used or tested as carriers to control the release of active molecules as pesticides. ZnO NM is also used as insecticide The Cu-based NMs allow for a decreased level of Cu application compared with Cu^{2+} salt. Multiwall carbon nanotubes (MWCTs) and CNT applications are proposed to control water uptake and limit the dessication of plants [5]. A particularly novel application of NMs is for the detection of plant pathogens [6]. The metal NMs actively being studied for this application include TiO_2–Nafion composites, zirconium phosphate, Cd–tellurium quantum dots, ZrO—graphite, and SWCTs. Nanomaterials may also enter an agricultural setting as residual nanoparticles that make their way into wastewater treatment plant sludges that are then applied to fields. Application of large quantities of sewage sludge in agriculture is significant in many countries. For example, in France and US the dry production of sludge (also referred to as biosolids) is approximately 20–23 kg/year/person. Of this the land application in France may be as high as 60–80% of the total sludge produced [7,8]. These sludges may also contain virus, antibiotics, and bacterial pathogens such as *Salmonella* and other undesirable compounds or pathogens [9,10]. In 2009 the EPA published a Statistical Analysis Report [11] of pollutant concentrations in treated sewage sludges. The various analytes were reported as organics, classical inorganic anions, steroids and hormones, pharmaceuticals and metals such as Ag, Ti, Fe, Cu, and Ni. The average concentration of Ag reported was approximately 10 mg/kg of dry biosludge. One of the first reports of the presence of nanomaterials in sewage sludge was published in 2010 [12] and indicated the presence of nanoparticles of Ag_2S as observed using TEM. Micro- and nano-TiO_2 originating from wear of exterior building facades have also been shown to make their way into urban wastewaters [13] where concentrations can approach 600 µg/l. A recent work aimed at updating the nanomaterial consumer product inventory (CPI), which started in 2005 [14], shows a large increase of total products from 54 to 1814 from the year 2005 to 2014 and coming from 622 companies and 32 countries. This analysis shows a fast penetration of nano-enabled products in the consumer products that have the potential of making their way into wastewater that may in turn yield ENMs in biosolids applied to agricultural lands.

19.2
Transport and Bioaccumulation by Plants

Whether ENMs are intentionally applied as nano-enabled agrichemicals, or if they are applied unintentionally through atmospheric deposition or application of biosolids, uptake by plants will require an initial transport step to the vicinity of plant tissues followed by bioaccumulation and possible translocation in the plant. Processes for biouptake, accumulation, and translocation have been considered in a recent review paper [15a] as well as implications for use in nanoagrichemicals [15b]. The bioaccumulation and translocation pathways of TiO_2, ZnO, CeO_2, SWCNT, MWCNT, Ag^0, Cu, Al, Fe, and Ni entering soils involve interactions with root tissues as well as the local root microbiome. The data are not totally confirmed for many NMs except may be for CeO_2, Cu, C60, and so on. Uptake via absorption onto leaves and stems has not yet been demonstrated conclusively [16,17]. Recent work examining effects on soil and rhizosphere microbiomes produced by CeO_2, CNT, and TiO_2 NMs at concentrations of 1 mg/kg of soil (clay–loam soil) with canola plants [18] showed that CeO_2 decreased microbial enzymatic activity in planted soil and altered the bacterial community structure in roots. Citrate-functionalized NMs lowered the impact on microbial enzymatic activities but triggered variability near the plant root. Double-wall CNTs were also investigated with different surface functionalizations: pristine or "raw" (rXDWCNTs), carboxylate-functionalized (fXDWCNTs), and DWCNTs coated with natural organic matter (NOM) (cXDWCNTs). fXDWCNTs and NOM increased Fe and Al contents in shoots and decreased Al, Cu, and Ni concentration in roots. cXDWCNTs altered bacterial microbiota in the rhizosphere (Figure 19.1).

The addition of different types of TiO_2 NPs (cubic and elongated anatase and elongated rutile) resulted in upregulation of enzymes associated with the scavenging of hydrogen peroxide when the TiO_2 was added as cubic anatase or rutile. In leaves, increased ascorbate peroxidase and cubic anatase increased the production of guaiacol peroxidase.

There are relatively few studies on the internalization and translocation of NPs in plants. One study reported that CNTs could penetrate thick seed coats and

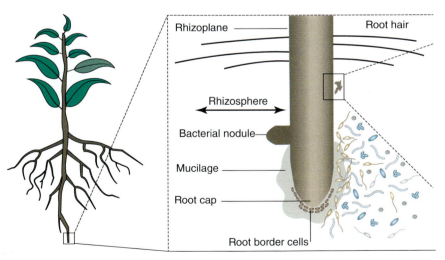

Figure 19.1 Rhizosphere with saprophytic and symbiotic bacteria and fungi associated. (With permission from Ref. [18].)

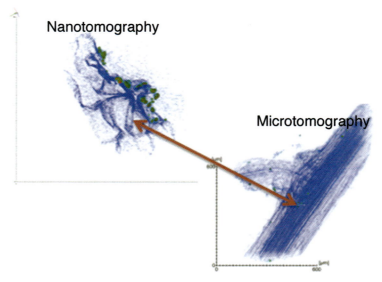

Figure 19.2 Ag_2S NPs accumulate on wheat roots at two scales: microtomography X (voxel = 0.6 μm) and nanotomography X (voxel = 50 nm). (With permission from V Vidal, CEREGE.)

increase water uptake inside tomato seeds [19]. The interactions between NPs and vascular plants have been reviewed [20]. This review cited studies on the bioaccumulation of SWCNT, MWCNT, CeO_2, CuO, Cu, Fe_3O_4, TiO_2, ZnO, and C70 in various plants. The majority of the NPs were observed to be associated with root cells, with C70 absorbed within the root cells and CuO NPs located in the root cells, intracellular space, cytoplasm, and nuclei of cortical cells, as well as in the xylem. Injection of Fe_3O_4 is associated with transfer in stems and leaves and Ag in the border cells, but some data do not show internalization and even no contacts, for example, in the case of MWCNT. The transformation of metallic and oxide NPs plays an important role in the bioaccumulation and transfer within the cells.

Micro- and nano-X-ray tomography is a powerful tool for visualizing the bioaccumulation and transfer of NPs within the cells. To date, no transfer within cells has been confirmed using these techniques. As an example shown in Figure 19.2 is the bioaccumulation of nano-Ag_2S on wheat roots.

Some studies have considered the internalization of TiO_2 and Ag^0 through leaves [21,22]. The data seem to confirm the internalization of TiO_2 and Ag^0 in the cuticle and penetration into the leaf tissue through the stomata.

19.3
Indirect Agricultural Application of NMs through Biowastes

As introduced earlier, biosolids coming from urban and industrial wastewater treatment plant (WWTP) represent an indirect agriculture application of NMs.

Table 19.1 Predicted exposure concentration in biosolids using the influent concentrations calculated using influent data from Ref. [29].

NPs	Biosolids concentration (g NPs/kg dry biosludges)
PVP–Ag 40 nm	2.87^{-2}
Transformation product	1.00
PVP–Ag 8 nm	2.87^{-2}
Transformation product	1.00
GA–Ag 25 nm	2.87^{-2}
Transformation product	1.00
GA–Ag 6 nm	2.87^{-2}
Transformation product	1.00
ZnO	1.5
Transformation product	592
TiO_2	247
Transformation product	247

In a recent report, Ref. [23] indicates that in the United States, 50% of the biosludges are used in agriculture, forestry, and parklands. In France, in 2014, approximately 40% of dry biosludges were applied in agriculture.[1] An OECD report in 2016 concerning the nanomaterials in waste streams [24,25] reported that dry biosolids contain on average 856 mg/kg of Ag, 4.5 g/kg of Ti, and 1.7 and 1.5 g/kg of Cu and Zn, respectively. It is unlikely that all of these metals can be traced to nanomaterials. However, a portion of Ag does appear to originate from nano-Ag^0 as nanoparticles of AgS were discovered within these sludges [26].

The concentrations of NPs susceptible to enter in a WWTP has been considered using probabilistic models [27] that derive estimates of the maximum expected concentrations from estimates of nanomaterial production taken to be 200, 47 700, and 11 700 tons/year for Ag, TiO_2, and ZnO, respectively.

Monte Carlo simulations [28] yield estimates of the concentration in g/kg of dry biosolids summarized in Table 19.1.

These estimates were obtained from bench-scale evaluations of nanoparticle removal and calculated values for the affinity of these NPs for the activated biosolids [30]. The ratio of NPs remaining in suspension to those in the settled sludge in primary and secondary settlers is referred to as the primary and secondary distribution coefficient. These distribution coefficients have been measured after 1 h of mixing followed by gravity separation in conditions simulating primary and secondary clarifiers where residence times are 2 and 5.5 h (Table 19.2) [31].

---

1) durable.gouv.fr/lessentiel/ar/272/0/diversite-sols-france.html.

Table 19.2 Primary and secondary distribution coefficients for various NPs in a wastewater treatment plant in the primary and secondary clarifiers.

NPs	Primary γ (l/mg)	Secondary γ (l/mg)
PVP–Ag 40 nm	0.0036	0.0040
PVP-Ag 8 nm	0.0026	0.0035
GA–Ag 25 nm	0.0014	0.00 064
GA–Ag 6 nm		
ZnO	0.0011	0.00 058
TiO$_2$	0.0088	0.0091
Pristine CeO$_2$ 8 nm	0.003	0.006
Citrate-coated CeO$_2$ 10 nm	0.001	0.003

Source: Modified from Refs [30,31].

Table 19.3 gives the values of affinity coefficients for NPs and activated sludge [30]. These data are in agreement with trends of removal reported by Hendren et al. (2013) [32] in batch studies, where the authors found that PVP-coated NPs were removed to a greater extent compared with GA–Ag NPs. Other researchers have reported high levels of removal of Ag NPs (99%) after 2 h of mixing without any impact of size [33–35].

Recently, a global approach of the life cycle releases of NMs [36] evaluated the release from WWTP to biosludges between 75 and 97%. The release of NMs to

Table 19.3 Percent removal dictated by γ and simulated α for the NPs studied.

NPs	C (mg/l)	Final removal	α calculated
GA–Ag 25 nm	10	71	0.0017
	50	73	0.0020
6 nm	10	71	0.0003
	50	65	0.0003
PVP–Ag 40 nm	10	95	0.0120
	50	94	0.0119
8 nm	10	88	0.0020
	50	90	0.0022
Pristine CeO$_2$ 8 nm	10	91	0.0064
	50	99	0.0088
Citrate-coated CeO$_2$ 10 nm	10	86	0.0023
	50	83	0.0018
TiO$_2$ 20 nm	10	95	0.0080
	50	99	0.0094
ZnO 30 nm	10	91	0.0088
	50	94	0.0090

Source: Modified from Ref [30].

the soils from different applications except agriculture represented approximately 1/3 of the global flux (soil, water, landfill) for TiO_2, Ag, ZnO, Fe and Fe oxides, Al_2O_3, and CeO_2. The study concerning TiO_2 has been made at full scale of a WWTP [37].

In conclusion, the concentrations of NPs associated with the biosludges used for agriculture application are important and can be added to the flux coming from other applications as textiles, coatings, paints, pigments, and cosmetics. Except for TiO_2 or CNT, the transformation of the other NPs within the WWTP can however greatly modify their reactivity.

19.4
Transformations of NPs in Soils after Application

CNTs and TiO_2 are insoluble for all practical purposes. However, many oxide and metal NPs have the potential to transform (e.g., redox state) or dissolve. ZnO and CuO NPs are very soluble. CeO_2 can be reduced and dissolved while Ag^0 may be both oxidized and dissolved. Very few papers have evaluated the transformation and resulting impact of these NPs.

19.4.1
Direct Application

Dissolution and transformation are very important pathways to be considered in evaluating the efficacy or ecotoxicity of NPs such as CuO, ZnO, Ag^0, and CeO_2. The importance of the solubility of ZnO in terms of toxicity for microalgae has been demonstrated [38]. The application of $CuSO_4$ to bamboo plants results in measurable concentrations of Cu in the leaves, stems, and roots. EXAFS measurements at the Cu K-edge show a total transformation of the initial cation in Cu(II)–malate, Cu(II)–histidine, Cu(I)–cyctseine, and Cu(I)S-inorganic after 70 days of contact in hydroponic growth condition and in true soil [39,40]. Following the addition of CeO_2 NPs to four plants – cucumber (*Cucumis sativus*), tomato (*Lycopersicum esculentum*), alfalfa (*Medicago sativa*), and corn (*Zea mays*), the XANES spectra of Ce LIII-edge do not show variation of the oxidation state, which is +4, irrespective of the presence of CeO_2 onto roots of the four plants [41]. Using a silver ion selective electrode in three different soils (two agricultural soils and one forest soil) leads to the release of Ag+ through dissolution of Ag NP after desorption. The pH and organic carbon content play an important role in the control of the speciation and mobility. Ag NPs are dissolved under normal soil conditions [42]. The bioactivity of Ag NPs and Ag+ is regulated by their interactions with Ca++ ions and dissolved organic carbon present in the soils. These cations significantly decrease the toxicity of both species against bacterial populations present in the microcosms close to the plants [43]. The soluble speciation Ag+ seems to be the most toxic. When Ag NPs are coated with polyvinylpyrolidone (hydrophilic) or oleic acid (amphiphilic) the dissolution in soils is reduced comparing with uncoated Ag NPs in solution. The toxicity to *Eisenia fetida* earthworms under an environmentally relevant exposure scenario in soil is very weak compared with $AgNO_3$ at the same concentration as Ag NPs used [44]. Nevertheless Ag NP solubility is well correlated with particle size as predicted by the modified Kelvin equation. Solubility of Ag NPs was not affected by the synthesis method and coating as much as by their size as shown in Ref. [45]. So the dissolution kinetics of Ag NPs are not affected by the coating but only by the size. The important parameter that could affect the reactivity of Ag NPs on a long term is the sulfidation that implies the precipitation of Ag_2S with a solubility constant around 10–50. The extent of sulfidation is affected strongly by the monodispersity of the Ag NPs that are more sulfidized than the polydisperse ones [46]. The initial properties of Ag NPs are important in assessing the properties for agriculture application and subsequently the environmental impact [47]. Concerning NPs as ZnO or CeO_2 [48], the use of EXAFS and XANES shows that ZnO is dissolved and Zn cations are associated with organic molecules similar to citrate ones with a translocation in soybean pods. A small part of Ce(IV) is reduced to Ce(III). The reduction of CeO_2 in contact with soy roots is not observed using the same techniques. Also, Zn is not in the ZnO form but associated with O similarly as in Zn–citrate or Zn–acetate. The CeO_2 NPs have been observed to be present in cucumber, alfalfa, tomato, and corn roots without reduction of Ce(IV) to Ce(III) [49,50]. The transformation of these NPs is evident as shown by some works and also organic molecules present in soils, bacteria cells, and the organs of plants play an important role for the transformations. Another parameter that is important is the sulfidation of Ag and Zn at low concentrations. The kinetic of sulfidation should be studied in soils in order to evaluate the lifetime of the initial NPs and the possible toxic effects on the plants and bacteria.

19.4.2
Indirect Applications from Biosludges

The presence of metals and NPs in soil amended by biosolids in agricultural sites has been evaluated, thanks to the analysis of long-term amended soils in Texas [51]. The amended soils showed higher concentration of Ag, Cu, and Zn, for example, in the upper layer of biosolid-amended soils. The data showed a low vertical mobility of these metals. Nanoparticles of TiO_2 were also detected in soil with a ratio of 0.07–0.2% of the total Ti. It was one of the first studies that seem to detect NPs in soils after application of biosludges. NPs as ZnO, CeO_2, and Ag present in biowastes coming from WWTP is important in terms of possible bioavailability and possible translocation within the roots, shoots, and so on. The majority of the studies concern Ag NPs. These papers show that Ag^0 NPs are oxidized and sulfidized in Ag_2S. The transformation to Ag_2S occurred in the nonaerated WWTP tank in less than 2 h [52]. In another work, the sulfidation kinetic of coating Ag NPs with PVP and citrate of 10 and 100 nm size, respectively, depends strongly on their size. The structure (polycrystallinity) of the

Ag NP may influence the kinetics of the sulfidation. Also, depending on the levels of sulfide availability and residence times in the sewer and the WWTP, Ag NP will be sulfidized to a variable degree. The stability of the Ag_2S NPs depends on the proportion between crystalline–amorphous structure under oxic conditions and further transformations such as crystallization during the digestion process. This last part of the WWTP condition, the speciation of Ag is applied for agriculture because nearly 100% of Ag is present in the biosludges [53]. Nevertheless, the amendment at environmentally relevant concentrations of bioslurry containing Ag NPs and $AgNO_3$ to terrestrial mesocosms planted with five plant species (*Carex lurida, Juncus effusus, Lobelia cardinalis, Microstegium vimineum*, and *Panicum virgatum*) showed important difference between the speciation of Ag after 30 days. The effect of the Ag NP treatment was as large as or larger than that of the treatment with $AgNO_3$ added at a fourfold higher concentration particularly in the N_2O fluxes, changes in microbial community composition, biomass, and extracellular enzyme activity, as well as effects on aboveground plant biomass [54]. Concerning ZnO (Table 19.1), the transformation of ZnO in Zn is quasi-total and for CeO_2 NPs coated and pristine the reductive dissolution and transformation in $CeHPO_4$ is around 11 and 33%, respectively, after 1 day of mixing [55].

The last phenomenon is the study concerning the transfer to the groundwater. The first study concerning the transfer of NPs has been made with C60 and a porous sandy soil. The distance of transfer was evaluated at maximum 10 m [56]. Another study assessed the transfer of TiO_2 through the same sandy soil [57]. Simultaneous particle aggregation and deposition nanoscale particle dimensions may favor aggregation kinetics, thus altering the transport and retention of these materials in saturated porous media. When surface chemistry favors nanoparticle–nanoparticle attachment (α_{pp}) over nanoparticle–collector attachment (α_{pc}), the rate of particle aggregation within pores may be comparable to that of deposition at ratios of collector to nanoparticle surface areas as high as 40. Aggregation of NPs in the porous media enhances NP deposition, however, aggregates that are not removed will sample a smaller portion of the available pore network within the column due to size exclusion. In a sandy soil the effect of tannic acid (TA) on the transfer of TiO_2 shows that TiO_2 nanoparticles and sand display very different affinities for the TA molecules, with preferential adsorption of TA onto sand rather than onto TiO_2 nanoparticles (Figure 19.3) [57] (from J. Labille personal communications and unpublished data).

These interactions strongly modify the respective surface properties and alter the fate of the nanoparticles in the porous medium. Depending on pH and ionic strength (I), TA may accelerate or delay the transfer of the nanoparticles. At pH $\approx \text{IEP}_{TiO2}$ or $I > \text{CCC}_{TiO2}$, bare TiO_2 nanoparticles self-aggregated and exhibited decreased breakthrough, the coated nanoparticles remained well dispersed in water and were more mobile in the medium. At pH $> \text{IEP}_{TiO2}$ and $I < \text{CCC}_{TiO2}$, the bare nanoparticles were well dispersed and mobile, the coated nanoparticles underwent new attractive interactions with the sand via TA bridging and deposited more heavily (Figure 19.4). The major difficulty of such studies is to evaluate

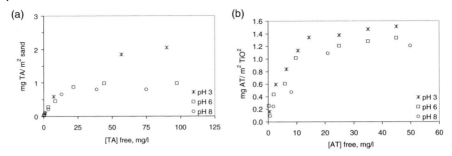

Figure 19.3 Adsorption isotherms of the TA on the sand (a) and TiO_2 NPs (b). Surface area of sand and TiO_2 NPs per batch = 3 and 2.5 m², respectively.

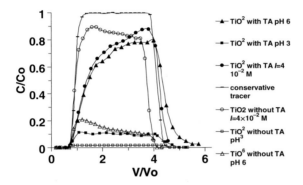

Figure 19.4 TA-induced facilitated TiO_2 NPs transfer in sand column.

the transfer in true soil and analyzing the components (organics, particulates as clays, oxides, carbonate, and so on) on their role in the transfer or blockage. This has been recently studied [58] and the authors show that the distribution of coated ZnO in sandy soil and loam sandy soil bounded to carbonate, Fe–Mn oxides, organic matter, or dissolved species is very similar, and it is very different for uncoated ZnO that shows a larger quantity of ZnO bound to carbonate. They conclude that the transfer is very limited due to the affinity for soil materials and at short time the dissolution of ZnO is very low.

19.5
Conclusion

The life cycle of ENMs used directly or indirectly for agriculture applications has not been studied extensively. Few studies to date do not consider the role of NP transformations and transport. The majority of the publications concern the effects on the plants and few the possible transformations and transfer. The need of studying the life cycle in relation to the efficiency of application of such materials as fertilizer, pesticides, and so on. is very important and certainly terrestrial mesocosms using real different soils will be very efficient. Studies using conditions that replicate soil conditions and plants to be encountered in agricultural applications are essential to understanding both the efficacy and the effects of nanomaterials in agricultural applications.

Acknowledgments

This chapter is based partly on works supported by the French ANR (Agence Nationale de la Recherche) through projects: Labex SERENADE (ANR-11-LABX-0064) funded by the Government through "'Investissement d'avenir'" A*Midex project (No. ANR-11-IDEX-0001-02) and GDRI I-CEINT (CNRS-DUKE Univ).

References

1. Gogos, A., Knauer, K., and Bucheli, T. D. (2012) Nanomaterials in plant protection and fertilization: current state, foreseen applications, and research priorities. *J. Agric. Food Chem.*, **60**, 9781–9792.
2. Thomas, J., Kumar, K., and Praveen, C. (2011) Synthesis of Ag doped nano TiO2 as efficient solar photocatalyst for the degradation of endosulfan. *Adv. Sci. Lett.*, **4**, 108–114.
3. Baruah, S. and Dutta, J. (2009) Nanotechnology applications in pollution sensing and degradation in agriculture: a review. *Environ. Chem. Lett.*, **7**, 191–204.
4. Gottschalk, F., Sonderer, T., Scholz, R.W., and Nowack, B. (2009) Modeled environmental concentrations of engineered nanomaterials (TiO2, ZnO, Ag, CNT, fullerenes) for different regions. *Environ. Sci. Technol.*, **43**, 9216–9222.
5. Kim, H., Kang, H., Chu, G., and Byun, H. (2008) Antifungal effectiveness of nanosilver colloid against rose powdery mildew in greenhouses, in *Nanocomposites and Nanoporous Materials VIII*, vol. **135** (ed. C.K. Rhee), Trans Tech Publications Ltd, Stafa, Zurich, pp. 15–18.
6. Servin, A., Elmer, W., Mukherjee, A., De la Torre-Roche, R., Hamdi, H., White, J. C, Bindraban, P., and Dimkpa, C. (2015) A review of the use of engineered nanomaterials to suppress plant disease and enhance crop yield. *J. Nanopart. Res.*, **17**, 92.
7. Peccia, J. and Westerhoff, P. (2015) We should expect more out of our sewage sludge. *Environ. Sci. Technol.*, **49**, 2015.
8. Tripathi, S., Sonkar, S. K., and Sarkar, S. (2011) Growth stimulation of gram (Cicer arietinum) plant by water soluble carbon nanotubes. *Nanoscale*, **3**, 1176–1181.
9. Khot, L. R., Sankaran, S., Maja, J. M., Ehsani, R., and Schuster, E. W. (2012) Applications of nanomaterials in agricultural production and crop protection: a review. *Crop. Prot.*, **35**, 64–70.
10. Peccia, J. and Westerhoff, P. (2015) We should expect more out of our sewage sludge. *Environ. Sci. Technol.*, **49**, 2015.
11. USEPA (2009) Targeted National Sewage Sludge Survey Sampling and Analysis. Technical report *EPA-822-R-08-016*. Available at http://water.epa.gov/scitech/wastetech/biosolids/upload/2009_01_15_biosolids_tnss-tech.pdf.
12. Kim, B., Park, C.-S., Murayama, M., and Hochella, M. Jr. (2010) Discovery and characterization of silver sulfide nanoparticles in final sewage sludge products. *Environ. Sci. Technol.*, **44**, 7509–7514.
13. Kaegi, R., Ulrich, A., Sinnet, B., Vonbank, R., Wichser, A., Zuleeg, S., Simmler, H.,

Brunner, S., Vonmont, H., Burkhardt, M., and Boller, M. (2008) Synthetic TiO$_2$ nanoparticle emission from exterior facades into the aquatic environment. *Environ. Pollut.*, **156**, 233–239.

14 Vance, M. E., Kuiken, T., Vejerano, E. P., McGinnis, S. P., Hochella, M. F. Jr., Rejeski, D., and Hull, M. S. (2015) Nanotechnology in the real world: redeveloping the nanomaterial consumer products inventory. *Beilstein J. Nanotechnol.*, **6**, 1769–1780.

15 (a) Schwab, F., Zhai, G., Kern, M., Turner, A., Schnoor, J. L., and Wiesner, M. R. (2015) Barriers, pathways and processes for uptake, translocation and accumulation of nanomaterials in plants – Critical review. *Nanotoxicology*, **10**, 257–278. (b) Servin, A., Elmer, W., Mukherjee, A., De la Torre-Roche, R., Hamdi, H., White, J. C., Bindraban, P., and Dimkpa, C. (2015) A review of the use of engineered nanomaterials to suppress plant disease and enhance crop yield. *J. Nanopart. Res.*, **17**, 92.

16 Rico, C. M., Majumdar, S., Duarte-Gardea, M., Peralta-Videa, J. R., and Gardea-Torresdey, J. L. (2011) Interaction of nanoparticles with edible plants and their possible implications in the food chain. *J. Agric. Food Chem.*, **59**, 3485–3498.

17 Khot, L. R., Sankaran, S., Maja, J. M., Ehsani, R., and Schuster, E. W. (2012) Applications of nanomaterials in agricultural production and crop protection: a review. *Crop Prot.*, **35**, 64–70.

18 Hamidat, M. (2015) Relation entre l'éco-conception des nanomatériaux et leur impact sur l'environnement. PhD thesis, Aix-Marseille University.

19 Khodakovskaya, M., Dervishi, E., Mahmood, M., Xu, Y., Li, Z., Watanabe, F., and Biris, A. S. (2009) Carbon nanotubes are able to penetrate plant seed coat and dramatically affect seed germination and plant growth. *ACS Nano*, **3** (10), 3221–3227.

20 Miralles, P., Church, T. L., and Harris, A. T. (2012) Toxicity, uptake, and translocation of engineered nanomaterials in vascular plants. *Environ. Sci. Technol.*, **46**, 9224–9239.

21 Larue, C., Castillo-Michel, H., Sobanska, S., Trcera, N., Sorieul, S., Cécillon, L., Ouerdane, L., Legros, S., and Sarret, G. (2014) Fate of pristine TiO$_2$ nanoparticles and aged paint-containing TiO$_2$ nanoparticles in lettuce crop after foliar exposure. *J. Hazard. Mater.*, **273**, 17–26.

22 Larue, C., Castillo-Michel, H., Sobanska, S., Cécillon, L., Bureau, S., Barthès, V., Ouerdane, L., Carrière, M., and Sarret, G. (2014) Foliar exposure of the crop *Lactuca sativa* to silver nanoparticles: evidence for internalization and changes in Ag speciation. *J. Hazard. Mater.*, **264**, 98–106.

23 Center for Sustainable Systems (2015) U.S. Wastewater Treatment Factsheet. Publ. No. CSS04-14, University of Michigan.

24 OECD (2016) *Nanomaterials in Waste Streams: Current Knowledge on Risks and Impacts*, OECD Publishing, Paris.

25 US EPA (2009 Targeted National Sewage Sludge Survey Sampling and Analysis. Technical report *EPA-822-R-08-016*. http://www.epa.gov/waterscience/biosolids/tnsss-tech.pdf.

26 Kim, B., Park, C. S., Murayama, M., and Hochella, M. F. Jr. (2010) Discovery and characterization of silver sulfide nanoparticles in final sewage sludge products. *Environ. Sci. Technol.*, **44** (19), 7509–7514.

27 (a) Keller, A. A., McFerran, S., Lazareva, A., and Suh, S. (2013) Global life cycle releases of engineered nanomaterials. *J. Nanopart. Res.*, **15**, 1692–1709. (b) Keller, A. A. and Lazareva, A. (2014) Predicted releases of engineered nanomaterials: from global to regional to local. *Environ. Sci. Technol. Lett.*, **1**, 65e70.

28 Barton, L. E., Auffan, M., Durenkamp, M., McGrath, S., Bottero, J.-Y., and Wiesner, M. R. (2015) Monte Carlo simulations of the transformation and removal of Ag, TiO$_2$, and ZnO nanoparticles in wastewater treatment and land application of biosolids. *Sci. Total Environ.*, **511**, 535–543.

29 Keller, A. A., McFerran, S., Lazareva, A., and Suh, S. (2013) Global life cycle releases of engineered nanomaterials. *J. Nanopart. Res.*, **15**, 1692–1709.

30 Barton, L. E., Therezien, M., Auffan, M., Bottero, J.-Y., and Wiesner, M. R. (2014)

Theory and methodology for determining nanoparticle affinity for heteroaggregation in environmental matrices using batch measurements. *Environ. Eng. Sci.*, **31** (7), 421–427.

31 Barton, L. E., Auffan, M., Olivi, L., Bottero, J.-Y., and Wiesner, M. R. (2015) Heteroaggregation, transformation and fate of CeO_2 nanoparticles in wastewater treatment. *Environ. Pollut.*, **203**, 122–129.

32 Hendren, C. O., Badireddy, A. R., Casman, E., and Wiesner, M. R. (2013) Modeling nanomaterial fate in wastewater treatment: Monte Carlo simulation of silver nanoparticles (nano-Ag). *Sci. Total Environ.*, **449**, 418.

33 Kiser, M. A., Ryu, H., Jang, H., Hristovski, K., and Westerhoff, P. (2010) Biosorption of nanoparticles to heterotrophic wastewater biomass. *Water Res.*, **44**, 4105.

34 Kaegi, R., Voegelin, A., Ort, C., Sinnet, B., Thalmann, B., Krismer, J., Hagendorfer, H., Elumelu, M., and Mueller, E. (2013) Fate and transformation of silver nanoparticles in urban wastewater systems. *Water Res.*, **47**, 3866.

35 Kaegi, R., Voegelin, A., Sinnet, B., Zuleeg, S., Hagendorfer, H., Burkhardt, M., and Siegrist, H. (2011) Behavior of metallic silver nanoparticles in a pilot wastewater treatment plant. *Environ. Sci. Technol.*, **45**, 3902.

36 Keller, A. A., McFerran, S., Lazareva, A., and Suh, S. (2013) Global life cycle releases of engineered nanomaterials. *J. Nanopart. Res.*, **15**, 1692.

37 Westerhoff, P., Song, G., Hristovski, K., and Kiser, M. A. (2011) Occurrence and removal of titanium at full scale wastewater treatment plants:implications for TiO_2 nanomaterials. *J. Environ. Monit.*, **13**, 1195.

38 Franklin, N. M., Rogers, N. J., Apte, S. C., Batley, G. E., Gadd, G. E., and Casey, P. S. (2007) Comparative toxicity of nanoparticulate ZnO, bulk ZnO, and $ZnCl_2$ to a freshwater microalgae (*Pseudokirchneriella subcapitata*): the importance of particle solubility. *Environ. Sci. Technol.*, **41**, 8484–8490.

39 Collin, B., Doelsch, E., Keller, C., Cazevieille, P., Tella, M., Chaurand, P., Panfili, F., Hazemann, J.-L., and Meunier, J.-D. (2014) Evidence of sulfur-bound reduced copper in bamboo exposed to high silicon and copper concentrations. *Environ. Pollut.*, **187**, 22–30.

40 Collin, B., Doelsch, E., Keller, C., Panfili, F., and Meunier, J.-D. (2012) Distribution and variability of silicon, copper and zinc in different bamboo species. *Plant Soil*, **351**, 377–387.

41 Lopez-Moreno, M., de la Rosa, G., Hernandez-Viezcas, J. A., Peralta-Videa, J. R., and Gardea-Torresdey, J. L. (2010) X-ray absorption spectroscopy (XAS) corroboration of the uptake and storage of CeO_2 nanoparticles and assessment of their differential toxicity in four edible plant species. *J. Agric. Food Chem.*, **58**, 3689–3693.

42 Benoit, R., Wilkinson, K. J., and Sauvé, S. (2013) Partitioning of silver and chemical speciation of free Ag in soils amended with nanoparticles. *Chem. Cent. J.*, **7**, 75.

43 Calder, A. J., Dimkpa, C. O., McLean, J. E., Britt, D. W., Johnson, W., and Anderson, A. J. (2012) Soil components mitigate the antimicrobial effects of silver nanoparticles towards a beneficial soil bacterium, *Pseudomonas chlororaphis* O6. *Sci. Total Environ.*, **429**, 215–222.

44 Shoults-Wilson, W. A., Reinsch, B., Tsyusko, O. V., Bertsch, P. M., Lowry, G. V., and Unrine, J. M. (2011) Effect of silver nanoparticle surface coating on bioaccumulation and reproductive toxicity in earthworms (*Eisenia fetida*) *Nanotoxicology*, **5** (3), 432–444.

45 Ma, R., Levard, C., Marinakos, S.M., Cheng, Y., Liu, J., Michel, F.M., Brown, G.E. Jr., and Lowry, G.V. (2012) Size-controlled dissolution of organic-coated silver nanoparticles. *Environ. Sci. Technol.*, **46**, 752–759.

46 Reinsch, B.C., Levard, C., Li, Z., Ma, R., Wise, A., Gregory, K.B., Brown, G.E. Jr., and Lowry, G.V. (2012) Sulfidation of silver nanoparticles decreases *Escherichia coli* growth inhibition. *Environ. Sci. Technol.*, **46**, 6992–7000.

47 Levard, C., Matt Hotze, E., Lowry, G. V., and Brown, G. E. Jr. (2012) Environmental transformations of silver nanoparticles: impact on stability and toxicity. *Environ. Sci. Technol.*, **46**, 6900–6914.

48 Hernandez-Viezcas, J.A., Castillo-Michel, H., Andrews, J.C., Cotte, M., Rico, C., Peralta-Videa, J.R., Ge, Y., Priester, J.H., Holden, P.A., and Gardea-Torresdey, J.L. (2013) In situ synchrotron X-ray fluorescence mapping and speciation of CeO_2 and ZnO nanoparticles in soil cultivated soybean (*Glycine max*) *ACS Nano*, **7** (2), 1415–1423.

49 Lo Pez-Moreno, M.L., de la Rosa, G., Hernandez-Viezcas, J.A., Castillo-Michel, H., Botez, C.E., Peralta-Videa, J.R., and Gardea-Torresdey, J.L. (2010) Evidence of the differential biotransformation and genotoxicity of ZnO and CeO_2 Nanoparticles on soybean (*Glycine max*) plants. *Environ. Sci. Technol.*, **44**, 7315–7320.

50 Peralta-Videa, J.R. and Gardea-Torresdey, J.L. (2010) X-ray absorption spectroscopy (XAS) corroboration of the uptake and storage of CeO_2 nanoparticles and assessment of their differential toxicity in four edible viezcas. *Agric. Food Chem.*, **58**, 3689–3693.

51 Yang, Y., Wang, Y., Westerhoff, P., Hristovski, K., Jin, V.L., Johnson, M.-V.V., and Arnold, J.G. (2014) Metal and nanoparticle occurrence in biosolid-amended soils. *Sci. Total Environ.*, **485–486**, 441–449.

52 Kaegi, R., Voegelin, A., Sinnet, B., Zuleeg, S., Hagendorfer, H., Burkhardt, M., and Siegrist, H. (2011) Behavior of metallic silver nanoparticles in a pilot wastewater treatment plant. *Environ. Sci. Technol.*, **45**, 3902–3908.

53 Kaegi, R., Voegelin, A., Ort, C., Sinnet, B., Thalmann, B., Krismer, J., Hagendorfer, H., Elumelu, M., and Mueller, E. (2013) Fate and transformation of silver nanoparticles in urban wastewater systems. *Water Res.*, **47**, 3866–3877.

54 Colman, B.P., Arnaout, C.L., Anciaux, S., Gunsch, C.K., Hochella, M.F. Jr., Kim, B., Lowry, G.V., McGill, B.M., Reinsch, B.C., Richardson, C.J., Unrine, J.M., Wright, J.P., Yin, L., and Bernhardt, E.S. (2013) Low concentrations of silver nanoparticles in biosolids cause adverse ecosystem responses under realistic field scenario. *PLoS One*, **8** (2), e57189.

55 Barton, L.E., Auffan, M., Bertrand, M., Barakat, M., Santaella, C., Masion, A., Borschneck, D., Olivi, L., Roche, N., Wiesner, M.R., and Bottero, J.-Y. (2014) Transformation of pristine and citrate-functionalized CeO_2 nanoparticles in a laboratory-scale activated sludge reactor. *Environ. Sci. Technol.*, **48** 7289–7296.

56 Lecoanet, H.F., Bottero, J.Y., and Wiesner, M.R. (2004) Laboratory assessment of the mobility of nanomaterials in porous media. *Environ. Sci. Technol.*, **38** (19), 5164–5169

57 Solovitch, N., Labille, J., Rose, J., Chaurand, P., Borschneck, D., Wiesner, M.R., and Bottero, J.-Y. (2010) Concurrent aggregation and deposition of TiO_2 nanoparticles in a sandy porous media. *Environ. Sci. Technol.*, **44**, 4897–4902.

58 Zhao, L.J., Peralta-Videa, J.R., Hernandez-Viezcas, J.A., Hong, J., and Gardea-Torresdey, J.L. (2012) Transport and retention behavior of ZnO nanoparticles in two natural soils: effect of surface coating and soil composition. *J. Nano Res.*, **17**, 229–242.

Part Five
Governance of Nanotechnology and Societal Dimensions

20
The Politics of Governance: Nanotechnology and the Transformations of Science Policy

Brice Laurent

PSL Research University, MINES ParisTech, CSI, i3 UMR CNRS, 60 Bd St Michel, 75006 Paris, France

20.1
An Issue of Governance

Nanotechnology has become a topic of research of its own within the social sciences. Numerous publications have appeared, raising issues related to the ethical implications of research in nanomedicine (particularly if connected to "human enhancements"), to the appropriate ways of engaging the public of nanotechnology, and to the range of roles to be allocated to social scientists in commenting on, and possibly contributing to the definition of nanotechnology programs [1,2]. Such an interest is directly connected to the initial integration of social science research early in the formulation of the program supporting nanotechnology research. The American promoters of nanotechnology and first directors of the National Nanotechnology Initiative coedited a series of volumes meant to discuss the "societal implications of nanotechnology" [3–5]. These "societal implications" ranged from the health and safety issues to the ethical concerns raised by the application of nanotechnology for "human enhancement." There were many reasons for these initial concerns. The early connection of nanotechnology with long-term promises of revolutionary developments should not be underestimated. Nanotechnology became fashionable in the 1990s because futurist Eric Drexler argued for the coming revolution that it would create. Controversies then followed between proponents (such as Drexler) of a revolutionary vision for nanotechnology and "realists" who saw in nanotechnology concrete market opportunities that too broad revolutionary statements could hinder by frightening potential consumers [6]. The initial visions for nanotechnology and the subsequent discussions about what exactly the "revolution" was about (see also [7]) made the examination of the "societal implications" of nanotechnology an integral component of nanotechnology.

The interest for nanotechnology implications did not stay in the United States science policy world. Framing nanotechnology as the "next industrial revolution"

Nanotechnology in Agriculture and Food Science, First Edition. Edited by Monique A. V. Axelos and Marcel Van de Voorde.
© 2017 Wiley-VCH Verlag GmbH & Co. KGaA. Published 2017 by Wiley-VCH Verlag GmbH & Co. KGaA.

did not go unnoticed in the international scene. In Europe, science policy officials at the European Commission soon developed their own version of nanotechnology programs that also included careful consideration about the "implications" issue. In so doing, the European institutions were both attentive to the specificities of European concerns (e.g., by not taking the framing of "human enhancement" for granting) and eager to avoid what they construed as previous failures of science/society relationships, including previous attempts at developing markets for genetically modified organisms [8].

In both the United States and Europe, the way of governing these "implications," and of ensuring an acceptable development of nanotechnology was meant to be "governance." Thus, the director of the US National Nanotechnology Initiative, Mihail Roco, wrote a paper with Ortwin Renn, a well-known specialist of risk perception studies, in which the two explained that nanotechnology was in need of "a switch from government alone to governance" [9]. They argued that instead of "a top-down legislative approach which attempts to regulate the behavior of people and institutions in quite detailed and compartmentalized ways," what was required was a system in which "people and institutions behave so that self-regulation achieves the desired outcomes" [9]. In Renn and Roco's perspective, the governance system was to be composed of the examination of the health and safety risks of nanomaterials at an early stage in the development of nanotechnology products, of international initiatives to promote common standards able to ensure the safety of nanotechnology objects, and of permanent interrogations of nanotechnology's existing and future ethical issues through the mobilization of social scientists as well as dialogues with "the public." This required a "coordinated approach" comprising the standardization of nanotechnology products, training programs for both scientists and social scientists, measures of public perceptions of nanotechnology, and careful risk examination.

What Renn and Roco suggested was a synthetic version of the nanotechnology programs as they were being developed in Europe as well as in the United States. The "governance system" proposed by Renn and Roco covered the range of initiatives undertaken on both sides of the Atlantic to frame the development of the field: As "nano" chemicals were crafted and studied, standardization operations were launched in national, European, and international standardization bodies and risk research was included in science policy programs; social scientists were called to investigate the social "implications" or "aspects" of nanotechnology; prospective reflections about future developments and ways of realizing them were launched; and "public engagement" in nanotechnology became a central concern of nanotechnology proponents, possibly eager to avoid what some of them construed as a potential public backlash [10]. In Renn and Roco's words, "governance" covered the whole range of these initiatives. As nanotechnology was construed as a science policy program both relying on anticipations of future development and comprising a reflection about its own "implications," it appeared as a condition for nanotechnology to exist. It is through the "governance system" advocated by Renn and Roco that research programs could be organized, that "implications" could be identified and dealt with, and that objects (such as "nanomaterials") could be defined so that they could become controllable entities able to be studied in scientific laboratories and circulated on markets.

Consequently, a proposition such as Renn and Roco's ought to be considered as a symptom to interrogate rather than a ready-made solution to follow. "Governance," as defined in the "governance system" proposed by Renn and Roco, ought to be analyzed as an engine for the development nanotechnology. Following nanotechnology's proponents in their push for "governance," one can start thinking about the political effects of this engine. Thus, Mihail Roco sees in nanotechnology a need for turning "government" into "governance," that is, going from a "a top-down legislative approach that attempts to regulate the behavior of people and institutions in quite detailed and compartmentalized ways" to a system that "attempts to set the parameters of the system within which people and institutions behave so that self-regulation or the ecosystem achieves the desired outcomes" (Roco, 2006: 3; quoted in Ref. [11]). "Governance," here, points to a mode of action based not on the legal constraint but on devices aimed to benefit from the inputs of the concerned individuals and organizations. This general stance echoes other uses of the term "governance," which a large body of scholarly works has commented upon. For instance, development policies promoted by international organizations such as the World Bank or the IMF have used "governance" as a way of promoting transparency and accountability, while delegating previously state-owned operations to private actors, supposedly more able to deal with practical problems and economic constraints [12]. Accordingly, the call for "good governance" goes with a recomposition of the role of the state and, even more, the transformation of the state in experiments about their very nature [13,14]. Within the European institutions, the 2001 White Paper on governance of the European Commission [15] is also characteristic of an approach to democratic legitimacy based on the consultation of various stakeholders, on transparency, and on accountability.

This has several consequences. First, the broad diffusion of the word "governance," in policy and academic circles, means that the word can be defined in various ways. It is "popular but imprecise" as scholar Roderick Rhodes puts it [16]. Governance is best thought of as an approach, open for redefinition, and potentially different in various contexts. At the core this approach is the objective of activating political subjects rather than imposing centralized legal constraints on them that requires transparency and accountability. Then the ways and means of "governance" and the details of what would constitute, for example and in the context of our interest here, the "governance system" of nanotechnology ought to be examined in details. Second, governance is not a neutral term, but has political effects – in the sense that is goes with a certain style of government that defines how decisions have to be made, what the domain of action of the state is/should be, and what roles citizens are expected to play. This means that the ways of conducting the governance of nanotechnology impacts such crucial components of democratic life as the sources of political legitimacy, the modalities of the sovereign action of the state, and the exercise of citizenship.

For all these reasons, it is necessary to interrogate the politics of governance in making nanotechnology exist *and* organizing democratic order. Nanotechnology, thereby appears as an opportunity to reflect on the contemporary forms of governance, taken as a style of government whereby delegation, experimentation, assessment, the involvement of stakeholders, and transparency are considered crucial to ensure both democratic legitimacy and the efficiency of public regulation. What does this style of government do? How to envision a critical analysis of it? Answering these questions requires a detailed discussion of the various instruments meant to implement the "governance system" that people like Mihail Roco and Ortwin Renn hope nanotechnology can rely on.

20.2
Operationalizing the Governance of Nanotechnology

The governance of nanotechnology is operationalized by a series of instruments. They comprise initiatives related to the "ethical, legal and social implications" (ELSI) or "aspects" (ELSA) of nanotechnology, voluntary codes meant to turn researchers and industrialists into "responsible scientists," and participatory mechanisms meant to "engage" public in the development of nanotechnology.

20.2.1
ELSI and ELSA Projects

Examples of the first category comprise numerous research projects funded by science policy bodies and meant to examine the "implications" of nanotechnology, including ethical issues and safety risks. The way of doing so is not consensual, and takes various forms in different political institutions. For instance, the European Ethical, Legal and Social Aspects (ELSA) projects about nanotechnology have been characterized by an explicit will for the exploration of what the "European values" might be. Researchers funded by the European Commission have discussed forms of "lay ethics" that would ensure the possibility for various public to formulate their concerns and values [17]. By contrast, the problematization of nanotechnology's "legal and social implications" in the United States is debatable regarding the role of ethics in science policy and in what conditions this additional public expertise can provide relevant advice for policy making ([18,19]; see Chapter 5 of Ref. [20]). As such, the implications of scholars and researchers taking inspiration from the social studies of science can be seen as an attempt in demonstrating the epistemological and political interest of a renewed approach for technology assessment described as a "real-time technology assessment" [21]. These differences are not incidental. They show that as the examination of the "implications" of nanotechnology has become an integral component of the science policy programs expected to support the development of the field, the nature of these implications and the "values" according to which they should be assessed are not consensual. This relates to the very nature of the political institutions involved. In the United States, the way of making the exploration of values a matter of expertise is consistent with a recurrent trope in the American public administration of science, making objectivity the core source of legitimate public action [22,23]. The situation in Europe appears quite different, as the European institutions are in search for both the sources of their democratic legitimacy, and the common values that could define an European identity.

20.2.2
Voluntary Codes

Voluntary Codes have been produced by private companies, think tanks, and public institutions. One of the most interesting examples might be the European "Code of Conduct" introduced by the European Commission in 2007 that defined a series of principles to ensure the "responsible development of nanotechnology." The Code proposes that researchers and research institutions do not engage in projects aiming at "human enhancement," that they refrain from applications of nanotechnology that could harm humans or the environment, and that they engage in transparent dialogues about their works. Interestingly for our reflection here, the code was meant to "achieve governance goals whereas legislation c(ould) not" ([24]: 5). This statement is worth analyzing in details. In the previous quote, what "legislation" cannot achieve in the European Union refers to matters related to "values," such as issues pertaining to the ethical boundaries one may wish to impose on technological development. By virtue of the subsidiarity principle, this domain is left to member states to decide – yet it raises numerous difficulties for the European institutions when controversial technological developments are concerned. Take for instance human embryonic stem cells. Whether to use them for research purposes has been widely discussed, and a topic for a public debate about ethics. The European science policy dealt with it by adopting a distributed approach that imposed constraints for research projects to receive funding from the European research policy institutions. Among these constraints, the fact that the embryonic stem cells were not to be produced in Europe (but could be acquired from outside of Europe) was central [25]. The approach to the ethical issues related to nanotechnology (including its potential use for "human enhancement" purposes) ought to be understood in the same way. Critics considered the Code as a purely voluntary tool that said nothing about the actual practice of nanotechnology research in European laboratories. But what should be considered is the role of the Code as a tool for governance – namely, as an instrument through which the control exercised by political institutions is distributed, and rely more on the will of individuals (such as researchers) and organizations (such as research institutions) than on the legal constraints. Accordingly, abiding to the Code became a condition for research projects to get European money, as evaluated during the review process of the research proposals sent to the Directorate General for Research of the European Commission.

Another interesting aspect of the Code relate to the previous section about the European ELSI projects. As said above, these projects were also meant to explore what the "European values" according to which one could evaluate nanotechnology's implications could be. In the Code, the reference to value is not as determined as one would expect in the case of, for instance, American bioethics. Take for instance the consultation document submitted by the European Commission for public comments when crafting the Code. The consultation paper was based on a series of "core European principles": "precaution," "inclusiveness," and "integrity" were mentioned, and complemented by "better and constant vigilance," "realizing societal benefits" through "societal debate," "credibility and trust" through "open and transparent public dialogue" and "protection of fundamental rights." This long list can be seen as too heterogeneous to be consistent. Yet, this diversity is a sign of an European unease about the definition of the values expected to characterize a common acceptable approach to technological progress. The Code, for that matter, appears as an instrument expected to govern this situation, as it has become one of the main components of the European approach to the ethics of nanotechnology, regularly presented by DG Research officials as the most visible attempt to define a "European approach to ethics," which would "promote dialogue" while "not imposing some forms of ethics rather than others."[1] The Code was expected to be used

> as an instrument to encourage dialogue at all governance levels among policy makers, researchers, industry, ethics committees, civil society organisations and society at large with a view to increasing understanding and involvement by the general public in the development of new technologies.[2]

For a part, "encouraging dialogue" had been the product of the very process that produced the Code that comprised preparatory consultations and regular revisions based on the consultation of interested stakeholders. But the Code is also expected to be taken up at a wider level. Together with the Council, the European Commission recommended that member states

> be guided by the general principles and guidelines for actions to be taken, set out in the Code of Conduct for Responsible Nanosciences and Nanotechnologies Research (. . .), as they formulate, adopt and implement their strategies for developing sustainable nanosciences and nanotechnologies (. . .) research.[3]

Hence, in a situation where subsidiarity rules make ethics and value a matter for member states (and not the European Commission) to act, the Code of

---

1) These are René von Schomberg's words during an academic conference (Darmstadt, September 28, 2010).
2) "Commission recommendation on a code of conduct for responsible nanosciences and nanotechnologies research & Council conclusions on Responsible nanosciences and nanotechnologies research 2009": 11–12.
3) "Commission recommendation on a code of conduct for responsible nanosciences and nanotechnologies research & Council conclusions on responsible nanosciences and nanotechnologies research 2009": 3

Conduct is a way for the European public bodies to act without sidestepping member states. As such, the Code is a perfect illustration of the connection between the instruments of governance, a practical organization of a political system based on will rather than constraint, and an imagination of a would-be European space that would be both efficient and legitimate.

20.2.3
Public Engagement

Third, "governance" goes with a concern for public participation. As soon as Roco formulated his concern for "governance," "public engagement" was heralded as a key theme in nanotechnology programs. In Europe and the US, this signaled an unprecedented mobilization of science museums for the sake of the public display of nanotechnology. Public display was accompanied by many original devices meant to ensure the participation of public. In America, science museums were gathered within the NISE network (National Informal Science Education), where "informal science education" covered a broad range of initiatives aimed to turn visitors into active participants in nanotechnology programs, as enthusiastic consumers, supporting taxpayers, or informed electors. While there were some debates in American science museums about the possibility for, modalities of, and ultimate goal of deliberative meetings with lay people about nanotechnology issues, the focus of the NISE network's initiatives has been on the individual and on the variation of pedagogical intervention. For a person in charge of the NISE network in Washington, DC, this was a stark contrast with the European situation, where "policy-makers do listen."[4] In Europe, the public offices in charge of funding nanotechnology research are also in charge of communication matters. A recent report about "communicating nanotechnology" offers a telling illustration of the perspective adopted by the European Commission on the appropriate communication of nanotechnology [26]. In stark contrast with the American situation, the focus in this report is on the "scientific understanding of the public," meaning that rather than conveying scientific information to the European public in the hope of turning them into supporters of nanotechnology, the objective is to learn what their interests and concerns are, and adapt the programs of nanotechnology development accordingly. The 2010 report uses such telling expressions as the "monitoring of public opinions," of which the European science museums ought to be part – for instance, by organizing focus groups involving visitors of nanotechnology exhibits. This has consequences for the ways in which the European public is imagined. In this perspective, the European public is a collection of individual citizens, not necessarily knowledgeable about nanotechnology and not necessarily concerned about it, but whose contributions to the political debate about European policy choices are deemed important. Such a perspective can be controversial, since its side effect is that *other* imaginations of the legitimate European public (for instance

---

4) Interview, October 2009, Washington DC.

NGOs mobilized for the public regulation of nanomaterials) are then considered less legitimate. Another controversial domain relates to the ways in which "the ethical, legal and social aspects" of nanotechnology are dealt with within this concern for the "scientific understanding of the public": Are these issues to be collectively explored with scientists and public? Or given social scientific facts about which European public ought to have an opinion? The difference is visible when comparing some of the ELSI projects funded by the European Commission and focusing on "lay ethics" [17] (which would be situated in the first case), and a report such as the one about communicating nanotechnology (which would be in the second).

20.3
The Constitutional Project of Governance

20.3.1
The Politics of Responsible Research and Innovation

The instruments of nanotechnology governance are not neutral. They define particular roles for citizens, experts, and policy makers. They distribute roles and responsibility among member states and political institutions – and, more than that, they shape political subjectivities and organizations. They have central roles in the organization of markets. Eventually, and as the comparison between European and American initiatives shows, they shape different political spaces across the Atlantic. The examples discussed above show that (at least for science policy) "governance" marks less disappearance of the state than a recomposition of its mode of action. They are significant of a more general evolution of science policy. Historians of science have proposed grant narratives, sketching the evolution of science policy. For instance, Dominique Pestre critically accounts for an evolution from a model based on the autonomy of science, publicly funded research, and a linear mode of reasoning from basic and autonomous research to industrial applications, to a science policy model whereby economic and research interests come together, and the connection of scientific evolution with the interests of various stakeholders is a central element [27]. Such a narrative can be read as a continuation, only under a critical guise, of the well-known description of the opposition between two "modes" of research production: first one based on the autonomy of science, and the second one where the constant interactions between academic research, policy institutions, and private companies are central [28].

These narratives are helpful to situate nanotechnology research. Indeed, the three types of governance instruments described above can be described as elements of "mode 2" of research production. Research on the social implications and greater public participation implies that the concerns of various stakeholders are hopefully taken into account when nanotechnology programs end up producing goods that can meet demands on the market. Voluntary Codes are ways for public bodies to act on researchers without constraining them, and coordination devices between academic research, public bodies, and private companies. Throughout all these examples, the scientific concern about the quality of research, the economic issue of the development of markets, and the democratic question related to the representation of concerned public are conflated.

But the grant narratives of the evolution of science policy are also simple, and poorly equipped to account for the variety in the enactment of nanotechnology governance. What emerges from the various attempts at introducing governance instruments for nanotechnology are various political constructs that, in some cases, restabilize existing political institutions and in other cases introduce new constructions. More generally, the connection among the sites where nanotechnology is problematized as a democratic issue (and where, accordingly, democracy itself is problematized) is a way of accounting for the construction of wider political spaces [20]. Our concern here for the governance of nanotechnology can lead to adopt such an approach about the sites where the instruments expected to implement governance are crafted and used.

Thus, one can see emerging in Europe a concern for "responsible research and innovation" (RRI), which follows the various experiments with the governance of nanotechnology. In a report written in 2013 for the DG Research, a group of scholars and officials who had worked on nanotechnology ELSA described the governance instruments experimented with nanotechnology as the early steps in the development of RRI, which was to become a central component of the European research policy.[5] RRI is an expression that has become widely used, and that gathers a wide range of initiatives, all related to either the examination of the ELSA of technological developments, voluntary Codes of Conducts, and public participation. Examining the European governance instruments for nanotechnology, one is then led to analyze its eventual outcome, namely, a research policy based on RRI. The actual transformation of scientific practice because of RRI can be (and is) debated. One could argue that increased public scrutiny on research could hinder promising breakthroughs, or, symmetrically, that the call for RRI is but an empty gesture hardly concealing the fact that European research is expected to provide engines for future economic growth [29]. More interesting for our concern here is what the current emphasis on RRI says about the politics of the European governance of science and technology. After nanotechnology, technological development appears, within RRI, as a collective project within which controversies are anticipated, while responsibility is distributed, thanks to nonconstraining instruments (such as Codes of Conduct, ethics review processes, etc.), where the long-term objectives of democratic legitimacy and economic competitiveness come together. This implies a close integration of the reflection about the objectives and practices of responsibility within the technical components of the European research policy. Officials at the DG Research of the

---

5) European Commission, 2013, Options for Strengthening Responsible Research and Innovation. Report of the Expert Group on the State of Art in Europe on Responsible Research and Innovation. KI-NA-25-766-EN-C.

European Commission see RRI as a "governance concept,"[6] and the expression is particularly interesting here. It makes "governance" a central component of the current European research policy, which, under the guise of RRI, expects to ensure the legitimacy of both technological developments and the interventions of the European institutions.

The politics of governance is then most apparent when one situates RRI within the recent evolutions of the European research policy. The 2000 Lisbon strategy, which hoped to make Europe a knowledge-based economy, defined target levels of public and private investment in R&D (namely, 3% of GDP) for member states. Evaluated in the early 2010s, it was considered a failure since the majority of member states never reached this target share. By contrast, the new forms of the European research policy after 2010, notably within the *Europe 2020* strategy, did not attempt to define minimal thresholds of investment, but target a limited number of objectives considered as priorities [30,31]. This evolution had a dual objective, namely, ensuring that these objectives (such as global warming or aging) correspond to the expectations and needs of the European citizens (an argument that has both democratic and economic resonances), and targeting a limited number of areas of interests in a situation where public spending are expected to be controlled. RRI, lying within this shift from the integration of the European research space to the careful choice of limited domains of interventions, thereby appears as a project whereby the European public is imagined in connection to a series of issues the European institutions have to solve for the sake of both democratic legitimacy and economic efficiency.

The European example shows that the study of the governance of nanotechnology can be a window into the analysis of the current evolutions of science policy, and, more generally, into discussions about the sources of legitimate public actions in contemporary democracies. The European case is particularly telling, as it is one of pervasive uncertainties about the very nature of the democratic space at stake: Who is the European public in the name of whom decisions ought to be made? What are the common values that are supposed to ground public decisions? RRI as a "governance concept" is an answer to these questions, and more than that, it proposes no less than a constitutional project, in that it offers a path for the construction of political subjects, while defining roles and responsibilities for the European institutions.

20.3.2
How to Think Critically about Governance

There are two approaches that we avoided so far. The first one would have taken "governance" at face value and wonder about how best to achieve it. This would have mirrored the position of nanotechnology's proponents. It would have prevented us for identifying the political effects of governance, the variety of instruments on which it is based, and the overall constitutional project it ends up

---

6) Presentation on "Responsible Innovation" by the Science in Society Unit, Vienna, June 2012.

shaping. But we also refrained from evaluating, even in a critical way, the "governance systems" that we saw emerging (particularly within the European institutions). How then to conceive of a critical approach to governance, understood as a constitutional project?

Nanotechnology has been targeted by critical groups, particularly in France, and their position is a useful reminder of the possibility for and the consequences of the radical critique of technological development. Since the early 2000s, a group called *Pièces et Main d'Oeuvre* (PMO) based in Grenoble, a city in the French Alps and a hub for nanotechnology research, has been vocal in the critique of nanotechnology development [20,32]. The interventions of PMO have taken the forms of carefully written anonymous pieces circulated on the Internet, as well as spectacular interventions during official events. PMO made national headlines in 2009 and 2010, when the group, accompanied by friends and followers across the country, successfully interrupted the public meetings organized as part of the national debate about nanotechnology commissioned by the French government. PMO criticizes nanotechnology as a program being conducted for the sole economic interests of the actors involved, which threatens the autonomy of science and will bring nothing but contestable developments, whether environmentally hazardous nanomaterials or useless (at best) and threatening (at worse) devices granting to technology even more control over humans. Throughout the course of its written production, PMO is careful not to be involved in nanotechnology programs. This implies a stark refusal to participate in any of the governance devices set up by public or private institutions. Hence, the refusal to participate in the national public debate alongside other stakeholders. This refusal is sometimes interpreted as an undemocratic attitude. Yet from PMO's viewpoint, it is entirely logical, when considering that governance mechanisms such as participatory devices are part and parcel of nanotechnology and contribute to make it exist as a technological program worthy of public interest.

The case of PMO is interesting because it directly echoes the position of the social scientist being asked to reflect on the governance of nanotechnology. As made explicit in this very text, the latter is bound to consider that "governance" is part and parcel of nanotechnology, and that the politics of it ought to be described. Yet, the way to do so might part away at this point. PMO's choice is to base the intervention on a critical distance expected to be absolutely effective. In participatory devices, there might be practical difficulties in maintaining it.[7] At the epistemological level, it prevents from considering in details the mechanisms whereby governance mechanisms acquire constitutional strength – a detailed examination that requires that one take seriously the objectives and motivations of instruments such as those examined in the previous section. The alternative then is to make the critical distance an epistemological and a political

---

7) Thus, activists questioned the ways in which they should voice their opinions during the public meetings and this proved more challenges than expected, as the organizers of the public debate sought to integrate all the opinions that were voiced, including the most critical. To the activists, such an inclusion was not acceptable.

question. For the researcher engaged in the critical reflection on governance, it is an epistemological question because it deals with the type of research endeavor that one undertakes. For that matters, the choice to focus on certain sites rather than others, and connect among them to make broader constructs explicit, is crucial. We have already discussed in this respect the current evolution of the European science policy. Distance is also a political question, since it is a matter of experimenting with forms of engagement, namely, participating in a project devoted to the "ELSA" of nanotechnology, intervening in a participatory devices, or being a member of a standardization committee. This list (which is far from exhaustive) also shows that the epistemological and the political questions are two sides of the same coin. It is through the variety of engagements with governance mechanisms that one can circulate among sites and connect among them.

Thus, the exploration of the variety across governance mechanisms helps identify several ways of problematizing nanotechnology and democracy in the Western world. We have already sketched the importance of RRI as a successor of the European governance initiatives targeting nanotechnology that construes the sources of democratic legitimacy differently than previous attempts at ensuring the European integration through research policy. In doing so, we have turned governance into a problematic approach, potentially contested and prone to various types of practical enactment. This has allowed us to identify not only new concerns for science policy institutions (such as the anticipation of potential controversies) but also the current recombinations of the European political economy.

References

1 Bennett, I. and Sarewitz, D. (2006) Too little, too late? Research policies on the societal implications of nanotechnology in the United States. *Sci. Cult.*, **15** (4), 309–325.

2 Macnaghten, P., Kearnes, M., and Wynne, B. (2005) Nanotechnology, governance and public deliberation: what role for the social sciences? *Sci. Commun.*, **27** (2), 268–287.

3 Roco, M. and Bainbridge, W. (eds) (2001) *Societal Implications of Nanoscience and Nanotechnology*, Springer, Dordrecht, The Netherlands.

4 Roco, M. and Bainbridge, W. (eds) (2003) *Societal Implications of Nanotechnology: Individual Perspectives*, National Science Foundation, Arlington, VA.

5 Roco, M. and Bainbridge, W. (2005) Societal implications of nanoscience and nanotechnology: maximizing human benefit. *J. Nanopart. Res.*, **7**, 1–13.

6 Laurent, B. (2010) *Les Politiques des Nanotechnologies. Pour un Traitement Démocratique d'une Science Émergente*, Charles Léopold Mayer, Paris.

7 Berube, D. (2006) *Nano-Hype: The Truth Behind the Nanotechnology Buzz*, Prometheus Books, New York.

8 Laurent, B. (2016) Perfecting European democracy: science as a problem of political and technological progress, in *Perfecting Human Futures: Technology, Secularization and Eschatology* (eds B. Hurlbut and H. Tirosh-Samuelson), Springer, Dordrecht, pp. 217–237.

9 Renn, O. and Roco, M. (2006) Nanotechnology and the need for risk governance. *J. Nanopart. Res.*, **8**, 153–191.

10 Rip, A. (2006) Folk theories of nanotechnologists. *Sci. Cult.*, **15** (4), 349–365.
11 Kearnes, M. and Rip, A. (2009) The emerging governance landscape of nanotechnology, in *Jenseits von Regulierung: Zum politischen Umgang mit der Nanotechnologie* (eds S. Gammel, A. Lösch, and A. Nordmann), Akademische Verlagsgesellschaft, Berlin, pp. 97–121.
12 Williams, D. and Young, T. (1994) Governance, the World Bank and liberal theory. *Polit. Stud.*, **42** (1), 84–100.
13 Harrison, G. (2004) *The World Bank and Africa: The Construction of Governance States*, Routledge, London.
14 Ehrenstein, V. and Laurent, B. (2015) State experiments with public participation: French nanotechnology, Congolese deforestation and the search for national public, in *Rethinking Participation. Science, Environment and Emergent Publics* (eds J. Chilvers and M. Kearnes), Routledge, London, pp. 123–143.
15 European Commission (2001) European Governance. A White Paper, Brussels, p. 428.
16 Rhodes, R.A.W. (1996) The new governance: governing without government. *Polit. Stud.*, **44.4**, 652–667.
17 Davies, S., Macnaghten, P., and Kearnes, M. (eds) (2009) *Reconfiguring Responsibility: Lessons for Public Policy, Part 1 of the DEEPEN Report*, Durham University, Durham.
18 Evans, J. (2000) A sociological account of the growth of principlism. *Hastings Cent. Rep.*, **30** (5), 31–38.
19 Evans, J. H. (2002) *Playing God? Human Genetic Engineering and the Rationalization of Public Bioethical Debate*, The University of Chicago Press, Chicago.
20 Laurent, B. (2017) *Democratic Experiments: Problematizing Nanotechnology and Democracy in Europe and the United States*, The MIT Press, Cambridge, MA.
21 Guston, D. and Sarewitz, D. (2002) Real-time technology assessment. *Technol. Soc.*, **24**, 93–109.
22 Jasanoff, S. (1990) *The Fifth Branch, Science Advisers as Policy-Makers*, Harvard University Press, Cambridge, MA.
23 Jasanoff, S. (1992) Science, politics and the renegotiation of expertise at EPA. *Osiris*, 7, 192–217.
24 von Schomberg, René; (2009) Organizing collective responsibility: on precaution, codes of conduct and understanding public debate. Keynote Lecture at the Meeting of the Society for the Study of Nanoscience and Emerging Technologies, Seattle, WA, September 11.
25 Jasanoff, S. (2005) *Designs on Nature: Science and Democracy in Europe and the United States*, Princeton University Press, Princeton.
26 Bonazzi, M. (2010) *Communicating Nanotechnology: Why, to Whom, Saying What and How*, European Commission, Brussels.
27 Pestre, D. (2008) Challenges for the democratic management of technoscience: governance, participation and the political today. *Sci. Cult.*, **17** (2), 101–119.
28 Gibbons, M., Limoges, C., Nowotny, H., Schwartzman, S., Scott, P., and Trow, M. (1994) *The New Production of Knowledge: The Dynamics of Science and Research in Contemporary Societies*, Sage, London.
29 de Saille, S. (2015) Innovating innovation policy: the emergence of 'Responsible Research and Innovation'. *J. Res. Innov.*, **2.2**, 152–168.
30 European Commission (2010) Europe 2020. A strategy for smart, sustainable and inclusive growth, Communication from the Commission, Brussels, March 3, 2010.
31 Lundvall, B.-A. and Lorenz, E. (2011) From the Lisbon strategy to Europe 2020, in *Toward a Social Investment Welfare State? Ideas, Policies and Challenges* (eds N. Morel and B. Palier et Jakob Palme), Policy Press, Bristol, pp. 333–351.
32 Joly, P.-B. and Kaufman, A. (2008) Lost in translation. The need for upstream engagement with nanotechnology on trial. *Sci. Cult.*, **17** (3), 225–247.

21
Potential Economic Impact of Engineered Nanomaterials in Agriculture and the Food Sector

Elke Walz, Volker Gräf, and Ralf Greiner

Max Rubner-Institut, Federal Research Institute of Nutrition and Food, Department of Food Technology and Bioprocess Engineering, Haid-und-Neu-Straße 9, 76131 Karlsruhe, Germany

21.1
Introduction

Nanotechnology is one of the key technologies of the twenty-first century. Compared to larger structures of the same material, the different physical, chemical, and biological effects of nanoparticle structures result in interesting and novel approaches in all areas of science and technology. The agrifood sector is certainly not one of the main potential users of nanotechnology, however, it can benefit from innovative ways to improve quality and sustainability of products and processing. The potential of this technology has been extensively described in scientific literature and the practical implementation, as well as the economic significance have been predicted in market forecasts.

Regarding the market volume of €731 billions in 2014 of the 26 world leaders in food production [1], a worldwide (only FAO available data are considered) use of pesticides of approximately 250 000 tons [2], and a consumption of 167 million tons of fertilizers (nitrogen, potash, phosphate; total nutrients) [3], respectively, the use of nanotechnology is becoming increasingly important in terms of optimization of production processes, improvement of food quality and safety, and resource-saving production in the agrifood sector. Accordingly, the following current and projected nanotechnology applications with impact on the agrifood sector have been summarized in Ref. [4]:

- Nanostructured food ingredients
- Nanodelivery systems for nutrients and supplements
- Organic nanosized additives for food, health food supplements, and animal feed applications
- Inorganic nanosized additives for food, health food supplements, and animal feed applications
- Food packaging applications

Nanotechnology in Agriculture and Food Science, First Edition. Edited by Monique A.V. Axelos and Marcel Van de Voorde.
© 2017 Wiley-VCH Verlag GmbH & Co. KGaA. Published 2017 by Wiley-VCH Verlag GmbH & Co. KGaA.

- Nanocoatings on food contact surfaces
- Surface functionalized nanomaterials
- Nanofiltration
- Nanosized agrochemicals
- Nanosensors for food labeling
- Water decontamination
- Animal feed

21.2
Potential and Possible Applications of Nanomaterials in the Food Sector and Agriculture

There has been a lot of publications showing the potential of nanomaterials in the food and agricultural sector in the last 10 years, as shown in Refs [4–44], but it is difficult to determine which products are still under development or already commercially available.

Figure 21.1 shows the number of publications per year containing the terms nanomaterials, nanoparticles, nanotechnology, nanocapsules, or nanoemulsions and food or agriculture for four different literature databases in the last 16 years (ScienceDirect searched within "title, abstract, keywords"; Web of Science within "topic"; PubMed within "title/abstract"; Scopus within "title, abstract, keywords"). It has to be mentioned that the term "nano" in these publications was defined by the authors themselves. Therefore, no uniform definition was used to identify a material as a nanomaterial and quite often the material described in the publication was significantly bigger than 100 nm.

In addition, there are reports questioning the success of nanomaterials in the food sector (e.g., Ref. [45]), and even eight years after this report it is still difficult to predict the future market penetration of nanomaterials in the agrifood sector.

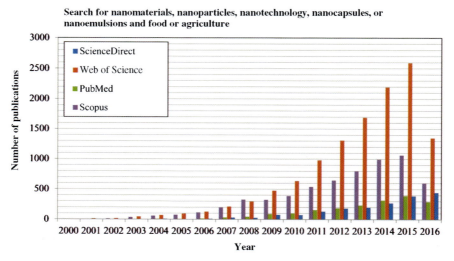

Figure 21.1 Number of publications dealing with nanomaterials in the agrifood sector.

Nanotechnology is intended to be used in foods (nano-inside) and in food contact materials, as well as in the analysis and the traceability of products (nano-outside). It is expected that this technology will result in improvements in production and process technology as well as optimized food quality in terms of taste, texture, and nutritional properties. In agriculture, the focus is on a more effective use of fertilizers, pesticides, and water treatment [9,11,12]. In addition, nanomaterials offer interesting aspects for the potential use of supplements present in animal feed [13].

So far, nanoapplications in the agrifood sector are mostly found in contact materials, particularly in the packaging field as well as in food additives and supplements, in form of metals and metal oxides, clays, and nanoformulations (nanocapsules, carrier systems) [16]. Accordingly, most products listed in the following nanoproduct registers/databases belong to these product groups:

- Inventory of nanotechnology applications in the agricultural and feed and food sectors [14].
- The project on emerging nanotechnologies: consumer products inventory – an inventory of nanotechnology-based consumer products introduced on the market [46].
- Nanotechnology in our food: an interactive database of consumer food products containing nanomaterials [47].
- The nanodatabase by the DTU Environment, the Danish Ecological Council and Danish Consumer Council [48].
- ANEC/BEUC inventory of products claiming to contain nano-silver particles available on the EU market [49].
- www.nanowatch.de – Die Nanoproduktdatenbank des BUND [50].

Reference [14] identified 55 types of nanomaterials and lists 633 nanotechnology applications for the food/feed/agri sector in 2014. A total of 276 nanomaterials were confirmed to be available on the market.

Regarding the products listed in all of the inventories mentioned above, it should be noted however, that many products disappear from the market shortly after their release, or the "nano-claim" cannot be verified and therefore, the products are removed from the databases. Actually, the products inventory of "the project on emerging nanotechnology" lists only seven food products. Two years ago 96 products were listed within the food category. The category of "supplements" however grew from 67 to 70 listed products (Table 21.1) [46].

Table 21.1 Number of "nanoproducts" listed in Ref. [46].

	May 2014	June 2015	July 2016
Food and beverage	204	117	118
Cooking	16	15	15
Food	96	7	7
Storage	20	20	20
Supplements	67	69	70

21.3
Nanotechnology: Market Research and Forecasts

21.3.1
Methodology of Market Research and Forecasts

To assess the economic potential of nanotechnology, for example, to create market forecasts for new technologies, different methods of market research are applied. They are usually based on data of former market analyses that are updated for subsequent years using mathematical methods.

For an overview of the current market as well as market changes, data are collected by primary (field research) and/or secondary (desk research) market research. Secondary data sources are referred to as existing data from different origins. So, particularly in the case of new technologies, patent and literature searches are used. However, information from companies (business data) is also applied, such as annual reports, press releases, stock market information, as well as data from concentrated data sources such as Factiva, Hoovers, and Bloomberg databases, which are reevaluated for the specific issue. The existing database in secondary methods is juxtaposed with new survey data in primary methods. Secondary market research initially serves to identify key players (e.g., industry experts, experts from research institutions, manufacturers, suppliers, distributors). Therefore, experts can be specifically selected to develop primary data sources. Secondary market research methods (surveys, interviews) are either qualitative or quantitative surveys. Qualitative approaches, such as group discussions, expert interviews, or in-depth interviews (expert workshops and interviews) are exploratory forms of market research. They are used to collect information that can subsequently be substantiated by quantitative methods. Qualitative approaches are well suited to inquire about attitudes and social expectations on the subject. On the other hand, results of quantitative inquiries (telephone, online interviews based on questionnaires, etc.) can be analyzed statistically.

Especially with long-term nanotechnology forecasts (10–20 years), various assumptions have to be made in terms of sales figures, for example, regarding the economic and political stability of regions considered over the forecast period, as well as on exchange rates in the currency market reports [51]. Different market reports will therefore differ in their forecasts. Among others the following factors will also contribute to the observed differences:

- Different assumptions by suppliers regarding the market success of nanomaterials and products consisting of or containing nanomaterials.
- The use of different definitions for nanomaterials, with respect to, differences in the interpretation of the definition for nanomaterials even when the same definition is applied.
- Differences in the product used for predicting potential sales figures (e.g., the nanomaterial itself or the product consisting of or containing this nanomaterial) [52].
- Differences in the estimation of the added value of a product consisting of or containing nanomaterials compared to the same product without nanomaterials.

21.3.2
Market Forecasts of Nanotechnology in Food and Agriculture: Publicly Available Data

Table 21.2 shows publicly available data from a selection of global market forecasts for nanotech-enabled products in food and agriculture (in US million dollars). iRAP predicted a compound annual growth rate (CAGR) of 11.65% for the nano-enabled food and beverage packaging sector from 2009 to 2014. The growth of intelligent packaging was predicted to be 18.7% for the same period. In another market report (persistence market research) CAGRs for the packaging sector are noted to be 12.7% from 2014 to 2020 with a predicted market volume of USD15 billions in 2020. However, it has to be mentioned that some of these very optimistic market forecasts were not fulfilled. In 2004, Helmut Kaiser Consultancy (HKC) [52] predicted a nanomarket volume of USD7 billions for 2006 and USD20.4 billions for 2010; some years later the consultancy firm conducted a new market research and shifted its prognosis approximately 10 years into the future [53]. This gives an impression how difficult and complicated it is to create market forecasts for the "nanofood and agriculture market."

21.4
Critical Considerations and Remarks Concerning Market Reports and Forecasts

Regarding the different results of market forecasts and to avoid misinterpretation of market data, some limiting conditions and basic parameters must be taken into account. For the economic potential of nanotechnology, these aspects have been discussed in Ref. [57].

As a matter of principles, it has to be clarified if the same nanomaterial definition was used in all interviews, databases, and by all respondents. This is an important requirement, because at present different definitions are applied for regulative and scientific purposes. While in most cases at least the size range is specified between 1 and 100 nm, there are still differences in details [58].

In the European food labeling regulation [59], for example, only engineered nanomaterials are considered, whereby "engineered" means intentionally produced. Instead of a defined size range only a size scale of 100 nm or less has been indicated. Understandably, these specifications give rise to different interpretations causing ambiguity on the term "intentionally." It is not clear if in this case "intentionally" signifies the intention to produce nanoscale material or if the term has a broader meaning. Furthermore, it is assumed that nanoscale organic materials must not be considered as nanomaterials (EU MEMO/11/704) within the scope of regulation (EC) No. 1169 [59].

Therefore, it is possible that enterprises are not aware of producing or providing nanomaterials, especially when considering that no standard analytical methods are available to detect and characterize nanomaterials in complex matrices. On the other hand, it can be assumed that the leeway given by the food labeling

Table 21.2 Selection of market forecasts (publicly available data) for nanotech-enabled products in food and agriculture (in million USD).

	2006	2007	2008	2009	2010	2011	2012	2013	2014	2015	2016	2017	2018	2019	2020
						Food and food packaging									
Helmut Kaiser Consultancy [52]	7000														
Helmut Kaiser Consultancy [53]					20 400					7000					20 400
Cientifica, 2006 (quoted in Ref. [4])	410						5800								
Food processing	100						1303								
Food ingredients	100						1475								
Food packaging	210						2930								
Food safety							97								
						Food, drink, agriculture									
Nanoposts, 2008, agriculture, food and drink (quoted in Ref. [54])		265								3210					
Nanosensors		2								360					
Encapsulation		3								320					
Nanocoatings		40								495					
Nanocomposites		180								1580					
Nanoporous membranes		40								455					

	Packaging for food and beverage industry												
iRAP, nano-enabled packaging for the food and beverage industry, 2009 [55]	4130	4210	4700[a]	5248[a]	5859[a]	6542[a]	7300						
Active packaging	2740	2790	3049[a]	3332[a]	3642[a]	3980[a]	4350						
Intelligent packaging	1030	1050	1246[a]	1479[a]	1756[a]	2084[a]	2470						
Controlled release packing	360	370	389[a]	410[a]	431[a]	454[a]	480						
Persistence market research, nano-enabled packaging market for food and beverages industry, 2014 [56]						6500	7321[a]	8250[a]	9298[a]	10479[a]	11810[a]	13310[a]	15000

a) Data calculated using denoted CAGRs.

regulation on the provision of consumer information [59] will be deliberately utilized. For this reason it is difficult to realistically portray the current market situation, which in turn is the basis of a market forecast.

Another source of deviations between different market reports and forecasts and between optimistic and pessimistic estimates of the market potentials is the use of sales figures related to different stages of the value chain. Huge differences in sales figures used for prediction have to be presumed when considering nanomaterial as produced on one hand and the final product consisting of or containing the nanomaterial on the other [60].

As already stated in Section 21.3.1, the challenge of market forecasts is to take into account the political and economic circumstances in the forecast period. Furthermore, it has to be considered that not all, but only a limited sample OR limited number of nanomaterial manufacturers, suppliers, or users are consulted in order to prepare the market forecast.

Last but not least, the market success of innovative products is hard to predict, because consumer perception of a novel technology and acceptance of the products derived from those technologies are of utmost importance.

Due to many uncertainties in the market prediction, different reports show divergent forecasts. However, the prediction of a steady increase of the market potential of nanomaterials in the agrifood sector is common to all of those market reports.

21.5
Obstacles Regarding Commercialization of Nanotechnologies in Food and Agriculture

There are several questions that have to be answered before new and innovative technologies or products make their way to the market:

- What is the consumer's perception regarding the new technology? Is there consumer acceptance of the new product? Are there convincing advantages of the new technology or the new product for an individual consumer?
- What are the advantages (e.g., quality, price, sustainability, health aspects) for consumers and industry to move from the conventional to the "nanoproduct?"
- Are the desired properties only achievable using nanomaterials?
- Is the production of the product consisting of or containing nanomaterials economic?
- What are the legal provisions/regulations for the use of the new technology or the commercialization of the products derived thereof [61]? Is nanolabeling of these products required and if so, is this labeling intended as pure information or as a risk assessment for consumers?
- Are there any potential risks originating from the use of the new technology and/or the new products? Is there an inappropriate distribution of risk and benefits, for example, between consumers and the industry?

Consumer acceptance is certainly one of the most important aspects with respect to the success of certain technologies or certain products. Throughout Europe, the use of nanomaterials in car paints and medical products is accepted. This acceptance declines the closer nanoproducts come to the human body [62]. Concerns were highest for products intended for human consumption [63,64]. Nevertheless, the Euro barometer survey 2010 showed that 41.1% of the EU consumers have a positive attitude toward nanotechnology and only 10% a negative one. However, it has to be considered that according to the survey 53.7% have never heard anything about nanotechnology. Thus, the awareness of the population related to this technology is low [65]. Perception of new technologies, among others, depends on emotional responses that are anchored in deep-seated cultural values and general attitudes. Therefore, acceptance of nanotechnology as well as nanotechnology-derived products in the agrifood sector varies from country to country and data from Europe are not simply transferable to other parts of the world.

Consumer acceptance can only be achieved if chances and risks are clearly communicated to the consumer. Therefore nanoproducts in the agrifood sector have to be actively promoted by companies. However, exactly this has been avoided by the industry because of concerns that this new technology might face similar rejection as gene technology. Labeling is another way to transparently inform about products consisting of or containing nanomaterials and to give consumers a free choice whether to use those products. This might contribute to generate trust in a population. As of December 2014, all ingredients present in food in the form of engineered nanomaterials need to be clearly indicated in the list of ingredients (Regulation (EU) No. 1169/2011) [59]. However, only very few – if any – products with nanolabeling could be found on the EU markets, even though there are databases and press reports indicating a widespread use of nanomaterials in agrifood products. This also seems to be true for other countries such as Australia and New Zealand [63,64]. As there is no internationally agreed definition for nanomaterials for the agrifood sector, and since the existing definitions in different countries are in general not really applicable, uncertainties exist if a certain product has to be labeled a "nanoproduct." In addition, the industry avoids labeling products as nano by making use of the leeway granted by the regulations and by reformulating products in a way that ingredients, which might be considered nanomaterials, are substituted due to the fear of a consumer backlash already mentioned above. This situation and practice does not obviously contribute in generating trust within a population. Last but not least, the economic competitiveness of such products needs to be evaluated prior to an extensive industrial engagement in this sector, even if scientific publications indicate a high potential for nanotechnology-derived products in the agrifood sector. In addition, reliable data on the benefits and risks of nanotechnology-derived products need to be generated before commercialization.

21.6 Conclusion

Without any doubt, engineered nanomaterials have enormous potential for the agrifood sector. Promising applications to improve product and process quality as well as for resource-friendly production have already been described in literature. However, because of the many uncertainties, engineered nanomaterials, particularly in the agrifood sector, might only be a flash in the pan, especially if regulatory challenges and consumer perception are not properly addressed. Trust is extremely important, and therefore a lack of transparency on the part of the industry will have a negative effect on consumer acceptance of nano-derived products.

References

1 FoodDrinkEurope (2015) Foodumsatz der führenden Agrar- und Lebensmittelkonzerne weltweit im Jahr 2014 (in Milliarden Euro). Available at http://de.statista.com/statistik/daten/studie/258495/umfrage/umsatz-der-weltweit-fuehrenden-lebensmittelkonzerne/ (accessed June 28, 2016).

2 FAO World use of pesticides (total) in 2013. Statistics Division FAOSTAT, Food and Agriculture Organization of the United Nations. Available at http://faostat3.fao.org/download/R/RP/E (accessed June 28, 2016).

3 FAO World consumption of fertilizers in nutrients in 2013. Statistics Division FAOSTAT. Available at http://faostat3.fao.org/download/R/RF/E (accessed June 28, 2016).

4 FAO/WHO (2010) FAO/WHO Expert Meeting on the Application of Nanotechnologies in the Food and Agriculture Sectors: Potential Food Safety Implications. Meeting report.

5 Duncan, T.V. (2011) Applications of nanotechnology in food packaging and food safety: barrier materials, antimicrobials and sensors. *J. Colloid Interface Sci.*, **363** (1), 1–24.

6 Bumbudsanpharoke, N. and Ko, S. (2015) Nano-food packaging: an overview of market, migration research, and safety regulations. *J. Food Sci.*, **80** (5), R910–R923.

7 Aschberger, K. et al. (2015) Nanomaterials in food: current and future applications and regulatory aspects. *J. Phys. Conf. Ser.*, **617**, 012032.

8 Chellaram, C. et al. (2014) Significance of nanotechnology in food industry. *APCBEE Procedia*, **8**, 109–113.

9 Fraceto, L. et al. (2016) Nanotechnology in agriculture: which innovation potential does it have? *Front. Environ. Sci.*, **4**. doi: 10.3389/fenvs.2016.00020.

10 Foss Hansen, S. et al. (2016) Nanoproducts: what is actually available to European consumers? *Environ. Sci. Nano*, **3** (1), 169–180.

11 Noruzi, M. (2016) Electrospun nanofibers in agriculture and food industry: a review. *J. Sci. Food Agric.*, **96** (14), 4663–4678.

12 Parisi, C., Vigani, M., and Rodríguez-Cerezo, E. (2015) Agricultural nanotechnologies: what are the current possibilities? *Nano Today*, **10** (2), 124–127.

13 Sekhon, B.S. (2014) Nanotechnology in agri-food production: an overview. *Nanotechnol. Sci. Appl.*, **7**, 31–53.

14 Peters, R. et al. (2014) Inventory of nanotechnology applications in the agricultural, feed and food sector. EFSA Supporting Publication 2014:EN-621.

15 Kumari, A. and Yadav, S. K. (2014) Nanotechnology in agri-food sector. *Crit. Rev. Food Sci. Nutr.*, **54** (8), 975–984.

16 Peters, R.J.B. *et al.* (2016) Nanomaterials for products and application in agriculture, feed and food. *Trend Food Sci. Technol.*, **54**, 155–164.

17 Gatos, K.G. (2016) Potential of nanomaterials in food packaging, in *Novel Approaches of Nanotechnology in Food* (ed. A.M. Grumezescu), Academic Press, pp. 587–621.

18 Kulkarni, A.S., Ghugre, P.S., and Udipi, S.A. (2016) Applications of nanotechnology in nutrition: potential and safety issues, in *Novel Approaches of Nanotechnology in Food* (ed. A.M. Grumezescu), Academic Press, pp. 509–554.

19 Chaudhry, Q. *et al.* (2008) Applications and implications of nanotechnologies for the food sector. *Food Addit. Contam.*, **25** (3), 241–258.

20 Momin, J.K., Jayakumar, C., and Prajapati, J.B. (2013) Potential of nanotechnology in functional foods. *Emir. J. Food Agric.*, **25** (1), 10–19.

21 Kim, D.-M. and Cho, G.-S. (2006) Nanofood and its materials as nutrient delivery system (NDS). *J. Appl. Biol. Chem.*, **49** (2), 39–47.

22 Weiss, J., Takhistov, P., and McClements, D.J. (2006) Functional materials in food nanotechnology. *J. Food Sci.*, **71** (9), R107–R116.

23 Livney, Y.D. (2015) Nanostructured delivery systems in food: latest developments and potential future directions. *Curr. Opin. Food Sci.*, **3**, 125–135.

24 García, M., Forbe, T., and Gonzalez, E. (2010) Potential applications of nanotechnology in the agro-food sector. *Ciencia e Tecnologia de Alimentos*, **30** (3), 573–581.

25 Liu, R. and Lal, R. (2015) Potentials of engineered nanoparticles as fertilizers for increasing agronomic productions. *Sci. Total Environ.*, **514**, 131–139.

26 Sozer, N. and Kokini, J. L. (2009) Nanotechnology and its applications in the food sector. *Trends Biotechnol.*, **27** (2), 82–89.

27 Neethirajan, S. and Jayas, D. S. (2011) Nanotechnology for the food and bioprocessing industries. *Food Bioprocess. Technol.*, **4** (1), 39–47.

28 Taylor, T.M. *et al.* (2005) Liposomal nanocapsules in food science and agriculture. *Crit. Rev. Food Sci. Nutr.*, **45** (7–8), 587–605.

29 Silva, H.D., Cerqueira, M.Â., and Vicente, A.A. (2012) Nanoemulsions for food applications: development and characterization. *Food Bioprocess. Technol.*, **5** (3), 854–867.

30 Valdés, M.G. *et al.* (2009) Analytical nanotechnology for food analysis. *Microchim. Acta*, **166** (1–2), 1–19.

31 Rashidi, L. and Khosravi-Darani, K. (2011) The applications of nanotechnology in food industry. *Crit. Rev. Food Sci. Nutr.*, **51** (8), 723–730.

32 Hannon, J.C. *et al.* (2015) Advances and challenges for the use of engineered nanoparticles in food contact materials. *Trends Food Sci. Technol.*, **43** (1), 43–62.

33 National Institute for Public Health and the Environment (RIVM) and Ministry of Health, Welfare and Sport (2010) Nanomaterials in Consumer Products : Update of Products on the European Market in 2010. RIVM Report 340370003/2010.

34 Dunford, N. (2005) Nanotechnology and opportunities for agriculture and food systems. *Nanotechnology*, **405**, 744–6071.

35 Handford, C.E. *et al.* (2014) Implications of nanotechnology for the agri-food industry: opportunities, benefits and risks. *Trends Food Sci. Technol.*, **40** (2), 226–241.

36 Rossi, M. *et al.* (2014) Scientific basis of nanotechnology, implications for the food sector and future trends. *Trends Food Sci. Technol.*, **40** (2), 127–148.

37 Boom, R.M. (2011) Nanotechnology in food production, in *Nanotechnology in the Agri-Food Sector*, Wiley-VCH Verlag GmbH, Weinheim, pp. 37–57.

38 Kampers, F.W.H. (2011) Packaging, in *Nanotechnology in the Agri-Food Sector*, Wiley-VCH Verlag GmbH, Weinheim, pp. 59–73.

39 Posthuma-Trumpie, G.A. and van Amerongen, A. (2011) Using nanoparticles in agricultural and food diagnostics, in *Nanotechnology in the Agri-Food Sector*,

40 O'Brien, N. and Cummins, E. (2011) Nano-functionalized techniques in crop and livestock production: improving food productivity, traceability, and safety, in *Nanotechnology in the Agri-Food Sector*, Wiley-VCH Verlag GmbH, Weinheim, pp. 75–87.

40 O'Brien, N. and Cummins, E. (2011) Nano-functionalized techniques in crop and livestock production: improving food productivity, traceability, and safety, in *Nanotechnology in the Agri-Food Sector*, Wiley-VCH Verlag GmbH, Weinheim, pp. 89–105.

41 Robinson, D.K.R. and Morrison, M. (2011) Nanotechnologies for improving food quality, safety, and security, in *Nanotechnology in the Agri-Food Sector*, Wiley-VCH Verlag GmbH, Weinheim, pp. 107–126.

42 Bugusu, B., Lay Ma, U.V., and Floros, J.D. (2011) Products and their commercialization, in *Nanotechnology in the Agri-Food Sector*, Wiley-VCH Verlag GmbH, Weinheim, pp. 149–170.

43 Chaudhry, Q., Castle, L., and Watkins, R. (2011) Nanomaterials in food and food contact materials–potential implications for consumer safety and regulatory controls, in *Nanotechnology in the Agri-Food Sector*, Wiley-VCH Verlag GmbH, Weinheim, pp. 191–208.

44 Dasgupta, N. *et al.* (2015) Nanotechnology in agro-food: from field to plate. *Food Res. Int.*, **69**, 381–400.

45 Busch, L. (2008) Nanotechnologies, food, and agriculture: next big thing or flash in the pan? *Agric. Human Values*, **25** (2), 215–218.

46 Project on Emerging Nanotechnologies: Consumer Products Inventory. Available at www.nanotechproject.org/cpi (accessed June 28, 2016).

47 Center for Food Safety Nanotechnology in our food – an interactive database of consumer products containing nanomaterials. Available at http://salsa3.salsalabs.com/o/1881/p/salsa/web/common/public/content?content_item_KEY=14112%20 (accessed June 28, 2016).

48 DTU Environment, the Danish Ecological Council and Danish Consumer Council The Nanodatabase. Available at http://nanodb.dk/en/about-us/ (accessed June 28, 2016).

49 BEUC (2013) ANEC/BEUC inventory of products claiming to contain nano-silver particles available on the EU market. Availabla at http://www.beuc.eu/publications/2013-00141-01-e.xls (accessed June 28, 2016).

50 BUND *Die Nanoproduktdatenbank des BUND*, BUND nanowatch.de. Available at http://www.bund.net/nc/themen_und_projekte/nanotechnologie/nanoproduktdatenbank/produktsuche/ (accessed June 28, 2016).

51 MarketsandMarkets (2015) Food Encapsulation Market: Global Trends & Forecast to 2020. Report Code FB 2207.

52 Helmut Kaiser Consultancy (2003) Study: Nanotechnology in Food and Food Processing Industry Worldwide 2003–2006–2010—2015: Tomorrow We Will Design Food by Shaping Molecules and Atoms Available at https://web.archive.org/web/20031209060614/http://www.hkc22.com/Nanofood.html (accessed June 16, 2016).

53 Helmut Kaiser Consultancy Study: Nanotechnology in Food and Food Processing Industry Worldwide 2011–2012–13–14–2015–2020—2025: Tomorrow We Will Design Food by Shaping Molecules and Atoms Available at https://web.archive.org/web/20130827111721/http://www.hkc22.com/Nanofood.html (accessed June 16, 2016).

54 Technology Strategy Board (2009) Nanoscale Technologies Strategy 2009-12; p. 26; Table 4 – Summary of technologies within market sector areas.

55 iRAP (2009) Global market for nano-enabled food and beverage packaging to cross $7 billion by 2014. Available at www.innoresearch.net/Press_Release.aspx?id=18 (accessed May 06, 2016).

56 Persistence Market Research (2014) Global nano-enabled packaging market for food and beverages industry will reach $15.0 billion in 2020. Available at http://www.persistencemarketresearch.com/mediarelease/nano-enabled-packaging-market.asp (accessed June 23, 2016).

57 Malanowski, N. and Zweck, A. (2007) Bridging the gap between foresight and market research: integrating methods to assess the economic potential of nanotechnology. *Technol. Forecast. Soc. Change*, **74** (9), 1805–1822.

58 Kreyling, W.G., Semmler-Behnke, M., and Chaudhry, Q. (2010) A complementary definition of nanomaterial. *Nano Today*, **5** (3), 165–168.
59 Commission, E. (2011) Regulation (EU) No. 1169/2011 of the European Parliament and of the Council of 25 October 2011 on the Provision of Food Information to Consumers. Official Journal of the European Union L 304/18.
60 Palmberg, C., Dernis, H., and Miguet, C. (2009) Nanotechnology: An Overview Based on Indicators and Statistics. STI Working Paper 2009/7, Statistical Analysis of Science, Technology and Industry.
61 Amenta, V. *et al.* (2015) Regulatory aspects of nanotechnology in the agri-feed-food sector in EU and non-EU countries. *Regul. Toxicol. Pharmacol.*, **73** (1), 463–476.
62 Federal Institute for Risk Assessment (2008) *Public Perceptions about Nanotechnology: Representative Survey and Basic Morphological–Psychological Study* (eds R. Zimmer, R. Hertel, and G.-F. Böl), BfR Wissenschaft, Berlin.
63 Food Standards Australia New Zealand (2016) Reports on the Use of Nanotechnology in Food Additives and Packaging. Available at http://www.foodstandards.gov.au/consumer/foodtech/Pages/Reports-on-the-use-of-nanotechnology-in-food-additives-and-packaging-.aspx (accessed June 28, 2016).
64 Food Standards Australia New Zealand (2016) Nanotechnology and food. Available at http://www.foodstandards.gov.au/consumer/foodtech/nanotech/Pages/default.aspx (accessed June 28, 2016).
65 Gaskell G., *et al.* (2010) *Europeans and Biotechnology in 2010: Winds of change?* (ed. Directorate-General for Research Science in Society and Food Agriculture, Fisheries, & Biotechnology), European commission, Luxembourg.

Conclusions

Monique A.V. Axelos[1] and Marcel Van de Voorde[2]

[1]INRA, UR1268 Biopolymères Interactions Assemblages, Rue de la Géraudière, 44316 Nantes, France
[2]Member of the Science Council of the French Senate and National Assembly, Rue du Rhodania, 5, BRISTOL – A, Appartment 31, 3963 Crans – Montana, Switzerland

To face global challenge on food nutrition security, it is necessary to increase food production while producing it more efficiently and with reduced detrimental environmental impacts. So, the food chain becomes critical to human well-being and survival. The development of future agricultural technologies is to assure sufficient healthy nutrition being a critical factor for the future of Europe and for countries around the world. We can expect that nanotechnology will certainly increase efficiency at all the steps, from production to consumption of food.

Since they first appeared in the early 1980s, nanotechnologies are no more in the infancy and concrete realities appeared in a large number of domains (electronic, medicine, materials, etc.). More recently, they also offer great benefits, from monitoring initial food production in agriculture and farming, through tailoring food for maximum nutrition and reduction in population malnutrition, cancer, and cardiovascular diseases, to reduction in waste products during transport and storage. As shown in this book, examples for the food chain are numerous, such as connected nanosensors to reduce the quantity of pesticides, insecticides, and chemical fertilizers used in the agriculture by releasing them at the right place, at the right time; biosensors and pathogen detectors for use in real time during production and storage for tracking contamination; nanocoating barriers to prevent spoilage and oxygen absorption in packaging; processed foods with tailored nanostructure for increasing food nutritional density; or nanoencapsulation to reduce the number of additives required in food products and offer a better protection for improving absorption of vitamins and micronutrients, minerals, and trace elements.

Nanotechnology offers promising new possibilities in the agriculture and food industry with some concrete application, but lack of awareness and uncertainty about the balance between potential benefits on one side and the potential risks

Nanotechnology in Agriculture and Food Science, First Edition. Edited by Monique A.V. Axelos and Marcel Van de Voorde.
© 2017 Wiley-VCH Verlag GmbH & Co. KGaA. Published 2017 by Wiley-VCH Verlag GmbH & Co. KGaA.

and possible long-term side effects on the other side are hesitating elements for the consumer. Consumer risk perception of health and safety concerns plays an important role in the uptake or rejection of this new technology. The number of questions not still well taken into account and financed on the identification of nanomaterials, on their short- and long-term behavior in human and in the environment, on their "Trojan horse" effect when mixed with other molecules, and so on has increased both citizen and scientist concerns but generates at the same time a lack of transparency in commercial communications and in institutional documents.

This vision on the high potential of nanotechnology benefits to meet the food and nutrition security global challenge will only be successful, if all nano- and microscopic processes in the food chain are fully understood, within a framework of research and development that determines the role of nanomaterials in agriculture and foods.

Research should be encouraged for ensuring (i) that nanoscience and nanotechnology concepts be explored to carry the potential to better understand and enhance the fertility of the soil and the yield of a crop from seed through to harvest. The nutritional and health-promoting qualities of a plant are of critical importance for food crops, as is optimizing; (ii) that nanodevices will help to develop cheap and highly integrated diagnostic systems of higher sensitivity, specificity, and reliability for chemical, biological, and medical analyses to play a major role in biodegradable packaging, protective coatings, and food monitoring; and (iii) that nanotechnology will provide a thorough nanoscale hazard and risk analysis before nanotechnology being exploited by both industry and consumers. Commercialization and marketing of nanofoods should only be done when nanospecific safe guidelines and laws and agreed nanospecific standard specifications and regulatory issues in food nanotechnology will be established involving the public in its decision-making. It is vital to establish a proper consumer strategy with respect to nanofoods. The fears and sensitivities with the public associated with health and safety are related to, for example, nano-enhanced additives in functional foods that improve taste or texture, risks to the consumer can be posed by release of nanoparticles from food packaging during handling or after disposal, or if nanoparticles migrate into foodstuffs. There is also potential for nanomaterial residues to accumulate in agricultural production of meat, milk, crops, and more generally in the environment (soil, water, animals, etc.).

Nanotechnology has been well developed in other domains: nanoelectronics, nanochemistry, nanomedicine, but the agriculture and food scientists and companies have not been very familiar with nanotechnology in the past. It is strongly recommended that governments stimulate innovation in this domain to the welfare of the agriculture and food industry and to safeguard the consumer from safe use. It is important to build up knowledge on safety aspects in parallel with scientific development so as not to risk regulatory delays and other hurdles to acceptance. This implies that independent research could be financed to evaluate and address the risk, to develop reliable and accessible information on real nanomaterials, to contribute to increase the level of awareness of the citizen, and to help decision policies.

This book highlights opportunities for research and development in nanoagriculture and nanofood products as well as for studies on nanotoxicology aspects of nanomaterials (eco- and human) and risk management. Governments, the FAO of UN, the OECD, the European Commission, research and consumer agencies, and industry should recognize the importance of this research and development technology and promote a breakthrough of nanotechnology in agriculture and food products so that commercialization and marketing of these new products can be realized in Europe and worldwide. Transparency is a key issue for the development of nanotechnology and we believe that precautionary principle and competitiveness are not two contrary concepts, rather they are mutually conducive.

Index

a

abiotic factors 118
acceptable daily intake (ADI) 73
– value 305
acetolactate synthase enzyme 21
acidification 50
activated sludge
– affinity coefficients for 338
active and intelligent materials (AIM) 187
active ingredient 116
– organic 123
active oxygen scavengers 168, 199
additives 199
adhesion 235
adsorption, distribution, metabolism, and elimination (ADME) profile 305
adsorption isotherms 342
aflatoxin B1 detection 21, 22
agglomerates 73
aging society 153
AgNPs. see silver nanoparticles (AgNPs)
agricultural industry 117
– food industry 49
– nanoparticles 13, 89
– productivity 7, 15
agricultural products 9, 15, 21, 303
– food production 377
agriculture sector, applications of nanotechnology in 18
– delivery of agriculture chemicals 19
– diagnosis and control of plant diseases 22
– and food sectors, challenges of using nanotechnology in 27
– high-value added products, waste reduction/production of 22
– nanosensors/nanobiosensors 11, 21
agrifood sector, nanotechnology applications 363

agrochemicals 8, 118
agrofood sector 100, 312
Ag_2S NPs accumulate on wheat roots 336
AI-nanocarrier
– bioavailability of 124
– combination 124
alginate beads 145, 148
– hydrogel beads, as Trojan Horse nanoparticle delivery systems for curcumin 146
allergy prophylaxis 313
aluminosilicates 22
aluminum, as food additive 77
American bioethics 354
aminolyzed/charged PET (A/C PET) 173
(3-aminopropyl)triethoxysilane (APTES) modification 123
amyloid fibrils 53
animal-derived proteins 141
animal feed 364
anomalous scattering 93
antibacterial agents 27, 121, 235
antibiotic resistance 121
anticaking agent 72
antimicrobials
– activities 198
– agents 22
– – released by delivered nanoparticles 248
– and antifungal surface coatings 118
– food packaging 179
– nanocomposites 179
– nanoparticle suspensions, as 118
– properties 202, 314
– silver nanocomposites 180
– substrate, formation impact 245
antisolvent precipitation 144
aquifers 5
aroma analysis 284
artificial fusions 212
artificially built multidomain enzymes 212

artificial nanostructures 50
artificial organic nanostructures 50
– found in food, types and uses of 52
– – lipid nanostructures 60
– – polysaccharide nanostructures 55
– – protein nanostructures 52
– research 49
– utilized in agri-food industry 51
ascorbate peroxidase 335
Aspergillus flavus 127, 179
Association Internationale pour l'Etude des Argiles (AIPEA) 170
atomic force microscopy (AFM) 21
– tips 87
atomic number Z
– scattering length, variation of 92
ATP-dependent extrusion systems 315
ATP production associated proteins 22
automatic measurements 259
avidine 38

b

Babinet theorem 91
Bacillus subtilis var. niger 181
bacterial cellulose nanowhiskers (BCNW) 173
bacterial growth, CA principle
– BacLAB based on solute diffusion model 239
bacterial microcompartments (BMC) 16, 222, 223
– framework for enzymatic nanoreactors 223
– natural BMC, engineering of 224
bacterial surface contamination 235
bactericides 121
Bayspec portable hyperspectral camera, with IMEC sensor chip 290
beam sizes 108
bentonite 74
beta-lactoglobulin 89
bioaccumulation 334, 335, 336
– of nano-Ag_2S on wheat roots 336
bio-based organic components 210
biocatalysts 215
biochemical sensors 272
– E-tongue systems 272
– technological trends for 272
biocompatibility 120
– compostable materials 167
– electronics 264
– nanosensors, for temperature and moisture monitoring 118
biodegradation 20
– polymers 172

biodiagnosis 221
biodiversity 6
bioenergy expansion 7
biofilms 235
– antimicrobial substrate, formation impact 245
– antimicrobial surfaces 244
– antimicrobial treatment of
– – antimicrobial agents released by delivered nanoparticles 248
– – by targeted drug release 246
– fluid interaction 236
– fluid interface 239
– fluid mixture 244
– formation 10, 236
– governing system of equations 247
– mathematical modeling 236
– models 237, 239
– – hybrid discrete–continuum models 239
– – individual-based modeling 17, 242
– – multidimensional continuum models 240
– – properties 243
– prevention 237
– properties 243
biofuels 23
biofunctionalization 201
bioimaging 221
biomarkers 264
biomass 341
– degradation 212
– derived compounds 22
biomimetic approaches 210
biomolecules, adsorption of 316
biomolecules-nanomaterial complex 317
bionanocomposites 300
bionanotechnology 22
biopolyester-based multilayers 172
biopolymers
– as building blocks to form hydrogel beads 140
– fibrils 146
– nanogels 138
biorecognition 265
bioresorbable electronics 264
biosensors 10, 117, 235, 377
– based on optical fiber technology 285
– design 22
– laccase/tyrosinase phenoloxidase enzymes, bi-immobilization of 22
– technologies 283, 285
biosludges 341
biosystems

– spatially constrained functional coupling 210
biotechnology 8
biotic factors 118
biowastes
– NMs, indirect agricultural application of 336
Bipolaris sorokiniana 126
BlueMoonGoods™ 180
Bombyx mori 20, 128
bovine serum albumin 35, 36
Bragg scattering 103

c

Caco-2 cells 318
Ca^{++} ions 339
calcium 23, 76
calcium carbonate 76, 77
calcium chloride 70, 77
Campylobacter jejuni 181
cancer 6, 377
Candida albicans 126
capillary electrophoresis 200
carbamate 201
carbonic anhydrases (CA) 223, 224
– biofilm model, combined differential–discrete 238
– 2D CA model 239
carbon nanomaterials (CN) 333
carbon nanotubes (CNTs) 3, 170
– double-wall 335
carboxysomes 223
cardiovascular diseases 6, 377
Carex lurida 341
carotenoids 135
Casein 35, 42
casein micelles 35
– in cow milk 105
– schematic structure 105
κ-casein protein fibrils 52
CcmM proteins 225
CcmN proteins 225
β-CDd nanoparticles, embedded in vitreous ice
– cryo-EM images of 59
Cd-tellurium quantum dots 334
cell death 121
cellular automaton (CA) model 238
cellular lipid-based organelles 210
cellulose nanocrystal 22, 125
cellulosomes 214
cementite nanowires 3
ceramic 198
channel geometry 240
checklist, for scattering 107

chemical/biological agents 201
chemical contamination 262
– detection of 263
chemical fertilizers 20
chemical properties 123
chemicophysical interactions 124
chemoautotroph microorganisms 223
chemotaxis 240
chitosan films 120, 181
chitosan nanoparticles 181
citrate-functionalized NMs 335
citric acid 57
Citrobacter freundii 224
Clay Minerals Society (CMS) 170
clays 198
– based nanofillers 168
click chemistry 123
climate conditions 9
CMOS-based technologies 279
– cameras 265
Code of Conduct 355, 357
coenzyme Q10 135
CO_2 indicator 185
cold sterilization 10
Collab4Safety report 304
colloidal silica 72
coloring agents 75
communication 5, 13, 24
– nanotechnology 355
complex coacervation 144
composite material, definition of 168
compound annual growth rate (CAGR) 367
computational tools 212
constitutional project, of governance 356
consumer acceptance 182, 371
consumer product inventory 334
consumer risk 378
contaminations 262
– measurement 262
contrast strategies 92, 94
conventional packaging materials 177
coordinated approach 350
copper 76
– copper oxide NPs 199
– derivatives 121
– nanoparticles 3, 334
– – synthesized by polyol method 181
– in plant protection 121
covalent strategies 123
CRISPR/Cas9 complexes 215
crops
– based food supply 8
– pathogens 117

crystalline phases 102
cubic anatase 335
Cucumis sativus 339
cupric salts 121
curcumin 55, 135, 146
– loaded micelles 148
– loaded nanoemulsions 147
cyanobacteria 224
cyclic peptides 212
cyclodextrins 57
cytochrome P450 212
cytotoxicity 200

d
dairy industry
– list of specific items to be tested in 262
dairy products 158
dairy sector, implementation of microsystems in 261
Danish Environmental Protection Agency 313
data acquisition platforms 294
data aggregation 293
data analysis 293, 294
– software 294
data storage 24
DDT pesticides 201
dehydrogenases 215
democratic legitimacy 353
dendrimers 19, 24
denitrification 8
dense confluent biofilms 238
designing particle characteristics 138
– composition 138
– particle charge 139
– particle size 139
– particle structure 140
deuterable polymer 102
diabetes 6, 135
2,4-dichlorophenoxyacetate 129
dielectric constant 35
dietary requirements 10
diffusion 9, 20
– path, tortuosity 171
diffusion-limited aggregation (DLA) model 237
diffusive wave scattering (DWS) 106
digestion models
– *in vitro* 316
digestion process
– enzymes 312, 315, 316, 317
– in'vitro models of 316
Directorate General for Research of the European Commission 353
discrete–continuum model 243

DNA-based scaffolds 215
DNA cleaving catalyst 215
DNA degradation 199
DNA modification 215
DNA specific affinity domain 212
D_2O buffer 100
double-slit experiment 90
double-tail lipids 103
drained foams
– SANS (reflectivity) 106
drug delivery 170
dry food, moisture uptake 185
drying 50
dynamic light scattering (DLS) 106, 318

e
economic growth 305
ecosystems services 6
educational programs 5, 11, 294
Eisenia fetida 340
elastic modulus 169
electrochemical sensing 202
electronic noses 283
electronic tongue 16, 301
electrospinning 145, 162
electrostatic complexes 144
electrostatic-driven encapsulation 221
electrostatic interactions 33, 146
emissions 21
– delivery system 25
emulsions 9, 103
– SANS results 104
encapsulation 19, 302
endocytosis 248
engineered inorganic particles in food products, techniques used to characterize 81
engineered nanomaterial (ENM) 177, 188, 311, 334
– dietary exposure 313
– inorganic nanomaterials 69
– – characterization of 78
– – as manufactured 79
– – as present in food matrices 80
– types of 313
enzyme hydrolysis, of proteins 160
Escherichia Coli 181, 244
esterification reaction 123
ethical issues 350
ethical, legal and social aspects (ELSA)
– projects 352
ethical, legal and social implications (ELSI) 352
– European projects 352, 353
– – funded by 356

ethylcellulose 185
ethylene 199
– nanosensor 187
ethylene vinyl alcohol (EVOH) possesses 172
EU regulation
– biocidal product regulation (EU 528/2012) 27
– regulation (EU) No 1169/2011 27
– regulation (EC) No 450/2009 on active and intelligent materials 27
– regulation (EC) No 1333/2008 on food additives 27
– regulation (EU) No 2283/2015 on novel foods and novel food ingredients 27
– regulation (EU) No 10/2011 on plastic materials 27, 188
European approach to ethics 354
European Commission 70, 188, 352
– on appropriate communication of nanotechnology 355
– nanomaterial defined 16
European Food Safety Authority (EFSA) 26, 69
– nanoinventory 78
European food sector 259
European governance
– initiatives 360
– of science 357
European institutions 353
European integration through research policy 360
European "European political economy 360
European research policy 357
– after 2010 358
European science
– museums 355
– policy 353
European values 353
existing technologies
– in applications 282
– characteristics 282
– – compact, state-of-the-art technology 282
– – integration 283
– – standardization 282
– – user-friendly technology 282
exopolysaccharide (EPS) 236, 246
extracellular enzyme activity 341
extrusion techniques 145

f

ω-3 fatty acids 61, 135
Fe-Mn oxides 342

fermentation process 50, 265
ferritin 217
fertilizers 6, 8
Fertilizer technology 8
fibril formation 146
fibroblast cells 200
field flow fractionation (FFF) 200
flavonoids 135
flavors 10
Flory–Huggins free energy 246
fluorescence emission 201
fluorescence resonance energy transfer (FRET) 186
fluorescent compounds
– QDs, lanthanides complexes 202
5-fluorouracil 57
foams 105
FokI enzyme 212
folded nucleotide domains (aptamer)
– 3D 215
follicle-associated epithelium covering 319
food 5, 15
– additives 72, 73, 76, 197
– – nanosized fraction of nanoparticles 71
– analysis 197, 200
– authenticity 26
– – and integrity 278
– availability 7
– nutrition security
– – global challenge 378
Food and Agriculture Organization of the United Nations (FAO) 15
Food and Drug Administration (FDA) 188, 294
foodborne illnesses 277, 304
foodborne pathogenic bacteria 202
food chain 8, 9, 10, 11, 27, 377
food choices/requirements drivers, of elderly 155
food coloring 72, 74, 305
food companies 301
food components 317
food consumption 377
food contamination, detection 197
food delivery system 25
food demand 8
food design 156
food grade
– fatty acids 77
– polysaccharides, used for manufacturing nanoparticles 56
– titanium dioxide and silica powders, methods used to characterize 80

food grown worldwide 11
food industry 259, 301
– new nanoelectronics-based technologies 18, 279
food loss 8, 12
food market 11
– microbial spoilage, electrochemical nanosensor 187
FoodMicroSystems 259, 269
FoodMicroSystems project 261
food nanogels 161
food nanotechnology 302
– regulations for 306
food nutritional density 377
food nutrition security
– global challenge 377
food packaging 8, 11, 25, 50, 182, 189, 197, 199, 303
– applications 178, 363
– materials 24
– polymers 167
– potential safety issues and current legislation 187
– preservative-modified atmosphere 177
– utilizing nanotechnology 52
food preservation
– nanomaterials in active packaging 178
– techniques 50
food processing 8, 12, 300
food products 26, 259, 305
– nanoparticles 13, 89
food quality 7, 19, 260, 278, 279, 304, 363
– assurance 277
– safety, direct nano-enabled indicators 186
food quality/safety
– indirect nano-enabled indicators of 184
food requirements 8
food safety 12, 116, 259, 262, 277, 301, 304
– risk 277
foods, and the elderly 153
food science 203
food sector 11, 261
– applications of nanotechnology 23
– – delivery of active compounds 24
– – food packaging 25
– – other applications 26
food security 5, 6, 12, 116
– sustainability 5
food spoilage micro-organisms growth 184
food storage 198
food supply 8, 259
food traceability 275

food value chain sustainability 12
food waste 8
Fourier transform 97
FOx diagnostics platform 288
FOx Diagnostics spin-off company 288
free energy 246
Fresherlonger™ 180
fullerenes 70
functional assembly
– through engineering of natural/synthetic complexes 212
– through natural/synthetic fusions of protein domains 212
functional coupling, through scaffold-independent structures 211
fungal hydrophobin rodlets 52
fungicides 121, 334
funnel toward sector or community oriented validation model 280
Fusarium graminearum 126
Fusarium oxysporum 127

g
GA-Ag NPs 338
galactokinase 212
β-galactosidase 212
gas absorbing 184
gas barrier 25
gas emissions 26
gaseous ammonia 187
gaseous emissions 167
gas flush 184
gasification 22
gas sensing devices, and systems 268
– E-nose instruments 269
– microchromatographers 270
gas sensors 21, 26
gastrointestinal environment 318
gastrointestinal tract (GIT) 312
– on ingested nanomaterials characteristics 314
– in vitro models of 19, 315
GC/MS systems 283
gelation 156
gelling agents 156
gel microparticle 162
genetically modified organisms (GMOs) 262
genetic transformation 18
genome engineering 212
genosensors 265
GFWS platform 8
GHz-THz sensor 291
gliadin 144

global food production 6, 16
global food security 15
globular proteins 146
glucose oxidase 55
gold 76
– monodisperse spheres form factor 96
– nanoparticles 3, 96, 184
– nanorods 21
– poly(vinyl pyrrolidone) core-shell nanocomposites 185
– uses in confectionary for decorations 77
governance 358
– good 351
– mechanisms 360
– system 350, 351, 352
grafting ontostrategy 123
grape-like aggregates 102
greenhouse gas emissions 8
Green Revolution 116
guaiacol peroxidase 335
guar gum/nanocrystalline cellulose film (GG/NCC) 60
guide RNA 215
Guinier plot 94
Guinier regime 94
gut models
– *in vitro* model 323

h

hazard characterization 306
hazard identification 306
health care systems 15
heat resistance 25
Helmut Kaiser Consultancy (HKC) 367
herbicides 20, 21, 129
heterogeneous catalysts 23
heterogeneous mosaic biofilms 238
high moisture extrusion cooking (HMEC) 50
high-performance liquid chromatography (HPLC) 200
homogenization 161
horse meatcrisis 260
HT29-MTX cells 319
human digestion process 315
human-edible crops 7
human enhancements 349
human gastrointestinal tract 136
human gut
– ex vivo models 318
human health 5
hybrid artificial nanostructures 50
hybrid discrete–continuum biofilm model 240

hydrocolloid gel particles 161
hydrogels 22, 198
– beads, fabrication methods for 143
hydrophobicity 212
– alumina silicate nanoparticles 20
– polyphenols 55
hydroxide pigments 75
hygroscopic nature 172
hyperspectral cameras 286
hyperspectral imaging 285
– new reflection-based camera technologies 285
hyperspectral sensor 291
hypertension 135

i

i-FAST project 290, 292
immunobooster properties 313
immunosensors 265
income growth 7
information technology 16
ingested nanomaterials
– characterization techniques 318
– digestive enzymes and food matrices 316
– pH/ionic strength 316
innovation 5, 11
innovative sensors use, during daily operations
– four-step model 279
– – awareness 279
– – implementation of new technologies 282
– – platform creation 280
– – validation of new technologies 281
innovative technologies 11
inorganic material, intrinsic properties of 197
inorganic nanoagrochemicals 8
instrumental analyses 278
intelligent food packaging
– easy-to-read 183
intelligent nanoagrochemicals 9
Intelligent packaging. *See also* food packaging
International Organization for Standardization (ISO) 70
Internet of Things (IoT) 293
intestinal barrier
– *in vitro* models of 318
ionic strength 35, 36, 341
iron hydroxide 70
iron nanoparticles 179
iron oxides 70, 75, 199, 200
isoelectric point (pI) 34
iStrip 184

j

Juncus effusus 341

k

Kaolin 122
Kelvin equation 340
kiwifruit bacterial canker 118
knowledge 13
– acquisition 5
– based decisions 10
Kraft Foods 301

l

lab-on-a-chip systems 283, 285
α-lactalbumin (α-La) 12, 35, 37, 40
– protein nanotubes 52
lactoferrin 35, 38
β-lactoglobulin (β-Lg) 35, 40, 159
– nanoparticles 55
leaching 8, 21
lactoperoxydase 39, 40, 41
lecithin 156
legal constraints 353
length-to-thickness ratio 171
light scattering 91
lignocellulosic materials 120
lipids
– compartments 216
– double-tail 102
– long-range structures 104
– nanomaterials 120
– organelles 210
– peroxidation 199
lipophilic bioactive molecules 137
lipophilic biomolecules 60
lipophilic ingredients 60
lipophilic nutraceuticals 137
liposomes 25, 61
liquid lipid nanoparticle (LLN) 138
Listeria monocytogenes 181, 187
livestock 15
Lobelia cardinalis 341
Lolium perenne 126
lumazine/riboflavin synthase complexes (LRSC) 217
Lycopersicum esculentum 339
lysozyme 35, 37, 101, 102

m

magnesium oxide nanoparticles 23
magnetic fluctuations 107
magnetic nanoparticles 183

major vault protein (MVP) 222
malic acid 57
malnutrition 7, 377
market forecasts
– nanotechnology, in food and agriculture
– – products 368
– – publicly available data 367
– reports, critical considerations and remarks concerning 367
mechanical stability 25
Medicago sativa 339
metabolons 210
metal detectors 277
metal ions 21
– pesticide formulations 116
metallic aluminum 77
metallic nanoparticles 9
metal nanomaterials 117
metal nanoparticles 21, 121
metal oxides 26, 200
– based NPs 197, 199, 200
methanol 199
metsulfuron-methyl 21
micelles 24, 70
microalgae cells 23
microbial/chemical analyses 277
microbial community composition 341
microbial contamination 24, 186
microbial food safety 300
microbiological contamination 262
microbiota exposure 312
micro electromechanical systems (MEMS) 271
microemulsions 25, 136, 137
microgels 158, 161
micronutrients 49
– deficiencies 7
microorganism detection 273
micro/smart systems 259
Microstegium vimineum 341
microsweets 73
microsystem technology 259, 260, 261
– areas 266
– detection methods 266
– – chemical and biochemical electronic sensors and systems 268
– – dielectric sensor 267
– – direct sensing, with electronic nose technology 267
– – hyperspectral imaging 266
– – imaging techniques 266
– – magnetic resonance imaging (MRI) 267

– – mid-infrared spectroscopy 266
– – near-infrared spectroscopy (NIRS) 266
– – process viscometer 267
– – ultrasound imaging 267
– – x-ray scanning 267
– food/beverage sector, implementation of 265
– meat sector, implementation of 264
milk proteins 186
millimeter wave sensors 18, 288
miniaturized hyperspectral camera systems 285
– IMEC 287
– types 286
miniaturized Raman spectroscopy systems 283
mixed-culture biofilm (MCB) 237
modern agricultural practices 116
molecular gastronomy 156
molybdenum 76
montmorillonite (MMT) 122
– hydrophilic and hydrophobic clays 122
– nanolayers
– – reinforcing effect of 170
MST devices 265
Mucor plumbeus 127
multidomain architectures 212
multidomain proteins 211
– composition, modulation of 211
multienzymatic complexes 210, 217
multifunctional nanomaterials 9
multiscale organization 33
multisensing platforms 259
multisite complexes 213
multiwall carbon nanotube (MWCT) 334
mushroom-shaped biofilm 239
mutations, in scaffold 213
mycotoxines 22, 201
myricyl palmitate 61

n

nanoagrichemicals 335
nanoagrochemicals 8
nanobarcodes 21, 26
nano-based kits 117
nanobiopolymers 123
nano(bio)sensors 16, 21, 26
– metal NPs 201
nanocapsules 20, 118
– improve dispersion, bioavailability, and absorption of nutrients 118
– production of 20
nanochemistry 378
nanocid®-based pesticides 129
nanoclays 122, 170, 199
– nanofilms, as barrier materials to prevent spoilage and oxygen absorption 118
nanocoatings
– barriers 377
– on food contact surfaces 364
– functionalized ligand 9
nanocomposites materials 169, 202, 203
– with antimicrobial properties 179
– with antioxidant properties 178
– definition of 168, 170
nanocrystalline cellulose 60
nanocrystalline TiO_2/polymer films
– UV illumination of 179
nanodelivery systems, for nutrients/supplements 363
nanodevices 378
nanoeffect 169
nanoelectronics 279, 378
nanoemulsions 9, 11, 20, 25, 61, 137
ω-3 nanoemulsions 61
nano-enabled agrichemicals 334
nano-enabled formulations 9
nano-enabled intelligent food packaging systems
– overview of 183
nano-enabled intelligent packaging
– stakes and challenges of 181
nano-enabled products 11
nano-enabled sensing
– principles of 183
nanoencapsulation 11, 20, 377
– of fertilizers 20
– flavor enhancers 118
– nutraceuticals
– – for targeted delivery 118
– pesticides 20
– techniques 20
nanofertilizers 21
nanofibrils 159
– of perylen-based fluorophore 186
nanofillers, in polymer-layered nanocomposites 170
nanofilms 118
nanofiltration 364
nanofoods 19, 299, 302, 378
– applications 301
– benefits of 307
– consumption of 305
– contact materials 10
– defined 300
– safety 304
nanoformulation 124

– agrochemicals 235
nanoindicator systems 184
nanoinsecticides 129
nano issue 4
nanolayers
– reinforcing effect of 170
nanoliposomes 61, 137
nanomaterial (NM) 3, 10, 15, 23, 115, 129, 188, 311, 334
– agrifood, publications dealing with 364
– applications for plant pathogens/pests 125
– – bacteria 14, 125
– – fungi 126
– – insects 14, 127
– – virus 128
– as color enhancers 118
– food and feed applications and food contact materials 119
– food sector and agriculture, potential and possible applications of 364
– functionalization 14, 123
– indication in list of ingredients 4
– indirect agricultural application of 336
– inorganic 119
– noble metals 120
– solutions absorbance/fluorescence of 320
nanomedicine 378
nanometer-sized particles 311
nanoparticles (NPs) 16, 145, 215
– affinity coefficients for 338
– analyte system 201
– for antimicrobial and antifungal surface coatings 118
– applied in food sector 24
– cell interaction 236
– colloidal delivery systems 136
– – biopolymer nanoparticles and nanogels 138
– – microemulsions 136
– – nanoemulsions 137
– – nanoliposomes 137
– – solid lipid nanoparticles 137
– control crop pests and plant pathogens 116
– CuO 336
– designed and different particle properties 139
– detection in food 200
– encapsulated 246
– endocytosis into cell 250
– foodborne pathogens
– – detect chemicals of 118
– gold 77
– inorganic 200

– – applications 202
– – for food science application 197
– metal (Ag, Cu) and metal oxide (ZnO, CuO, TiO_2) 198
– models for intercellular/surface delivery 248
– percent removal dictated by γ and simulated α 338
– physical-chemical properties 183
– for selective binding and removal of chemicals and pathogens from food 118
– sensors 201
– structure factors 99
– suspensions, as antimicrobials 118
– TiO_2 detected in soil 340
– trapped inside hydrogel beads to form Trojan horse nanoparticle delivery systems 141
– zinc oxide (ZnO) 198
nanopesticides 8, 116, 120
nanopores 186
– fibers 9
– foams 9
– materials 20
nanoprobes 87
nanoprocessing 21
nanoproducts 8, 365
nanoreactors 211
nanoscale 188
– adsorbents 10
– colloidal delivery systems 136
– delivery systems 19, 20, 129
nanoscience 15, 16
– fundamental of 16
nanosensors 9, 10, 182, 301, 312, 377
– for food labeling 364
– nanobiosensors for smart packaging 25
nanosilver 22, 27, 300
– particles 22
nanosizes
– additives, inorganic food 363
– agrochemicals 364
– in food products 69
– *in situ* 87
nano-smart dust 21
nanostructure
– food ingredients 363
– layers 172
– materials 3, 70
nanostructured lipid carrier (NLC) 61, 62
nanosurfactant spreading, on cell membrane 251
Nanotech Consortium 301
nanotechnologies 12, 188, 235, 349, 363
– agricultural and food sectors 15, 116, 117

– applications 9, 116
– – different industrial and agricultural fields 119
– – in food science and technology 23
– – modern agriculture 19
– benefits of 307
– defined by US Environmental Protection Agency 117
– design and development
– – top-down and bottom-up approach 117
– food and agriculture, obstacles regarding commercialization of 370
– in food science 33
– in food sector, promising benefits of using 25
– and foods for elderly 157
– FSAI, global applications of 300
– gene sequencing 10
– governance 357
– human enhancement 349
– market research and forecasts 366
– – methodology of market research and forecasts 366
– nanomaterials 119
– – combined organic/inorganic nanomaterials 122
– – inorganic nanomaterials 120
– – organic nanomaterials 120
– risk/benefit factors 4
nanotechnology
– governance of 352
– societal implications 349
Nanotechnology Consumer Products Inventory 312, 313
nanotubes and nanoparticles, as gelation and viscosifying agents 118
nano zinc oxide 300
National Informal Science Education 355
National Informal Science Education (NISE)
– initiatives 355
– network 355
National Nanotechnology Initiative (NNI) 3
$Na_2Ti_3O_7$ nanotubes 185
natural biochemical reactions 210
natural/engineered nanocompartments
– encapsulated biosystems 16, 216
natural organic matter (NOM) 335
Navier–Stokes equation 239
near-infrared spectroscopy (NIRS) 270
– state-of-the-art benchtop NIR devices 270
networked food production supply 12
neutrons activation 108, 109
new technology developments 18, 285
– transmission-based technology 18, 288
Newtonian fluids 241
Nicotiana benthamiana 224
N-isopropylacrylamide (NIPAM) 198
nitrogen 21
noble metal NPs 201
N_2O fluxes 341
nonbiodegradable packaging materials 180
noncovalent associations 215
noncovalent interactions 34
noncovalent surface modifications of nanomaterials 123
nonpolar bioactive molecules 137
nonpolar tails 136
nonpolymerase chain reaction (PCR)-based method
– microdevices/components, needed for micro-PCR 263
– μPCR device 274
nontoxicity 120
nuclear magnetic resonance imaging (NMRI) 292
– benchtop NMR analyzer 291
– low-resolution 291
nucleic acid-peptide interactions 216
nucleic acids 209
nucleic acids-based scaffolds 215
numerical simulations 240, 242
nutraceutical compounds 312
nutraceuticals 135, 136
nutrient delivery systems 9
nutrient fortification 9
nutritious food 12
nylon-clay composite materials 169

O

obesity rates 7
oil-in-water microemulsions 136
oleic acid 339
oleosome 33
omega-3-fatty acids 302
optical detection 201
optical fiber biosensor technology 288
optical microscopy 87
orders of magnitude 91
organic active compounds 130
organic dyes 202
organic fatty acid chlorides 123
organic matter 342
organic nanomaterials 119, 120
organic nanosized additives for food 363
organic nanostructures 49
organophosphorus 201

oscillations 211
osmotic pressure 20, 241, 246
osteopontin 39–41
ovalbumin 35, 38, 50
ovotransferrin 35, 38, 39
oxygen (O_2) 178
– absorber 190
– scavenging 202

p

packaging 11, 167
– foil 186
– intelligent 177
– materials 300
palladium 76, 199
Panicum virgatum 341
particle sizes 9, 27, 74, 77, 278, 306
– distribution 72, 75
Pdu compartments 224
pectin 120, 145
– lysozyme coacervates 103
Pedersen–Schurtenberger model 97
PEGylation 221
Penicillium expansum 126
PE/nylon packages 179
peptides 135
Percus–Yevick function 97
pesticides 20, 27, 116, 129
– formulations 116
pests
– plant pathogens control, development of novel nanoformulations 123
– resistance 129
petroleum reserve 116
pharmaceutical delivery systems based on biodegradable polymers 124
pharmaceutical formulations 124
pharmoacokinetics 305
phase distribution 173
phase separation 100
pheromone methyl eugenol 130
Phoma glomerata 126
phospholipid bilayer 212
photoactivated indicator 184
photodegradation 57
photolithographic approaches 209
phyllosilicates 171
physical barriers 199
physical sensors 265
physicochemical processes 89
physicochemical properties 4
Phytophtora infestans 127
Pièces et Main d'Oeuvre(PMO) 359

plant-based agriculture
– agricultural production 116
– nanotechnology 19
plant-derived proteins 141
plant diseases 116
plasma membrane 22
plastic packaging materials 168
plastic polymers 25
platinum 76, 199
Pleospora herbarum 126
politics of governance 358
polycationic polysaccharide 59
polyelectrolytes
– complexation 58
– nanoparticles 59
– scattering from proteins 101
polyethylene glycol (PEG) 128
polyhydroxyalkanoates (PHAs) 172
polylactic acid (PLA) 172
poly(DL-lactide-*co*-glycolide) (PLGA) 125
polymer chains, form factors 97
polymer-clay nanocomposites 120
polymer film 20
– carbon coated copper nanoparticles 185
polymeric packages 167
polymerization process 72
polymer-layered nanocomposites 170, 171
– with improved flexibility 25
polymer–nanoparticle composites 235
polymer silicate nanocomposites 25
polymers incorporated nanomaterials
– active packaging 25
polyphenols 55
polypropylene 199
polysaccharides 142, 173
– nanoparticle preparation, mechanisms for 58
– nanostructures 55
polystyrene nanoparticles (PS-NPs) 317
polyvinylalcohol 199
polyvinylpyrolidone 339
population
– growth 7
– percentage, in age groups over 60 and over 80 years old, by country 154
pore size 100
Porod law 95
portable aroma systems 284
portable hyperspectral camera technology 290
potassium ferrocyanide 74
preservative-modified atmosphere packaging 177

Index | 393

printed technology 262
probing oxygen sensivity 185
process analytical technologies (PAT) 278, 294
prolamin 300
propylene films 199
protein oxidation 199
proteins
– engineering strategies 221
– fusions 212, 213
– gel, cross-linked 50
– nanocompartments 216
– – structural organization of 221
– – structural properties and biotechnological applications of 218
– nanomaterials 55, 120
– nanostructures 34, 35
– polysaccharide complexes 159
– protein interactions 213
– scaffolds 213
– transport 216
– used as nanoparticle carriers, for range of bioactives 54
Pseudomonas aeruginosa 125
Pseudoperonospora cubensis 125
pyrolysis 22
pyruvate dehydrogenase 217

q

quality control 7, 290
quality-directed screening systems 278
quantum dot (QD) 127, 201

r

radiation facilities 109
radio frequency identification (RFID) 184, 293
– enabled sensor 187
Raji B cells 319
reactive oxygen species (ROS) 305
reflectivity 107
remote sensing system 10
resource-intensive food 6
resource-saving production 363
responsible research and innovation(RRI) 357
resveratrol-loaded stearic acid 62
Rhizopus stolonifer 127
rhizosphere 335
– microbiomes 335
Rhyzopertha dominica 128
risk acceptability 5
risk assessment
– approache 4
– food containing nanomaterials 306
– need for 306

risk management 379
– approache 4
RNA-protein hybrid scaffolds 215
RNA scaffolds 215
rural development 12
ruthenium polypyridyl complexes 185

s

safer food 49
safety
– assessment 278
– monitoring 279
– regulations 12
– tests 10
saliva lubricates food 315
Salmonella enterica 181, 224
salting transform food structures 50
sample damage 109
sample–detector distances 91
sample environment 109
sample sizes 13, 108
SAS profile 93
– advanced form factors 95
– aggregation 98
– form factors 93, 96
– hard spheres 97
– intermediate regime 95
– natural systems 102
– phase separation 100
– polymer interpenetration 98
– proteins/polymer 13, 100
– structure factors 13, 97
scaffold
– and compartment-based confinement mechanisms for coupled enzymatic reactions 214
– structure 213
scanning tunneling microscope 3
scattering
– length density (SLD) 92, 107
– principles 90
– techniques 88, 89, 106
– vector 90
science policy programs 350
scientific-based knowledge 10
selenium 76
self-assembling polysaccharide nanoparticles 59
self-organization mechanisms 209
self-regulation 19
semiconductors quantum dot 201
sensing food quality 183
sensors 265

– chips 279
– nanoparticle 201
– technology 278
sensory analysis 278
serum albumin 36
setup sensor configuration Aquantis 289
shelf life 49, 199, 260
– of food product 50
shell-dependent nanocompartments 221
shell formation 221
shell-independent nanocompartments 217
shell-restricted compartments 211
signal-to-machine noise ratio 109
silica 73
– NPs doped with metal NPs 202
silicates 73, 74
silicon-based nanostructures 210
silicon dioxide (SiO_2) 200
silver 76, 307
silver ions 121, 198
silver nanocomposites, antimicrobial activity of 180
silver nanoparticles (AgNPs) 20, 22, 76, 179, 313
– agglomeration 317
– as antibacterial agents 197
– colloidal solutions 321
– consumer products, categories of 314
– cytotoxic effect of 198
– cytotoxicity assessment and application 320
– cytotoxicity studies 321
– embedded packaging material 314
– exposure 320
– incubation 321
– loaded in zeolites 198
– risk assessment of 306
– zeta-potential 316
single particle - inductively coupled plasma mass spectrometry (SP-ICP-MS) 318
single-walled carbon nanotube (SWCNT) 21
SiO_2 nanoparticles 89
Sitophilus oryzae 122, 128
small-angle neutron 88
small-angle scattering spectrometer 108
smart packaging 189
smart phones 264
SNEDDS formulation 104
SnO_2 nanobelt functionalized 186
societal debate 354
socioeconomic development 7
soft and nutritious gel particles, rational design of 155

– building-up healthy gels with soft textures 158
– molecular gastronomy 15, 156
– structure and food properties 155
soft foods, elderly and food-related issues 15, 153, 155
soil microorganisms 21
soil organic matter 8
solar energy conversion 115
solid lipid nanoparticle (SLN) 61, 137
solubility 55, 141
– of animal and plant proteins derived from various origins 142
sonication 50
spectral fingerprinting 284
spectral systems 283
spectrometer setups 13, 109
Staphyloccccoccal enterotoxin B 187
Staphylococcus aureus 181, 198
state-of-the-art benchtop NIR devices 284
static light scattering (SLS) 106
statistics-based screenings 277
Stokes equation 241
succinic acid 57
sulfidation 340
sulfur-containing NMs 333
supplements 8, 16, 199
surface biocides 11
surface chemistry 341
surface-enhanced Raman scattering 201
surface functionalized nanomaterials 364
surface plasmon resonance 201
surfactants 123, 136, 137
sustainability 12
synthetic biology 209, 210
synthetic genetic mechanisms 211
synthetic multidomain enzymes 212
synthetic pesticides 116
synthetic polymer 100
synthetic silica (SAS) 72
synthetic supramolecular assembly 212
synthetic transcription factor 213
system biology 8

t
TA-induced facilitated TiO_2 NPs transfer in sand column 342
tannic acid 341
tartaric acid 57
tastes 10
Taylor cone 145
technologies
– constraints 260

– innovation 305
– to produce food nanostructures and microstructures 160
– and suppliers, existing, examples of 283
terrestrial biodiversity 5
(2,2,6,6-tetramethylpiperidine-1-oxyl, TEMPO)-mediated oxidation of nanomaterials 123
texture measurements 278
texture-modified (TM) food 155
– bottom-up planning/design of soft gel particles for 159
thermal resistance 198
thermal treatments 160
thermodynamic incompatibility 143, 144
thermoplastic polymers 179, 198
thermoplastic/thermosetting polymer matrix 115
ticaprin 61
time-temperature integrators 26
titanium 76, 307
titanium oxide (TiO_2) 70, 74, 200, 305
– Nafion composites 334
– nanocoatings 26
– nanoparticles 26, 317, 341
– – in management of plant diseases 20
TM foods
– nanostructures uses 157
– technological alternatives for design of 160
toxicity
– of fertilizers 21
– of titanium dioxide 200
– Trojan-horse mechanism of 322
toxicological/safety assessments 306
tracking and tracing systems 275
training programs 350
transcription, of reporter gene 213
transesterification 22
transferrins 38
transformations
– of metallic and oxide NPs 336
– of NPs, in soils after application 339
– – direct application 339
– – indirect applications from biosludges 340
translocation pathways 335
transport and bioaccumulation, by plants 334
triacylglycerols 61
Tribolium castaneum 128

triethanolamine 27
tripolyphosphate (TPP) 58
Trojan horse 378
– delivery systems to deliver curcumin-loaded lipid nanoparticles 147
– nanoparticle delivery systems 140
tunneling 211
turbidity 148

u

ultra-performance liquid chromatography (UPLC) 200
urbanization 6
US Department of Agriculture (USDA) 16
UV-activated indicator 184
UVA light 179
UV light activation 184

v

validation of cleaning actions 277
validation procedures 10
validation process combine technical evaluations with 289
van der Waals forces 123
Verticillium dahliae-infested soil 127
vitamin C 61
vitamin K 315
volatile analysis 278
volatilization 8
Voluntary Codes 353, 356

w

waste reduction 24
wastewater treatment plant
– primary and secondary distribution coefficients for NPs 338
water decontamination 364
wave-based technologies 278
waxes 61
whey protein isolate (WPI) 55, 143
wireless monitoring system 9
World Bank 351
World Health Organization 6

x

x-ray fluorescence (XRF) 283
x-rays 107
– crystallography 35
– 3D 292
– inspection 277
– scattering 88
– tomography 292
Xylella fastidiosa 125
xyloglucan chain 97

y
Young slits 90

z
Zea mays 339
zein 144, 300
zeolites 198
zinc 76, 307

zinc oxide (ZnO) 70, 78, 198, 199
– bound to carbonate 342
– nanoparticles 181
– NPs, antibacterial activity of 198
– TiO$_2$ nanocomposites 186
zirconium phosphate 334
ZrO—graphite 334